Analysis of Wildlife Radio-Tracking Data

Analysis of Wildlife Radio-Tracking Data

Gary C. White
Department of Fishery & Wildlife Biology
Colorado State University
Fort Collins, Colorado

Robert A. Garrott
Department of Fisheries & Wildlife Conservation
University of Minnesota
St. Paul, Minnesota

Academic Press
San Diego New York Boston
London Sydney Tokyo Toronto

This is an Academic Press reprint reproduced directly from the pages of a title for which type, plates, or film no longer exist. Although not up to the standards of the original, this method of reproduction makes it possible to provide copies of book which would otherwise be out of print.

Find Us on the Web! http://www.apnet.com

Copyright © 1990 by Academic Press
All Rights Reserved.
No part of this publication may be reproduced or transmitted in any form or by any means, electronic or mechanical, including photocopy, recording, or any information storage and retrieval system, without permission in writing from the publisher.

ACADEMIC PRESS
A Division of Harcourt Brace & Company
525 B Street, Suite 1900
San Diego, California 92101-4495
United Kingdom Edition published by
Academic Press Limited
24–28 Oval Road, London NW1 7DX

Library of Congress Cataloging-in-Publication Data

White, Gary C.
   Analysis of wildlife radio-tracking data / Gary C. White,
     Robert A. Garrott.
     p. cm.
   Includes bibliographical references.
   ISBN 0-12-746725-4 (alk. paper)
    1. Animal radio tracking.   I. Garrott, Robert A. II. Title.
QL60.4.W45 1990
599'.0028--dc20   89-17884
                                                                      CIP

Contents

Preface xi

CHAPTER 1
Preliminaries. *1*
    Map Coordinate Systems *2*
    Entry of Data for Computer Processing *6*
    Summary *10*
    References *10*

CHAPTER 2
Design of Radio-Tracking Studies *13*
    Radio-Tracking Studies and the Scientific Method *13*
    Treatments, Controls, and Replicates *17*
    Sampling and Statistical Considerations *19*
    Field Considerations *21*
    Final Thoughts *22*
    Summary *23*
    References *23*

Contents

CHAPTER 3
Effects of Tagging on the Animal 27
    Design of Experiments to Detect Effects of Transmitters on Animals 28
    Effects of Transmitters on Animals 35
    Summary 38
    References 38

CHAPTER 4
Estimating Animal Locations 41
    Nontriangulation Location Techniques 42
        Homing in on the Animal 42
        Aerial Tracking 42
        Satellite Tracking 45
    Triangulation Location Techniques 47
        Two Receiving Stations 47
        Three or More Receiving Stations 58
        Performance of Lenth's Estimators 69
        Data Quality Control and Censoring 72
    Summary 74
    References 75

CHAPTER 5
Designing and Testing Triangulation Systems 79
    Measuring Accuracy of Directional Bearings 80
    Special Considerations for Mobile Triangulation Systems 90
    Designing the Triangulation System 94
    Summary 110
    References 111

CHAPTER 6
Simple Movements 113
    Identifying and Correcting Data Errors 113
    Still Graphics 115
    Animated Graphics 119

Migration and Dispersal *121*
Measuring Fidelity *133*
Animal Association *137*
Sample Size *140*
Summary *141*
References *141*

CHAPTER 7
Home Range Estimation *145*
Independence of Observations *147*
Minimum Convex Polygon *148*
Bivariate Normal Models *155*
   Jennrich–Turner Estimator *155*
   Weighted Bivariate Normal Estimator *160*
   Multiple Ellipses *161*
   Dunn Estimator *161*
   Testing Bivariate Normality *162*
Nonparametric Approaches *166*
   Grid Cell Counts *166*
   Fourier Series Smoothing *168*
   Harmonic Mean *170*
   Problems Common to All of the Nonparametric Methods *172*
Computer Programs for Home Range Calculation *173*
Extension of Home Range Estimators *174*
Evaluating Home Range Estimators *174*
Similarity of Home Ranges *175*
Preferred Home Range Estimator *178*
Usefulness of the Home Range Concept *178*
Summary *179*
References *180*

CHAPTER 8
Habitat Analysis *183*
Availability *183*
Utilization *185*

Contents

Preference 186
   χ^2 Analysis 186
   Marcum–Loftsgaarden Analysis 191
   Heisey's Analysis 192
   Johnson's Analysis 193
   Friedman Test 194
   Differences between Animals 196
   Selecting a Test 197
Critical Habitat 198
Accuracy of Radio-Tracking Locations 200
Home Range Approach 201
Sample Size 202
Summary 203
References 204

CHAPTER 9
Survival Rate Estimation 207

Binomial Distribution 208
Parameter Estimation by Numerical Methods 213
Program SURVIV 215
   Example 220
   More Complex Applications of Program SURVIV 222
   Assumptions Required for Use of Program SURVIV 223
Methods Incorporating Time Until Death 226
   Program MICROMORT 230
   Kaplan–Meier Method 232
Medical Survival Analysis 242
   Cox Model 244
   Logistic Regression 245
   Discriminant Analysis 248
   χ^2 Analysis 248
   Censoring 249
   Example Analysis 249
Summary 251
References 252

CHAPTER 10
Population Estimation 255
   Capture–Recapture Estimation 255
      Combining Lincoln–Petersen Estimates 257
      Example Calculations 262
      Assumptions 262
      Preferred Estimator 264
      Sample Size Determination 267
   Line Transects 269
   Aerial Surveys 269
   Summary 269
   References 270

CHAPTER 11
Data Analysis System 271
   Need for an Analysis System for Radio-Tracking Data 271
   Data Analysis System Design 273
   Current Directions 274
   Summary 275
   References 275

Appendices
   1. Introduction 277
   2. FIELDS: Radio-Tracking Data Preprocessor 293
   3. BIOCHECK: Radio-Tracking Data Checking 295
   4. BIOPLOT: Radio-Tracking Data Plotting and Editing 299
   5. HOMER: Home Range Estimation 301
   6. PC SURVIV User's Manual Version 1.4 307
   7. SAS Home Range Estimation Procedures 343
   8. Survival Estimation Computer Listing 361
   9. SAS Monte Carlo Simulation of Capture–Recapture 367

   Index 373

Preface

Advances in technology, particularly miniaturization of electronic components, have allowed wildlife biologists to remotely monitor free-ranging animals while they pursue their normal movements and activities. Transmitters, each with a unique identifying frequency, are attached to the animals, and signals from these transmitters are received by the biologists to track the animals. Such use of radio-tracking equipment to obtain biological information about animals is known as radio tracking. The design of radio-tracking studies and the analysis of data collected from such studies is the subject of this book.

For nearly 20 years we have been continually involved in studies which have relied primarily on radio-tracking techniques for the collection of data. These studies have involved a variety of North American vertebrates, such as voles, lemmings, foxes, deer, and elk, and have addressed a diversity of topics, such as daily movements, migrations, habitat use, survival, and population estimation. Like most telemetry users, we have struggled with questions about study design, sample sizes, and appropriate analytical procedures. When conducting extensive searches of the published literature to find answers to these questions, we found an abundance of information on field techniques and radio-tracking equipment but relatively little on design and analysis. The information we did find was usually scattered through a wide variety of publications and did not always satisfy our needs.

Many telemetry users have had similar experiences. The potential to extract biological information from data bases developed with radio-tracking information is often never fully realized. We feel this problem is due primarily to the lack of a generalized text which consolidates the many analytical techniques and computer programs available into a single reference. Such references have been written for capture–recapture methods (cf. White et al., 1978; Arnason and Baniuk, 1977) and line transect density estimation methods (Laake et al., 1979; Gates, 1979) and have proven extremely useful. This book is an attempt to produce a similar reference for those using radio-tracking to investigate free-ranging animals.

We have organized the material in this text into a logical sequence of chapters, with each chapter introducing and discussing a specific telemetry topic or analytical problem. The first three chapters deal with designing radio-tracking studies and the mechanics of data collection, and we recommend that all readers, regardless of their research interests and experience, review these chapters. Chapters 4 and 5 discuss the topics of estimating an animal's position and triangulation and will be of interest to anyone concerned with spatial data. In Chapter 6 we discuss a number of topics dealing with animal movements, such as graphic presentation, migration, dispersal, fidelity, and animal association, which have received little attention with respect to the development of analytical tools. Chapters 7 through 10 are each devoted to discussing a particular analytical problem which has a relatively large body of literature associated with it, such as home range estimation, habitat utilization, and estimating survival rates and population size. In the final chapter (11) we identify topics we feel would be fruitful areas for future research and discuss the need for a portable, comprehensive software package for the analysis of radio-tracking data.

The recent advances in microcomputer technology have provided researchers and biologists the opportunity to obtain powerful machines at a minimum cost. Hence, most scientists and managers employing radio-tracking techniques also have access to a microcomputer, whereas five years ago, access to a mainframe was often unavailable. Therefore, throughout this text we have emphasized software that can be executed on IBM PCs and compatibles or Apple machines that have DOS software compatibility. Because of the quantitative nature of the material, the reader is expected to have had one or more introductory statistical courses and some skills in basic algebra. We have attempted, as much as possible, to tailor the text to those with such training. Occasionally, however, we have included technical presentations of analytical techniques which require relatively advanced training. We strongly encourage those without a strong quantitative background to skip over these sections.

Preface

We are grateful to the following individuals who have provided software for our use and modification: John Cary, James Dunn, Michael Samuel, and Michael Stüwe. Ken Sejkora, Eric Anderson, David Saltz, and Joel Schmutz assisted with workshops that helped develop many of the ideas presented in the text. A. William Alldredge, David R. Anderson, Richard M. Bartmann, I. Lehr Brisbin, Jr., Len H. Carpenter, John Cary, William R. Clark, R. Bruce Gill, Thomas E. Hakonson, and Michael E. Samuel also stimulated discussions that refined our thinking. Members of the FW696 class at Colorado State University and graduate students at the University of Minnesota also contributed to our enthusiasm and illuminated our thinking. I. Lehr Brisbin, Jr. and William R. Clark provided critical reviews of draft manuscripts. Ralph Franklin and D. Heyward Hamilton provided support for our research that kept us motivated to write this book. We are deeply grateful to all these people.

Neither of us could have completed this project without the ardent support of our wives, Liz and Diane. We owe them a debt for relinquishing their time to our demands.

Arnason, A. N., and L. Baniuk. 1977. User's Manual: POPAN-2, A Data Maintenance and Analysis System for Recapture Data. Department of Computer Science, University of Manitoba, Winnipeg, Canada.

Gates, C. E. 1979. LINETRAN User's Guide. Institute of Statistics, Texas A&M University, College Station, Texas. 47 pp.

Laake, J. L., K. P. Burnham, and D. R. Anderson. 1979. User's Manual for Program TRANSECT. Utah State University Press, Logan, Utah. 26 pp.

White, G. C., K. P. Burnham, D. L. Otis, and D. R. Anderson. 1978. User's Manual for Program CAPTURE. Utah State University Press, Logan, Utah. 44 pp.

CHAPTER

1

Preliminaries

Wildlife radio-tracking data are at least three dimensional, consisting of x and y coordinates in space and a t coordinate in time. Throughout this text an animal's location will be identified as the vector (x, y, or t). Occasionally, other associated information will be added to this vector, such as elevation above sea level (z), habitat type of the location, height above the ground (i.e., height of a bird in a tree), depth below the water surface (i.e., depth of a dive by a seal), or activity of the animal at the time of the location (feeding, resting, etc.). These attributes augment the basic data obtained by a wildlife tracking study, but often are the main objective of the study. Relating these variables to the x, y, or t vector is an aspect of the data analyses considered in this book.

The topic of this text is the analysis of data collected from tracking free-ranging animals. Our purpose is to provide methods to extract as much information as possible from the data. Before collection of the data begins, some basic decisions must be made as to how to attack the problem. First, a map coordinate system must be chosen for use in the study. Second, a system must be chosen for recording the data for computer processing. The selection of a software system requires that hardware be available to run it. In this chapter we discuss these questions, providing some possibilities to help make these decisions, plus making recommendations based on our experiences.

Map Coordinate Systems

The most familiar of the map coordinate systems is the system of latitude and longitude. This is a circular coordinate system, with any point on the globe delineated on the north–south axis by a latitude coordinate and on the east–west axis by a longitude coordinate. Both coordinates are measured in degrees, minutes, and seconds. The disadvantage of this coordinate system is that it is not rectangular; hence, the calculation of the distance between any two points identified by latitude and longitude is not straightforward. The advantage of this coordinate system with respect to radio-tracking studies is that it is continuous for the entire world. This is important for studies of mobile species which migrate long distances, such as pelagic birds, marine mammals, and many of the bird species which nest in temperate, boreal, and polar regions. Although radio-tracking studies of these species have been severely restricted in the past, due to technological limitations of transmitting and receiving systems, currently emerging satellite technology will undoubtedly result in an increase in investigations of these species. This is probably the best coordinate system to use for such studies which involve very large areas.

A second coordinate system, which is generally less familiar to many telemetry users, is the Universal Transverse Mercator (UTM) system (Edwards 1969, U.S. Army 1973, Snyder 1987), a worldwide rectangular coordinate system used to locate a point on the earth's surface. Because coordinates on a spherical surface cannot be on a Cartesian coordinate system without breaks or gaps, UTM coordinates are not continuous over the entire globe. Between 80° S latitude and 84° N latitude, the earth has been divided into 60 zones, each zone generally consisting of 6° of longitude (Fig. 1.1). Within each zone the coordinate system is continuous, but orientation of the coordinate systems between zones is slightly different, resulting in breaks in the coordinate system at each zone boundary. Each UTM zone is constructed around the prime meridian for the zone, that is, the longitude in the middle of the zone, with bounding meridians divisible by 6. Zones are numbered from 1 to 60, starting with zone 1 east of 180° W longitude (Fig. 1.1). Thus, the prime meridian for UTM zone 1 is 177° W longitude, and the bounding meridians are 180° to 174° W longitude. The prime meridian for UTM zone 12 is 6° × 12 zones *minus* 183°, or 111° W longitude, giving the bounding meridians of 114° and 108° W longitude. Note that longitudes west of 0° are negative, that is, 111° W longitude corresponds to −111° longitude. Because the width of a zone becomes 0 at the North and South poles, the UTM system is generally used only between 80° S latitude and

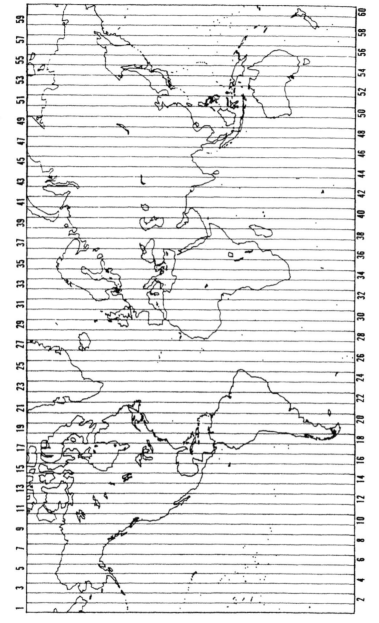

Figure 1.1 A Miller cylindrical projection map of the world illustrating the 60 universal transverse mercator zone designations.

84° N latitude. From latitudes 84° N and 80° S to their respective poles, the Universal Polar Stereographic (UPS) projection is used instead.

UTM coordinates consist of two 7-digit numbers, with units in meters. The numbers increase as one moves east and north. Within each UTM zone 500,000 m is added to the x coordinates, so that all coordinates are positive (Snyder 1987). Otherwise, all locations west of the prime meridian for the zone would have negative values. The y coordinate indicates the distance (in meters) from the equator, but, due to the zone breaks in the coordinate system, the x coordinate does not provide a similar reference. More or less precise designations are made by adding or subtracting digits. For example, the zone 13 coordinates (0291923, 4441087) would locate a square meter in northwestern Colorado. However, (0291, 4441) would be less precise and would locate only a square kilometer, because the last three digits of each coordinate have been dropped.

Most radio-tracking studies are regional in nature, involving relatively small areas which usually fall within one UTM zone. The UTM coordinate system is ideal for recording locations in this situation for two reasons. First, it is based on metric measurements, the universal standard for scientists. Second, and more importantly, the UTM system within a zone provides a continuous Cartesian coordinate system, allowing easy calculation of the distances between points and simplifying the calculations used in triangulation, when bearings from different locations are used to estimate an animal's location. A UTM grid can easily be superimposed over any study area covered by the U. S. Geological Survey 7.5-minute topographical maps. At the edge of each map there are small tick marks labeled with the UTM coordinates. By drawing north–south and east–west lines connecting tick marks with the same coordinates, a 1-km UTM grid is created over the map. Information on the UTM zone and declination of the UTM coordinate system from true north are printed at the lower left margin of the maps (Fig. 1.2).

With bad luck, one may end up with a zone change directly across the study area. The quick approach to solving this problem is to extend one of the zones to cover the entire study area. To do this, convert the coordinates from one zone into longitude and latitude, and then convert them into UTM coordinates for the zone of interest. However, if the study area is large, then serious errors can be introduced with this approach. A more complex but less error-prone approach is to select a meridian through the center of the study area as a baseline for a new UTM coordinate system.

A number of computer programs are available that will convert locations from one coordinate system to another. Snyder (1987) provides algorithms for

1 Preliminaries

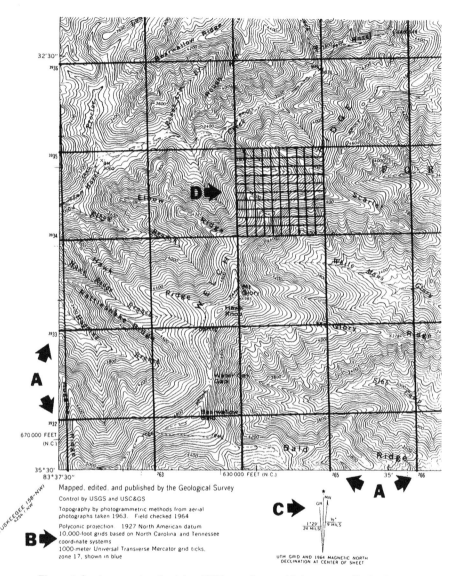

Figure 1.2 An example of a 1-km UTM coordinate grid drawn on a U. S. Geological Survey 7.5-minute topographical map. (A) Locations of UTM tick marks and their coordinates. (B and C) Locations of information on UTM zone and declination of the coordinate system from true north, respectively. (D) A 100-m grid on drafting film that can be used to obtain more precise locations.

converting between coordinate systems that are commonly used by the U.S. Geological Survey, with numerical examples of conversions provided in an appendix. Tucker and Campbell (1976) provide a set of FORTRAN subroutines for the conversion of locations between coordinate systems. Dodge et al. (1986) provide a program for laptop and personal computers (UTMTEL) that can convert latitude and longitude coordinates to UTM coordinates.

One commonly encountered system that cannot be converted is the township–range land-mapping system used in the United States. Because this system is implemented on the ground through surveys, errors have occurred. Hence, direct conversion to another system is not possible without a database to supply the corrections. Because of the survey errors inherent in the system, distances between locations cannot be reliably computed. Hence, we do not recommend that radio-tracking locations be recorded using the land survey system.

Entry of Data for Computer Processing

Preparing wildlife radio-tracking data for computer analysis requires that the information be entered into a computer database. Many database packages are available to perform this data entry. Before selecting a package, the user must be certain that the package is able to manipulate the data as necessary to perform the analyses described in this text, or at least that the package can easily export the database to a statistical package that can perform the manipulations. Some of the requirements for the database package are that dates and times can be easily entered and manipulated and that error-checking routines can be implemented to check the data as they are entered. The package must be able to handle the volume of data that will be entered; that is, database packages that maintain the entire mass of data in the computer's memory may not be adequate for a large radio-tracking study. Finally, some simple graphics capabilities are helpful in visually verifying data immediately after entry.

The x, y, and t variables are usually the most important values entered in a computer database. In addition, other variables that should be recorded are the animal's identity (usually the transmitter's frequency), age, and sex. Age and sex are often used as covariates or classification variables needed for analyses and do not require much storage capacity. A code which specifies the type of information recorded for the entry should also be included. For example, the place of initial capture should be differentiated from a routine location. The final location (e.g., the site of an animal's death) should be differentiated for other records. Another example is the last location recorded for an animal before the radio failed. The type of information code would be used to determine whether the records were to be used for home range estimation (routine locations) or survival rate estimation (chronologically, the first and last records).

Entry of the date and the time for computer processing is more complex than entry of the x and y coordinates. Although the Julian date is usually used for analysis, we do not recommend entering data as Julian dates because of the difficulty in visually interpreting values without a ready conversion table. Rather, we suggest that the date and the time be entered as *yymmdd.hhmm*, where *yy* is the last two digits of the year, *mm* is the month (01–12), *dd* is the day of the month (01–31), and the decimal point separates the day from the fraction of a day. The hour (*hh*) is recorded on a 24-hour clock (00–23), and the minute (*mm*) can have values from 00 to 59. An important reason for recording the date and the time in this format is that the file can be sorted chronologically by time using these 11 columns. The database package must be able to compute from these values the amount of time between two locations.

We advise that the following variables be placed in the file with this order:

Variable	Content
ID	Animal identity (usually radio frequency)
TIME	Date and time in *yymmdd.hhmm* format
XCOOR	UTM *x* coordinate (seven digits)
YCOOR	UTM *y* coordinate (seven digits)
AGE	Age of animal, usually an integer or letter code
SEX	Sex of animal, either a numerical or letter code
TYPE	Type of observation (initial capture, location, death, etc.), usually an integer or character code
OTHER(S)	Additional information pertaining to this record (cause of death, weak radio signal, etc.)

An ASCII file in this format can be sorted according to animal identity and time of location using the default SORT command available on most operating systems, although we would use the database system used for data entry for this task. Other information of importance can be entered on the remainder of the record. Specific types of information can be extracted, such as initial capture locations, routine locations of adults, or only locations of deaths of juveniles.

A special problem arises in radio-tracking studies in which the same radio is placed on more than one animal over the course of the study. In such a case a separate identification code might be used for each animal, and two variables are recorded in the file: the animal's identification code and its transmitter's frequency. Another complication is also resolved with this convention. An animal can be recaptured and its transmitter replaced with one of a different frequency. Thus, the animal's identity is not lost when the transmitter frequency changes. The need for this additional variable should be determined in the early stages of the study and the data file constructed accordingly.

In this text, we will emphasize the use of the SAS (SAS Institute Inc. 1985) PC system because of its flexibility and power for programming the various analyses discussed as well as the ability to perform separate analyses by individuals or other group categories. PROC SORT provides a convenient sorting mechanism to reorder the data file. SAS provides the computer functions necessary to convert dates and times into Julian dates and back again for printing in the output. Sorting of the file can be performed using the dates without special handling of the data. Time between two locations can be computed in a variety of units. SAS's programming language allows computing distances traveled between consecutive locations, or even the creation of a variable containing the total distance traveled during specified intervals. We use SAS because of the availability of high-quality graphics and a variety of statistical analyses. Often, example analyses will be demonstrated in the text with SAS code. SAS provides all of the necessary routine statistical analyses required to summarize wildlife tracking data, including the capability of summarizing the data by individual animals or categories of animals with the BY statement.

In addition, programming analyses are less complicated to do with the SAS DATA step programming language than with a computer language such as FORTRAN or BASIC, primarily because a flexible procedure is provided to input the data and the facilities to sort and manipulate the data are available to the programmer as part of the system. Utilities are also available for both line printers and higher quality output devices, which easily provide visual representations of data. The fundamental reason that we have selected SAS for use in this text over other data analysis systems is the power of the programming language. We are not aware of any other statistical package that provides the potential in its language to program the triangulation methods, home range estimators, or survival analyses presented in this text as SAS programs. SAS is also available on a variety of computers and runs under more than just the DOS operating system generally available on personal computers. Finally, SAS provides an adequate database management system to handle a large radio-tracking project.

Because of the specialized nature of many of the procedures presented,

1 Preliminaries

some other specialized programs have become popular, which we discuss as required. However, we do restrict ourselves to software that operates on a personal computer; that is, we do not discuss some of the older packages that are limited to a specific mainframe environment. A brief description of some of the software used in this text appears in Table 1.1. For other packages, the reader will need to acquire the latest version from the software's author or from

TABLE 1.1
Software Packages of Use in Radio-Tracking Data Analysis on a Personal Computer

Program	Purpose
PC SAS	General statistical analysis system that allows complex programming as well as graphical and statistical analyses (SAS Institute Inc. 1985)
TRIANG	FORTRAN (also BASIC) program to convert bearings to a radio signal from two or more known locations to an estimate of the radio's location (White and Garrott 1984)
XYLOG	BASIC program to convert bearings to a radio signal from two or more known locations to an estimate of the radio's location (Dodge et al. 1986)
UTMTEL	Laptop computer BASIC program for calculating UTM coordinates from latitude/longitude (Dodge et al. 1986)
BIOPLOT	Interactive plot of spatial data on screen for error checking (described in text)
BIOCHECK	Error detection via distances moved and/or speed limit (described in text)
FIELDS	Prepare data for input to BIOPLOT, BIOCHECK, and HOMER (described in text)
HOMER	Home range estimation (described in text)
McPAAL	Home range estimation (available from Michael Stüwe, Smithsonian Institute, Washington, D.C.)
DC80	Home range estimation (available from John Carey, University of Wisconsin, Madison, WI)
TELEM/PC	Home range estimation (available from Steve Sheriff, Missouri Department of Conservation, Columbia, MO)
HOME RANGE	Home range estimation (available from E. O. Garton, University of Idaho, Moscow, ID)
SURVIVE	Survival rate estimation, comparing groups (White 1983)
MICROMORT	Survival rate estimation, comparing rates by cause of death (Heisey and Fuller 1985)
LOTUS 1-2-3	Plotting radio-tracking data (available commercially)
Quattro	Plotting radio-tracking data (available commercially)
FREELANCE	Publication quality graphics of data from Lotus 1-2-3 or Quattro plots (available commercially)

an electronic bulletin board. One bulletin board that specializes in wildlife- and fishery-related software is SESAME, located at Raleigh, North Carolina.

Summary

1. Radio-tracking data are three dimensional, consisting of two spatial coordinates (x and y) and one time coordinate (t).
2. The UTM projection system is preferred for representing the spatial coordinates and is assumed throughout the text.
3. All of the analyses presented can be performed on an IBM personal computer or an IBM-compatible computer with the DOS operating system.
4. The structure of the data file to record wildlife radio-tracking data should include at least the animal's identity, age, and sex; the date and time of the observation; the type of observation; and the UTM coordinates of the animal's location. Other information might also be required, depending on the objectives of the study.
5. The SAS programming, statistics, and graphics packages are used extensively throughout the text to perform the analyses, and we provide an example code.
6. Additional programs are discussed that perform specific tasks for the analysis of wildlife radio-tracking data.

References

Dodge, W. E., D. S. Wilkie, and A. J. Steiner. 1986. UTMTEL: A laptop computer program for location of telemetry "finds" using LORAN C. Massachusetts Cooperative Wildlife Research Unit, Amherst. 21 pp.

Edwards, R. L. 1969. Archaeological use of the Universal Transverse Mercator grid. Am. Antiquity 34:180–182.

Heisey, D. M. and T. K. Fuller. 1985. Evaluation of survival and cause-specific mortality rates using telemetry data. J. Wildl. Manage. 49:668–674.

SAS Institute Inc. 1985. SAS® Language Guide for Personal Computers, Version 6 Edition. SAS Institute Inc., Cary, NC. 429 pp.

Snyder, J. P. 1987. Map projections—a working manual, Prof. Pap. 1395. U.S. Geological Survey, Washington, D.C. 383 pp.

Tucker, T. C. and L. J. Campbell. 1976. CATCH: Computer assisted topography, cartography and hypsography. Part 2. MAPPROJ: A subroutine package for a number of common map projections, ORNL/TM-3790. Oak Ridge Natl. Lab., Oak Ridge, TN.

U.S. Army. 1973. Universal Transverse Mercator Grid, TM 5-241-8. Headquarters, Department of the Army, Washington, D.C. 64 pp.

White, G. C. 1983. Numerical estimation of survival rates from band recovery and biotelemetry data. J. Wildl. Manage. 47:716–728.

White, G. C. and R. A. Garrott. 1984. Portable computer system for field processing biotelemetry triangulation data. Colo. Div. Wildl. Game Inf. Leafl. 110:1–4.

CHAPTER

2

Design of Radio-Tracking Studies

Radio-tracking provides a useful technique for studying the mechanics of wildlife populations. Movements (Chapter 6) provide information on how animals use the environment, migration patterns, dispersal, and activity patterns. Home range estimates (Chapter 7) quantify the area used by an animal. Habitat use studies provide information on habitat preference and, if properly defined, can provide information on the need for various habitat types (Chapter 8). Survival studies provide estimates of mortality rates (Chapter 9), and population estimation (Chapter 10) studies estimate the number of animals in the population. However, whether radio-tracking should be used in a study depends on the objectives of the study, the type of data to be collected, and the constraints put on the investigator regarding funding, field conditions, equipment limitations, and the species under study. The purpose of this chapter is to discuss the design of wildlife radio-tracking studies, with particular emphasis on conducting experiments to demonstrate cause-and-effect relationships.

Radio-Tracking Studies and the Scientific Method

A key step in the implementation of any radio-tracking study is careful design. Although this may seem obvious to most readers, it is an extremely important point which is occasionally overlooked or ignored by telemetry users. Sargeant (1980:58) stated, "It is likely that more money and effort have been wasted on ill-conceived radio-tracking studies than on the use of any other field technique." It appears that the reason for these "ill-conceived" studies is the general attitude of some investigators that, by placing transmitters on a handful of animals and "tracking" them, one is guaranteed to obtain good biological data. This fact could not be further from the truth. Radio-tracking is nothing more than a specialized technique available to the investigator along with hundreds of other techniques for collecting information. A similar opinion has been expressed by Lance and Watson (1979:113):

> . . . the purely descriptive study based on no apparent hypothesis and in which the objective is merely general information. Such a use of radio-tracking might be legitimate for exploring a wholly unknown subject, but these are few in ecology nowadays, and miscellaneous data-gathering is a poor reflection of the technique's true potential.

To avoid this pitfall when planning a radio-tracking study we suggest applying the five steps of the basic scientific method. The first step is defining the scope of the problem to be addressed. Some studies may have very specific objectives and, hence, a detailed definition of the problem can be formulated at the very outset of the planning process. Often, however, the scope of a study is quite broad, leading to a very general definition. The second step is to study existing information that pertains to the defined problem. This can be a tedious and time-consuming process if a large body of literature has been published in the field of interest, but the time is well spent. Reviewing the results of previous studies allows one to become familiar with the current "state of knowledge" and aids in the formulation of specific hypotheses, the third step in the scientific method. This is an important step in the planning process, as explicitly defining each hypothesis to be tested *before* the study begins allows the investigator to design data collection procedures and statistical tests that will optimize one's chance of obtaining conclusive results. Once methodology and sample size decisions have been made, the fourth step, data collection, is initiated. The final step in the scientific method is careful analysis of the data, allowing the investigator to draw conclusions about rejecting the hypotheses tested.

Wildlife radio-tracking studies can be divided into three basic conceptual designs: descriptive, correlational, and manipulative. The procedures outlined in the scientific method can be applied in all three designs, but are routinely used only in manipulative studies. *Descriptive studies* use radio-tracking to observe natural behavioral processes, usually with no attempt at hypothesis formulation or testing before the data are collected. Such studies are very common among telemetry users and include general investigations of home range size and shape and seasonal and daily movements. These studies are useful in learning more about the natural history of a species, but are limited to learning what an animal does, not why the animal is doing it (Sanderson 1966). To illustrate this point, consider a study of habitat utilization. A group of instrumented ungulates is located at random times throughout the day to determine the habitats used. Thus, the primary result of this descriptive study is the proportion of time an individual animal spends in each habitat type. The investigator now knows the animal's location, but not why the animal used a particular habitat type or what it was doing while in that habitat.

A more informative approach would be to formulate one or more hypotheses and design a *correlational study* to determine whether the behavior of the monitored animals supports this hypothesis. For example, the instrumented animals have available three habitat types, a grass–forb community, a low shrub community, and a closed-canopy forest community. One possible hypothesis is that as the ambient air temperature increases, the ungulates select habitats with a vegetative structure that provides thermal cover (i.e., reduced temperatures). To test this hypothesis, the investigator monitors the microclimate temperatures in each of the habitat communities while simultaneously tracking the ungulates. Results of the study may support the hypothesis, in that the temperatures within the forest community were consistently lower than the temperatures in the grassland and shrub communities. In addition, as the temperatures rose in the nonforested communities, the animals tended to avoid these habitats and occupy the forested community. In this example, a correlation is developed between elevated air temperatures and use of the forest habitat type, providing the investigator with more information than a descriptive study. However, the investigator must be aware that in a correlational study, a relationship between two or more variables does not imply cause and effect. In other words, the data do not indicate that higher air temperatures caused the animals to move from the grassland and shrub communities to the forest community, although a correlation is present. Other factors not monitored during the study may have been responsible for the observed response. In our temperature–ungulate example, lack of harassment from parasitic insects, rather than air temperature, may cause the animal to use the forest habitat. Similarly, evidence that a habitat was selectively used (i.e., occupied in higher proportion than its availability) does not imply that the habitat is "critical" to the animal's well-being.

Manipulative studies require that the system be perturbated. In order to conclusively demonstrate that a particular habitat type is needed to insure the welfare of a population, that habitat must be made unavailable to the population and the decline in the individual's health and/or the fitness of the population demonstrated. Thus, the study design requires an experimental approach, with appropriate manipulations (treatments), controls, and replicates. One experimental approach to the ungulate habitat example would be to monitor microclimate temperature in each of the three habitats on four study areas while simultaneously tracking instrumented animals. After the preliminary data are collected, showing a correlation between increased air temperatures and use of the forest habitat, two of the study areas are randomly selected for treatment. Ideally, the treatment should only involve altering the variable of interest, in this case, artificially increasing the air temperature within the forest habitat. A more practical treatment, however, may be removal of the forest canopy. Data on air temperature and habitat use would again be collected on all four areas and results compared between times for each study area and between the treatment and control areas. If the data from the controls were similar between monitoring periods, there were no effects due to time, and any changes in the treatments could be attributed to the perturbation.

The above example is an over-simplification, but it illustrates the importance of manipulative studies. While descriptive and correlational studies may provide insights into why animals behave in a particular manner, manipulative experiments, including both time and geographic controls, are needed to obtain conclusive results about why animals behave as they do.

Consider another, even simpler, example of the importance of experimental manipulations as opposed to correlational studies. You are watching television, when the screen begins to scroll. Two possible explanations (hypotheses) are: (1) your television failed or (2) the broadcasting station failed. How do you decide which of these hypotheses is incorrect? Without hesitating, you switch channels, performing the manipulative experiment to distinguish between the two possible hypotheses. Without the manipulation of changing channels, you have no information to decide between the two explanations. Even in everyday life, we use experimental manipulations to collect information!

Let's extend the example even further. Suppose your TV screen begins to scroll, and you hear an airplane passing overhead. An approach based on purely observational methods would result in the conclusion that the airplane caused the problem. By switching channels, you could determine whether your TV was the cause of the problem, but not whether the airplane was the ultimate cause. Only by manipulating the airplane could this cause-and-effect mechanism be established.

Treatments, Controls, and Replicates

Both temporal and spatial controls are needed to demonstrate that a treatment has an effect (see Green 1979 for a good review of this process). Consider the hypothetical experiment described in Fig. 2.1. The purpose of this diagram is to stress the danger when only a treatment area is being monitored, that is, no spatial control is included in the study. Two scenarios are depicted—one in which the treatment has a significant effect on the survival rate (Fig. 2.1A), and the second in which there is no significant effect of the treatment relative

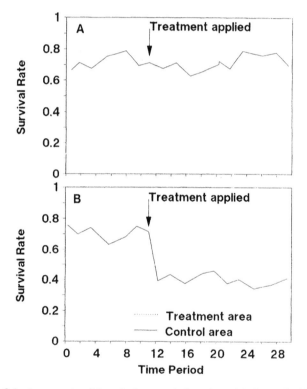

Figure 2.1 An example of the mistaken conclusions that might be reached if an experimental design lacks a spatial control. (A) The two study areas have about the same survival rate prior to the treatment. After the treatment, the survival rate on the treatment area drops, while the control remains the same, suggesting a treatment effect. (B) The treatment had no effect, but due to the drop in survival on both the treatment and control areas from an extraneous factor, the researcher would have concluded that the treatment did affect survival, had the control area not been measured.

to the control (Fig. 2.1B). However, had the control area not been monitored in the second scenario, the investigator would have concluded that a significant effect had taken place.

Now consider the alternate problem, that is, a spatial control but no temporal controls, depicted in Fig. 2.2. By applying the treatment to one area without prior monitoring for differences, the experimental protocol suffers from the possibility that the two areas are not identical except for the treatment effects. Thus, in Fig. 2.2A, if the researcher did not monitor survival on the control area prior to the treatment, and only monitored the differences between

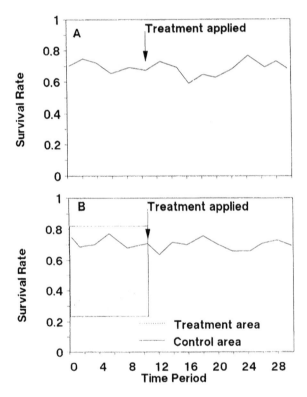

Figure 2.2 An example of the mistaken conclusions that might be reached if an experimental design lacks a temporal control. (A) The treatment and control areas have differences prior to the treatment. These differences are greater after the treatment, suggesting a treatment effect. (B) The treatment had no effect. Due to differences between the treatment and control areas, the researcher would have concluded that the treatment affected population survival if measurements in the shaded area were not taken prior to the treatment.

the treatment and control area after the treatment is applied, he would conclude that differences existed because of the treatment effect. In fact, the differences were apparent between the two study areas prior to the treatment, which had no effect. In contrast, Fig. 2.2B shows a situation in which there is a difference between the two study areas, but the treatment effect enlarges this difference.

In both Figs. 2.1 and 2.2, only a single pair of areas is shown, with the pair consisting of a treatment area and a control area. In fact, a single pair of areas only allows inferences to be drawn about the one pair of areas, not inferences about the general effect of the treatment to all possible areas. We refer the reader to Hurlbert (1984) and Stewart-Oaten et al. (1986) for excellent discussions of statistical problems associated with the design of perturbation studies. As a general rule, without replication of treatment–control pairs of study areas or of populations, the researcher can only draw inferences to the specific areas or populations studied. Given the difficulties of replicating entire studies, it is not surprising that few experimental field studies are conducted with adequate replicates. Much of the replication of areas and populations, therefore, will have to take place in the literature through publication of numerous studies of similar nature. Eberhardt (1988) suggests that one way of obtaining adequate replication in the field of wildlife management and research is to use the extensive number of management units within a state as individual sampling units. Whether or not adequate replication is possible for a given study, we feel strongly that, at a minimum, experimental designs must include both temporal and spatial controls so that proper inferences about at least the specific areas or populations can be drawn.

Sampling and Statistical Considerations

Another aspect which must be considered when designing radio-tracking studies is sampling. Design of a proper sampling scheme causes the investigator to consider the definition of the statistical population being studied, the size of the sample to be taken, and the probability that the experiment will return a correct conclusion (i.e., the power and Type I error rate of the experiment).

Because few automated systems are available, most investigators spend large amounts of time in the field manually collecting data. The intensive fieldwork required to collect data limits the number of animals that can be tracked and the frequency with which data can be collected. Hence, it is essential for the investigator to carefully plan the sampling scheme for the study. If variability between sex and/or age classes can be expected for the parameters to be estimated, and only a limited number of animals can be tracked, the investigator should consider instrumenting only one sex–age class. Instrumenting the first animals captured often results in studying small numbers of each sex–age class, making definitive conclusions difficult. A solid conclusion based on an experiment with high power may be possible if only a segment of the animal population is studied. In contrast, if all segments are studied at once, with the available sampling effort distributed accordingly, no one portion of the population may be adequately studied.

Appropriate statistical procedures should be selected prior to beginning the field portion of a wildlife radio-tracking study, in order to calculate the sample sizes needed to gain reasonable statistical power and/or precision. The power of an experiment is the probability that a false null hypothesis will be rejected. That is, given that the null hypothesis is false, we want to be sure that the statistical test we will be using is going to have adequate data to reject the null hypothesis. The power of a statistical test is a function of the sample size (amount of data). Thus, determining how much data must be collected requires that the statistical test be identified and the power evaluated for different sample sizes. An experiment with low power or a statistic with a large confidence interval provides little or no conclusive information.

If the data from the instrumented animals are to be extrapolated to the total population or a larger group of animals, the instrumented animals should represent a random sample from the larger group. That is, the statistical population to be studied must be defined and a representative sample of the animals drawn from this overall population. True random samples are usually difficult to achieve, but the investigator should strive to sample animals in as unbiased a manner as possible. For example, a random sample of the statistical population must be collected for attaching transmitters. Most trapping methods take what comes to the trap, not a particularly random sample of the population. O'Gara and Harris (1988) demonstrated differences in the condition and age of mule and white-tailed deer killed by automobiles and by predators. Garrott and White (1982) showed differences in age and sex ratios of mule deer, depending on how the Clover trap was baited and the height of the door.

Next, the investigator must sample the animals with radios over time. Most analytical procedures also require that individual observations be independent, which is a reflection of sampling frequency. In many types of radio-tracking studies, independence of observations can usually be assumed. However, in studies in which data are collected intensively, such as monitoring daily movements, activity, and habitat use, the investigator must try not to artificially

2 Design of Radio-Tracking Studies

inflate sample sizes by sampling so frequently that independence of the observations becomes questionable. All of these factors must be balanced in order to determine a sampling scheme that will satisfy the requirements for a sound study, while recognizing financial and logistical limitations.

Field Considerations

Once the objectives, conceptual design, statistical procedures, and sampling scheme of the study have been outlined, a radio-tracking system must be designed that will meet the requirements of the study. At this point in the planning, there are a multitude of technical decisions to be made that will directly affect the performance of the system and, hence, the success of the study. If the investigator is not well versed in the technical aspects of radio-tracking equipment, he should seek advice from experienced telemetry users as well as vendors of this equipment. An initial decision must be made on the frequency of the radio-tracking system which, in turn, affects many of the system's physical and operating characteristics. Brander and Cochran (1971), Cochran (1980), Sargeant (1980), Mech (1983), and Kenward (1987) provide informative discussions on this topic, and Kolz (1983) discusses regulations governing the use of frequencies. Basic transmitter characteristics that must be considered include whether the generated signal will be continuous or pulsed (if pulsed, what will be the optimum pulse width and length?); transmitter life; weight, volume, and shape of the unit; how the transmitter will be housed to protect the components from the environment; frequency separation between individual transmitters; and the type of antenna to be used. In addition to the basic transmitter, circuits can be incorporated which can measure a variety of environmental (e.g., motion, temperature, pressure, pH, light) and physiological (e.g., heart rate, body temperature) parameters. In addition to specifications for the transmitter itself, the method of attaching the transmitter to the animal must be decided. The unit may be carried externally on collars, harnesses, ear tags, and similar devices, or the transmitter may be carried internally via implantable or ingestible packages. Throughout the entire process of designing the transmitter-attachment unit, care must be taken to minimize the potential impacts of the instrument package on the animals (see Chapter 3).

While a wide variety of factors must be considered in designing the transmitter package, only two factors are generally important when designing the receiving system: accuracy and cost. The most commonly used radio-tracking receiving systems are hand-held, mobile, and manually operated fixed towers.

Less commonly used systems include automated fixed towers (Cochran et al. 1965, Deat et al. 1980, Lemnell et al. 1983), grids (Chute et al. 1974, Cunningham et al. 1983), and satellites (Buechner et al. 1971, Kolz et al. 1980, Jennings and Gandy 1980, Fancy et al. 1988). The abilities of these receiving systems to accurately locate a transmitter within a given study area vary tremendously. There is also wide design latitude within each type of system that results in additional variation in accuracy. So, in order for the investigator to design a receiving system appropriate for the study, the range of acceptable accuracy must be defined. For example, if the objective of the study is to determine the rates and the causes of mortality in a population of game birds, a hand-held receiving system would, in all likelihood, be sufficient to monitor the status of the animals and locate dead birds.

If, however, the objective of the study is to determine how the birds utilize the available habitats, a more accurate receiving system, such as fixed towers, may be required in order to assign a habitat type to each animal location. All too often, the receiving system that is most convenient is employed rather than first determining what receiving system would provide the accuracy needed to gain meaningful data. Unfortunately, designing a receiving system to provide a specified accuracy is not easily accomplished, because the accuracy is usually unknown until it is actually tested on the study area. General guidelines are available, however, that will give an investigator some indication as to the type and the complexity of the receiving system needed to meet specified requirements (see Chapter 4). Cederlund et al. (1979) provide a good comparison of hand-held, mobile, and fixed tower receiving systems used on a common study area, and numerous papers are available in the literature which discuss the accuracy (precision and bias) of specific receiving systems (Heezen and Tester 1967, Springer 1979, Hupp and Ratti 1983, White 1985, Lee et al. 1985, Garrott et al. 1986).

Final Thoughts

The process of designing a radio-tracking study is iterative. The objectives suggest the population to be sampled, which suggests the sampling procedures, which suggests the statistical analysis of the data, all of which are influenced by the field techniques. In reality, optimal experiments can only be designed when the answer is already known! In practice, the investigator works through the series of decisions to design a study numerous times, because the interrelationships between the components outlined in this chapter force this kind of process.

Summary

1. Cause-and-effect relationships can only be determined through manipulative experiments; that is, the ecological system must be perturbated in order to demonstrate conclusively that a cause-and-effect relationship exists between two variables. Wildlife tracking techniques lend themselves particularly well to manipulative experiments, but unfortunately have seldom been used for such research.
2. The statistical analysis of manipulative experiments is not a simple task. Numerous pitfalls await the investigator who is unaware of the problems of replication, randomization, and the proper construction of the hypothesis to be tested.
3. To properly sample a wildlife population, the population must be rigorously defined. Given that the population has been defined, a representative (presumably random) sample must be taken. In reality, investigators may have difficulty in delineating a wildlife population from which to draw a statistical sample of animals and, further, can seldom take a random sample of the members of the population with available capture techniques.
4. Prior to beginning the field portion of a wildlife tracking study, the hypothesis to be tested and the power of the statistical procedure to test this hypothesis should be determined.
5. Often, the statistical design of a wildlife tracking study will determine the field techniques used. A wide range of field equipment and procedures exists, so that usually an appropriate protocol is obtainable.

References

Brander, R. B. and W. W. Cochran. 1971. Radio-location telemetry. Pages 95–105 *in* R. J. Giles ed. Wildlife Management Techniques, 3rd ed. The Wildlife Society, Washington, D.C.

Buechner, H. K., F. C. Craighead, Jr., J. J. Craighead, and C. E. Cote. 1971. Satellites for research on free-roaming animals. BioScience 21:1201–1205.

Cederlund, G., T. Dreyfert, and P. A. Lemnell. 1979. Radiotracking techniques and the reliability of systems used for larger birds and mammals. Swedish Environ. Protection Board, Solna, pm 1136. 102 pp.

Chute, F. S., W. A. Fuller, P. R. J. Harding, and T. B. Herman. 1974. Radio tracking of small mammals using a grid of overhead wire antennas. Can. J. Zool. 52:1481–1488.

Cochran, W. W. 1980. Wildlife telemetry. Pages 507–520 *in* S. D. Schemnitz, ed. Wildlife Management Techniques, 4th ed. The Wildlife Society, Washington, D.C.

Cochran, W. W., D. W. Warner, and J. R. Tester. 1965. Automatic radio-tracking system for monitoring animal movements. BioScience 15:98–100.

Cunningham, C. R., J. F. Craig, and W. C. Mackay. 1983. Some experiences with an automatic grid antenna radio system for tracking freshwater fish. Pages 135–149 *in* D. G. Pincock ed. Proc. 4th Int. Wildl. Biotelemetry Conf. Applied Microelectronics Institute and Technical Univ. of Nova Scotia, Halifax.

Deat, A., C. Mauget, R. Mauget, D. Maurel, and A. Sempere. 1980. The automatic, continuous and fixed radio tracking system of the Chizé Forest: theoretical and practical analysis. Pages 439–451 *in* C. J. Amlaner, Jr. and D. W. Macdonald, eds. A Handbook on Biotelemetry and Radio Tracking. Pergamon Press, Oxford, England.

Eberhardt, L. L. 1988. Testing hypothesis about populations. J. Wildl. Manage. 52: 50–56.

Fancy, S. G., L. F. Pank, D. C. Douglas, C. H. Curby, G. W. Garner, S. C. Amstrup, and W. L. Regelin. 1988. Satellite telemetry: a new tool for wildlife research and management, Resour. Publ. No. 172. U.S. Fish and Wildl. Serv. Washington, D.C. 54 pp.

Garrott, R. A. and G. C. White. 1982. Age and sex selectivity in trapping mule deer. J. Wildl. Manage. 46:1083–1086.

Garrott, R. A., G. C. White, R. M. Bartmann, and D. M. Weybright. 1986. Reflected signal bias in biotelemetry triangulation systems. J. Wildl. Manage. 50:747–752.

Green, R. H. 1979. Sampling Design and Statistical Methods for Environmental Biologists. Wiley (Interscience) New York. 257 pp.

Heezen, K. L. and J. R. Tester. 1967. Evaluation of radio-tracking by triangulation with special reference to deer movements. J. Wildl. Manage. 31:124–141.

Hupp, J. W. and J. T. Ratti. 1983. A test of radio telemetry triangulation accuracy in heterogeneous environments. Pages 31–46 *in* D. G. Pincock, ed. Proc. 4th Int. Wildl. Biotelemetry Conf. Applied Microelectronics Institute and Technical Univ. of Nova Scotia, Halifax.

Hurlbert, S. H. 1984. Pseudoreplication and the design of ecological field experiments. Ecol. Monogr. 54:187–211.

Jennings, J. G. and W. F. Gandy. 1980. Tracking pelagic dolphins by satellite. Pages 753–756 *in* C. J. Amlaner, Jr. and D. W. Macdonald eds. A Handbook on Biotelemetry and Radio Tracking. Pergamon Press, Oxford, England.

Kenward, R. E. 1987. Wildlife Radio Tagging. Academic Press, San Diego, CA. 222 pp.

Kolz, A. L. 1983. Radio frequency assignments for wildlife telemetry: a review of the regulations. Wildl. Soc. Bull. 11:56–59.

Kolz, A. L., J. W. Lentfer, and H. G. Fallek. 1980. Satellite radio tracking of polar bears instrumented in Alaska. Pages 743–752 *in* C. J. Amlaner, Jr. and D. W.

Macdonald eds. A Handbook on Biotelemetry and Radio Tracking. Pergamon Press, Oxford, England.

Lance, A. N. and A. Watson. 1979. Some strength and limitations of radio-telemetry for research on grouse. Pages 112–114 *in* T. W. I. Lorel ed. Woodland Grouse Symposium. World Pheasant Association, Suffolk.

Lee, J. E., G. C. White, R. A. Garrott, R. M. Bartmann, and A. W. Alldredge. 1985. Assessing the accuracy of a radiotelemetry system for estimating mule deer locations. J. Wildl. Manage. 49:658–663.

Lemnell, P. A., G. Johnsson, H. Helmersson, O. Holmstrand, and L. Norling. 1983. An automatic radio-telemetry system for position determination and data acquisition. Pages 76–93 *in* D. G. Pincock ed. Proc. 4th Int. Wildl. Biotelemetry Conf. Applied Microelectronics Institute and Technical Univ. of Nova Scotia, Halifax.

Mech, L. D. 1983. Handbook of Animal Radio-Tracking. Univ. of Minnesota Press, Minneapolis. 107 pp.

O'Gara, B. W. and R. B. Harris. 1988. Age and condition of deer killed by predators and automobiles. J. Wildl. Manage. 52:316–320.

Sanderson, G. C. 1966. The study of mammal movements—a review. J. Wildl. Manage. 30:215–235.

Sargeant, A. B. 1980. Approaches, considerations and problems associated with radio tracking carnivores. Pages 57–63 *in* C. J. Amlaner, Jr. and D. W. Macdonald eds. A Handbook on Biotelemetry and Radio Tracking. Pergamon Press, Oxford, England.

Springer, J. T. 1979. Some sources of bias and sampling error in radio triangulation. J. Wildl. Manage. 43:926–935.

Stewart-Oaten, A., W. W. Murdoch, and K. R. Parker. 1986. Environmental impact asessment: pseudoreplication in time? Ecology 67:929–940.

White, G. C. 1985. Optimal locations of towers for triangulation studies using biotelemetry. J. Wildl. Manage. 49:190–196.

CHAPTER

3

Effects of Tagging on the Animal

An underlying assumption of most radio-tracking studies is that the instrumented animals are moving through the environment, responding to stimuli, and behaving in a manner similar to noninstrumented animals. That is, the animals carrying transmitters are a representative (random) sample of the entire population, and the transmitters do not affect them in any way that makes their responses different from noninstrumented animals. It is reasonable, however, to expect that the researcher impacts the animals through the capture, handling, and attachment of the instrument package (Fuller 1987). Impacts may range from subtle behavioral changes which are manifested only temporarily to long-term changes that affect an animal's survival and reproduction. Whether or not these effects can be dismissed as inconsequential depends on the nature of the effects and the objectives of the study.

For example, female gallinaceous birds are commonly trapped on their nests and fitted with backpack or breast-mounted transmitters. The bird may become temporarily preoccupied with the newly fitted transmitter, spending much time in preening and comfort behaviors. This relatively subtle and short-term effect of instrumenting the bird would probably be of little consequence if the objective of the research is the determination of annual survival rates. However, if the study is to focus on reproductive success, the hen's preoccupation with the transmitter may interrupt incubation and result in decreased reproductive success. On the other hand, the transmitter package may be too heavy to be carried comfortably by the bird or the harness may be fitted poorly, resulting in decreased agility and mobility. The consequences of this effect may not appear during a short-term study of reproductive success, when the hen is relatively immobile during incubation, but could result in decreased long-term survival due to difficulty in escaping predators.

Design of Experiments to Detect Effects of Transmitters on Animals

Few investigators discuss the possible impacts of instrumenting the subject animals (cf. Fig. 3.1). If the topic is addressed, it is usually in the form of a subjective evaluation, such as "the animals appeared normal" or "the animals seemed to be unaffected." Another common approach is to assume that the transmitter package has negligible effects if the animals successfully complete

Figure 3.1 The importance of proper attachment of transmitters to animals is recognized when a case such as this male white-tailed deer (*Odocoileus virginianus*) is reported. The abrasions caused by a collar that is too tight may affect the behavior of this animal, and possibly its survival. (Courtesy of Kenneth E. Larson.)

3 Effects of Tagging on the Animal

biological or behavioral processes such as mating, establishing and maintaining a territory, or producing offspring. Such statements and rationale are weak, indicating only that the transmitter packages are not overtly deleterious to the well-being of the animals in question. In many cases, however, this is the best assessment of impacts that can be expected. It is difficult to perceive how a researcher can formally test the effects of a transmitter package used on wolverines (*Gulo gulo*) in the rugged wilderness of Montana or on giant pandas (*Ailuropoda melanoleuca*) in the jungles of China. For many species, however, measuring the effects of instrumentation may be more feasible, and in these cases some general principles may be developed.

Whether a formal study of instrumentation effects should be implemented depends on the characteristics of the investigation, such as the species to be studied, transmitter weight and configuration, attachment method, and research objectives. Placing a 400-g radio collar on an elk (*Cervus elaphus*) for the purpose of studying seasonal movement patterns could, in all likelihood, be implemented without testing the collar's effect. In contrast, placing a transmitter implant into the peritoneal cavity of voles (*Microtus* sp.) in order to study reproductive behavior would be a questionable practice unless the researcher demonstrated, through controlled experiments, that the instrumentation procedure did not affect the animals. Also, if many other investigators used the same size package on a particular species, it may be more generally assumed (possibly incorrectly!) that the impact is negligible.

The need to study the impacts of attaching a transmitter to an animal is clearly evident; however, it is often difficult to design an appropriate experiment. A well-designed experiment would compare the performance of a population of instrumented animals with a control population of non-instrumented animals. Frequently, however, radio-tracking is used to collect specific types of data which are difficult or impossible to obtain in any other way, making a control impossible. Without a control, it is difficult to demonstrate that attaching a transmitter to an animal has no effect. Thus, it is generally impossible to demonstrate that tagging an animal with a transmitter has no effect. Part of this dilemma is due to the logic of the situation. We can only demonstrate an effect. Hence, we conclude that there is no effect if we are unable to measure one. The null hypothesis (i.e., no effect) of the statistical test is accepted, so that the experiment does not allow the investigator to conclude that no effect is taking place, only that his experiment did not detect any effects. Presumably, a better designed experiment with greater power will always show an effect when previous experiments have not shown one.

As an example of the problem, we present a discussion of one of our attempts at testing for the effects of transmitters on mule deer (*Odocoileus hemionus*) (Garrott et al. 1985). High coyote (*Canis latrans*) predation rates on collared fawns during the winter caused us to hypothesize that collars might subject fawns to increased predation by providing coyotes a more secure hold on the neck, or by visually isolating collared animals, thus attracting the coyote's attention. We therefore instrumented 45 male mule deer fawns with the standard break-away radio-tracking collars we normally used and 46 with ear-tag transmitters which did not provide a secure hold or visually isolate the tagged fawn. Configurations were alternated between successive male animals captured with drop nets. We concluded no differences ($P = 0.67$) in mortality rate between ear-tagged versus collared fawns and "failure to detect a difference in mortality rates between these two groups of instrumented fawns indicates that these two mechanisms did not influence predation rates of radio-collared fawns."

However, closer scrutiny of this study suggests that the power is lacking to detect a meaningful difference. The test of the hypothesis of no attachment effect is performed with a 2×2 χ^2 test.

Attachment method	Coyote predation	Other fate
Ear-tag attachment	n_{11}	n_{12}
Neck-collar attachment	n_{21}	n_{22}

Although 91 instrumented animals sounds like a large study, was this level of effort adequate to detect a biologically important difference in predation rates? First, we must define what difference in predation rate is "biologically important." Based on data from Garrott et al. (1985), about 60% of the instrumented animals died from coyote predation. Thus, to start the process of designing a study, suppose we want to detect at least a difference of 0.3 in coyote predation rates; that is, we hypothesize that the coyote predation rate on neck-collared fawns is 0.6 and that ear-tag fawns would only have a 0.3 coyote predation rate. How many collars are needed to detect a difference of this magnitude with 80% certainty? The problem described is that of testing for differences between two percentages. Sokal and Rohlf (1981:766–767) provide an easy-to-understand description of the method to calculate the needed sample size, given that equal numbers of animals are used in each treatment group. Define the current coyote mortality rate on fawns tagged with the standard collar as $p_1 = 0.6$, and the expected decline for fawns tagged with ear tags as $p_2 = 0.3$. The calculation of the number of instrumented animals (n) needed for each group requires the computation of two intermediate quantities, $\bar{p}$ and A:

$$\bar{p} = (p_1 + p_2)/2,$$
$$= (0.6 + 0.3)/2.$$
$$= 0.45,$$

$$A = \left[t_{\alpha(x)}\sqrt{2\bar{p}(1-\bar{p})} + t_{2\beta(x)}\sqrt{p_1(1-p_1) + p_2(1-p_2)} \right]^2,$$
$$= \left[1.96\sqrt{2(0.45)(1-0.45)} + 0.842\sqrt{0.6(1-0.6) + 0.3(1-0.3)} \right]^2,$$
$$= 3.785,$$

$$n = \frac{A\left[1 + \sqrt{1 + 4(p_1 - p_2)/A}\right]^2}{4(p_1 - p_2)^2},$$
$$= \frac{3.785\left[1 + \sqrt{1 + 4(0.6 - 0.3)/3.785}\right]^2}{4(0.6 - 0.3)^2},$$
$$= 48,$$

where $1 - \beta = 0.80$ from the 80% certainty (80% power) of being able to detect a 0.3 difference or $\beta = 0.20$, the probability of not rejecting the null hypothesis when it is true (i.e., a Type II error), and $\alpha = 0.05$ from the typical 1:20 chance of making a Type I error. Thus, 48 collars are needed for each group of animals, for a total sample of 96. From these sample size estimates, we see that a total sample of 91 radios was not quite adequate to detect a doubling in predation rates by coyotes from ear-tag transmitters to the standard collar transmitters, that is, from a 0.3 predation rate to 0.6. A SAS program (SAS Institute Inc. 1985) to make these calculations and its output for a range of values are shown in Fig. 3.2. From this output, we see that, to detect a change from 0.6 to 0.5, 408 transmitters would be required for each group.

The above sample size calculations were computed from analytical formulas that apply specifically to a 2 × 2 χ^2 table. Often, analytical methods are not available to compute the power of a χ^2 table, so we present a "quick and easy" numerical method for computing power for any χ^2 table (Burnham et al. 1987:214–215), including unequal sample sizes between treatments. Assume, as before, that you can postulate the parameter values under the alternative hypothesis; that is, you know the values of p_1 and p_2. Then the SAS program in Fig. 3.3 will compute the power for a range of sample sizes.

```
title 'Calculation of sample size necessary';
title2 'to detect a difference in 2 proportions, p1 and p2';
data;
*  p1 > p2 is assumed, but absolute values are used in the calculations;
   input p1 p2;
   alpha=0.05 /* Type I error rate (Rejecting H0 when it is true) */;
   beta=0.20 /* Type II error rate (Accepting H0 when it is false) */;
   pbar=(p1+p2)/2;
   A=(tinv(1-alpha/2,1000)*sqrt(2*pbar*(1-pbar))+
      tinv(1-beta,1000)*sqrt(p1*(1-p1)+p2*(1-p2)))**2;
   n=int((A*(1+sqrt(1+4*abs(p1-p2)/A))**2)/(4*abs(p1-p2)**2)+0.5);
   cards;
0.6 0.59
0.6 0.58
0.6 0.55
0.6 0.5
0.6 0.4
0.6 0.3
0.6 0.2
;
proc print;
run;
```

Calculation of sample size necessary
to detect a difference in 2 proportions, p1 and p2

OBS	P1	P2	ALPHA	BETA	PBAR	A	N
1	0.6	0.59	0.05	0.2	0.595	3.79004	38100
2	0.6	0.58	0.05	0.2	0.590	3.80423	9610
3	0.6	0.55	0.05	0.2	0.575	3.84068	1576
4	0.6	0.50	0.05	0.2	0.550	3.88096	408
5	0.6	0.40	0.05	0.2	0.500	3.88454	107
6	0.6	0.30	0.05	0.2	0.450	3.78476	48
7	0.6	0.20	0.05	0.2	0.400	3.57991	27

Figure 3.2 SAS procedure and output to calculate the sample size necessary to detect a difference in two proportions.

```
*-------------------------------------------------------------*
|    SAS code to compute the power of a chi-square test for   |
|    a 2 × 2 table. The method can be extended to an r × c    |
|    table if necessary.                                      |
*-------------------------------------------------------------*;
title 'Computation of power of experiment for 2 × 2 table';
title2 'to detect a difference in 2 proportions, p1 and p2';
data power;
    alpha=0.05; p1=0.3; p2=0.6;
    do n=20 to 50 by 10;
        obs11=p1*n; obs12=n-obs11;
        obs21=p2*n; obs22=n-obs21;
        expect11=n*(p1+p2)/2; expect12=n-expect11;
        expect21=n*(p1+p2)/2; expect22=n-expect21;
        noncent=(obs11-expect11)**2/expect11
               +(obs12-expect12)**2/expect12
               +(obs21-expect21)**2/expect21
               +(obs22-expect22)**2/expect22;
        df=1;
        critchi=cinv(1-alpha,df);
        power=1-probchi(critchi,df,noncent);
        keep p1 p2 alpha n noncent critchi df power;
        output;
    end;
    label p1='Binomial probability for control group'
          p2='Binomial probability for treatment group'
          alpha='Type I error rate'
          n='Sample size for each row'
          df='Degrees of freedom'
          noncent='Non-centrality parameter'
          critchi='Critical chi-square value to reject H0'
          power='Probability of rejecting false H0';
proc print;
run;
```

Computation of power of experiment for 2 × 2 table
to detect a difference in 2 proportions, p1 and p2

OBS	ALPHA	P1	P2	N	NONCENT	DF	CRITCHI	POWER
1	0.05	0.3	0.6	20	3.63636	1	3.84146	0.47891
2	0.05	0.3	0.6	30	5.45455	1	3.84146	0.64638
3	0.05	0.3	0.6	40	7.27273	1	3.84146	0.76939
4	0.05	0.3	0.6	50	9.09091	1	3.84146	0.85432

Figure 3.3 SAS procedure and output to calculate the power of a χ^2 test to detect a difference in two proportions.

The program computes the expectation of the observed value for each cell under the alternative hypothesis. The values are computed in the variables *obs11*, *obs12*, *obs21*, and *obs22* shown in Fig. 3.3. Then the program computes the expectation of the expected value for each cell under the null hypothesis, which are the variables *expect11*, *expect12*, *expect21*, and *expect22* in Fig. 3.3. These eight values are then used to compute the χ^2 statistic for the table, which is really the noncentrality parameter of the corresponding noncentral χ^2 power curve for the test. That is, the noncentrality parameter specifies the distribution of the test statistic under the alternative hypothesis. Thus, we are interested in how often the test statistic will exceed the critical value of the test statistic (3.84 for 1 degree of freedom and $\alpha = 0.05$). Figure 3.4 illustrates this process.

The final step for the SAS code in Fig. 3.3 to complete the power computation is to compute the critical χ^2 value with the *cinv* function, and then the power with the *probchi* function. The *probchi* function actually computes the Type II error rate (β), so the value is subtracted from 1 to obtain power.

From the output in Fig. 3.3, we see that the sample for each row of the table must be between 40 and 50 animals, consistent with the value predicted

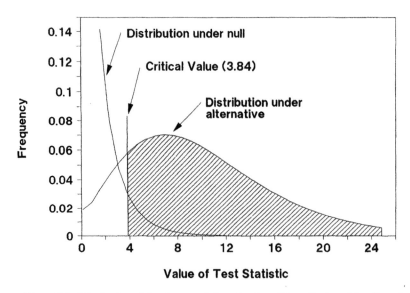

Figure 3.4 Distribution of the χ^2 test statistic under the null hypothesis and the alternative hypothesis for $p_1 = 0.3$ and $p_2 = 0.6$, with 50 transmitters for each group. The power of the test is the probability that the test statistic exceeds the critical value (3.84), and is represented by the shaded area in the graph.

from the analytical procedures used earlier. Generally, the quick method suggests that slightly smaller samples are needed than the analytical procedure. Given the usually poor guesses for p_1 and p_2, this slight difference is insignificant.

Effects of Transmitters on Animals

A common concern when instrumenting relatively small animals is the weight of the transmitter, particularly for flying species. One approach to testing for impacts from heavy transmitters is to fit animals with varying sizes of packages. Greenwood and Sargeant (1973) fitted captive mallards (*Anas platyrhynchos*) and blue-winged teal (*Anas discors*) with back-mounted transmitter packages in three weight ranges, with a fourth group of noninstrumented birds of each species serving as a control. Results of the study indicated that all treatment groups were impacted as these birds lost more weight than did the control groups. There was no correlation between body weight loss and weight of the transmitter packs in the mallards, but a correlation was evident in the teal. Lightweight transmitter packs had less effect on the weight loss of the teal than did heavier transmitter packs. Amlaner et al. (1979) conducted a similar experiment with pairs of breeding herring gulls (*Larus argentatus*). In order to eliminate the effects of capture and handling, a fourth group of noninstrumented birds was handled in a similar manner as the instrumented birds and was included in the study. Clutch survivorship during the first week was lower for gull pairs subjected to more severe treatments. In addition, renesting occurred only in the noninstrumented treatment group and in one pair of gulls carrying the lightest transmitter pack. A third example of the impacts of transmitter package weights is provided by Warner and Etter (1983), who concluded that the longevity of free-ranging hen ring-necked pheasants (*Phasianus colchicus*) equipped with transmitter packs decreased with increased weights.

Many investigators have noted that after attaching a transmitter to an animal there is a period of several days to 1–2 weeks during which the animals alter normal behaviors as they acclimate to the transmitter package. A commonly reported change is an increase in comfort movements such as preening and shaking (Greenwood and Sargeant 1973, Gilmer et al. 1974, Siegfried et al. 1977). Other temporary effects noted by various investigators include disruption of incubation behavior (Amlaner et al. 1979) and breeding behavior (Hirons and Owen 1982), changes in intraspecific interaction (Nenno and Healy 1979), avoidance of water (Greenwood and Sargeant 1973), abnormal levels of activity (Boag 1972, Hamley and Falls 1975, Leuze 1980, Webster and Brooks

1980, Perry 1981, Birks and Linn 1982), and decreased food intake (Boag 1972, Webster and Brooks 1980, Perry 1981).

A variety of effects has also been documented after the initial adjustment period. Transmitters have caused hair and feather wear and skin irritation (Bartholomew 1967, Hessler et al. 1970, Corner and Pearson 1972, Greenwood and Sargeant 1973, Perry 1981, Kenward 1982, Wywialowski and Knowlton 1983, Hines and Zwickel 1985, Jackson et al. 1985). Birds have become entangled in loosely fitted harnesses (Schladweiler and Tester 1972, Hirons and Owen 1982, Hines and Zwickel 1985), and trailing whip antennas may become entangled in vegetation or cause electrocution of birds that commonly perch on power lines (Dunstan 1977). Transmitters have impaired movements (Banks et al. 1975) and digging ability (Corner and Pearson 1972) in small mammals. Weight loss has been noted for waterfowl (Greenwood and Sargeant 1973, Schladweiler and Tester 1972, Perry 1981) and the breeding behavior of American woodcock (*Philohela minor*) has been disrupted (Ramakka 1972). Warner and Etter (1983) concluded that radio-marked hen ring-necked pheasants (*Phasianus colchicus*) experienced lower survival rates than did normal birds, and Johnson and Berner (1980) felt that 28-g transmitter packs may affect the survival of cock pheasants that weighed less than 897 g. Erikstad (1979) noted decreased chick survival rates for instrumented female willow ptarmigan (*Lagopus lagopus*) when compared to noninstrumented birds. Decreased survival has also been documented for meadow voles (*Microtus pennsylvanicus*) (Webster and Brooks 1980), black-tailed jack rabbits (*Lepus californicus*) (Wywialowski and Knowlton 1983), and may have occurred in a study of Greenland collared lemmings (*Dicrostonyx groenlandicus*) (Brooks and Banks 1971).

The reader may have noticed in the previous discussion that most studies of instrumentation effects have been conducted on birds. A bibliography of wildlife radio-tracking studies listed 163 avian studies and 351 mammalian studies (Patric et al. 1982), indicating that despite the fact that mammalian radio-tracking studies are twice as common as avian studies, little data are available on impacts of transmitters on mammals. The preponderance of impact studies on avian species is no doubt due to the relatively small body size of most birds and their dependence on flight. The body weight–transmitter weight ratio appears to be the factor which concerns most investigators. This concern is also illustrated in the mammalian literature, as most impact studies have been conducted on small mammals, such as voles (*Microtus* sp.) (Hamley and Falls 1975, Leuze 1980, Webster and Brooks 1980) and mice (*Peromyscus leucopus*) (Smith and Whitney 1977, Smith 1980), with little attention to the more frequently studied large predators and big game animals. Although the weight of the transmitter package may not be of great concern with larger animals, there is still potential for significant impacts. Researchers have suggested or demonstrated instrumentation effects on mountain lions (*Felis concolor*) (Garcelon 1977), mule deer (*Odocoileus hemionus*) (Goldberg and Haas 1978, Wenger and Springer 1981, Garrott et al. 1985), white-tailed deer (*Odocoileus virginianus*) (Clute and Ozoga 1983), caribou (*Rangifer tarandus*) (Pank et al. 1985), river otters (Melquist and Hornocher 1979), and sea otters (*Enhydra lutris*) (Garshelis and Siniff 1983).

A review of the available literature suggests five general recommendations for researchers planning to instrument wild animals:

1. Use the smallest possible transmitter package when instrumenting relatively small animals.
2. Transmitter packages to be placed on animals which may depend on cryptic coloration for survival should be as inconspicuous as possible.
3. Whenever possible, transmitters and their attachments should be tested on captive animals before they are placed on free-ranging animals. Test animals should be of the same sex and age as the animals for which the transmitters are intended in the wild.
4. Allow several days or up to 1 week for newly instrumented animals to acclimate to the transmitter before collecting data which will be considered indicative of normal behavior.
5. Whenever possible, avoid instrumenting animals during their reproductive period, as many species appear to be particularly sensitive to disturbance during this time.

These recommendations are only general guidelines and certainly do not apply to all species or research programs.

We do not want to give the reader the impression that the use of radio-tracking in the study of free-ranging animals is of questionable value. On the contrary, most investigators who have studied the effects of this technique have concluded that, although transmitter attachment may alter some aspects of an animal's behavior or physiology, the changes do not significantly affect the integrity of the research. This brief discussion is meant to familiarize the reader with many potential impacts which may result from instrumenting animals and to encourage further studies designed to investigate the consequences of attaching transmitters to wild animals. Consideration of this problem is crucial to the interpretation of all radio-tracking studies.

Summary

1. Detection of the effects of instrumentation on animals is difficult because of the logic of the problem: The investigator can never prove that the transmitter is not having an effect, only that the experiment failed to detect an effect. Therefore, the power of the experiment is what determines the validity of the test for an effect.
2. The transmitter weight–body weight ratio of flying animals is the factor that concerns most investigators.
3. Common sense should be used in placing transmitters on animals, to minimize potential impacts on the animal and thus maximize the chances of obtaining reliable data from the study. The smallest possible transmitter package, cryptic coloration, testing for transmitter effects on captive animals, time for the animal to adapt to the transmitter, and avoiding critical life history periods of the animal should be considered in designing a study.

References

Amlaner, C. J., Jr., R. Sibly, and R. McCleery. 1979. Effects of telemetry transmitter weight on breeding success in herring gulls. Pages 254–259 *in* F. M. Long ed. Proc. 2nd Int. Conf. Wildl. Biotelemetry. Univ. of Wyoming, Laramie.

Banks, E. M., R. J. Brooks, and J. Schnell. 1975. A radiotracking study of home range and activity of the brown lemming (*Lemmus trimucronatus*). J. Mammal. 56:888–901.

Bartholomew, R. M. 1967. A study of the winter activities of bobwhites through the use of radiotelemetry. Occas. Pap. C. C. Adams Cent. Ecol. Stud. No. 17:1–25.

Birsk, J. D. S. and I. J. Linn. 1982. Studies of home range of the feral mink, *Mustela vison*. Symp. Zool. Soc. London 49:231–257.

Boag, D. A. 1972. Effect of radio packages on behavior of captive red grouse. J. Wildl. Manage. 36:511–518.

Brooks, R. J. and E. M. Banks. 1971. Radio-tracking study of lemming home range. Commun. Behav. Biol. 4:1–5.

Burnham, K. P., D. R. Anderson, G. C. White, C. Brownie, and K. H. Pollock. 1987. Design and analysis methods for fish survival experiments based on release-recapture. Am. Fish. Soc. Monogr. 5:1–437.

Clute, R. K. and J. J. Ozoga. 1983. Icing of transmitter collars on white-tailed deer fawns. Wildl. Soc. Bull. 11:70–71.

Corner, G. W. and E. W. Pearson. 1972. A miniature 30-MHz collar transmitter for small mammals. J. Wildl. Manage. 36:657–661.

Dunstan. T. C. 1977. Types and uses of radio packages for North American Falconiform and Strigiform birds. Pages 30–39 *in* F. M. Long ed. Proc. 2nd Int. Conf. Wildl. Biotelemetry. Univ. of Wyoming, Laramie.

Erikstad, K. E. 1979. Effects of radio packages on reproductive success of willow grouse. J. Wildl. Manage. 43:170–175.

Fuller, M. R. 1987. Applications and considerations for wildlife telemetry. J. Raptor Res. 21:126–128.

Garcelon, D. K. 1977. An expandable drop-off transmitter collar for young mountain lions. Calif. Fish Game 63:185–189.

Garrott, R. A., R. M. Bartmann, and G. C. White. 1985. Comparison of radio-transmitter packages relative to deer fawn mortality. J. Wildl. Manage. 49:758–759.

Garshelis, D. L. and D. B. Siniff. 1983. Evaluation of radio-transmitter attachments for sea otters. Wildl. Soc. Bull. 11:378–383.

Gilmer, D. S., I. J. Ball, L. M. Cowardin, and J. H. Riechmann. 1974. Effects of radio packages on wild ducks. J. Wildl. Manage. 38:243–252.

Goldberg, J. S. and W. Haas. 1978. Interactions between mule deer dams and their radio-collared and unmarked fawns. J. Wildl. Manage. 42:422–425.

Greenwood, R. J. and A. B. Sargeant. 1973. Influence of radio packs on captive mallards and blue-winged teal. J. Wildl. Manage. 37:3–9.

Hamley, J. M. and J. B. Falls. 1975. Reduced activity in transmitter carrying voles. Can. J. Zool. 53:1476–1478.

Hessler, E., J. R. Tester, D. B. Siniff, and M. M. Nelson. 1970. A biotelemetry study of survival of pen-reared pheasants released in selected habitats. J. Wildl. Manage. 34:267–274.

Hines, J. E. and F. C. Zwickel. 1985. Influence of radio packages on young blue grouse. J. Wildl. Manage. 49:1050–1054.

Hirons, G. J. M. and R. B. Owen, Jr. 1982. Radio tagging as an aid to the study of woodcock. Symp. Zool. Soc. London 49:139–152.

Jackson, D. H., L. S. Jackson, and W. K. Seitz. 1985. An expandable drop-off transmitter harness for young bobcats. J. Wildl. Manage. 49:46–49.

Johnson, R. N. and A. H. Berner. 1980. Effects of radio transmitters on released cock pheasants. J. Wildl. Manage. 44:686–689.

Kenward, R. E. 1982. Techniques for monitoring the behaviour of grey squirrels by radio. Symp. Zool. Soc. London 49:175–196.

Leuze, C. C. K. 1980. The application of radio tracking and its effect on the behavioral ecology of the water vole. *Arvicola terrestris* (Lacepede). Pages 361–366 *in* C. J. Amlaner, Jr. and D. W. Macdonald eds. A Handbook on Biotelemetry and Radio Tracking. Pergamon Press, Oxford, England.

Melquist, W. E. and M. G. Hornocher. 1979. Development and use of a telemetry technique for studying river otter. Pages 104–114 *in* F. M. Long ed Proc. 2nd Int. Conf. Wildl. Biotelemetry. Univ. of Wyoming, Laramie.

Nenno, E. S. and W. M. Healy. 1979. Effects of radio packages on behavior of wild turkey hens. J. Wildl. Manage. 43:760–765.

Pank, L. F., W. L. Regelin, D. Beaty, and J. A. Curatolo. 1985. Performance of a prototype satellite tracking system for caribou. Pages 97–118 *in* R. W. Week and F. M. Long eds. Proc. 5th Int. Conf. Wildl. Biotelemetry. Univ of Wyoming, Laramie.

Patric, E. F., G. A. Shaughnessy, and G. B. Will. 1982. A bibliography of wildlife telemetry and radio tracking, Rhode Island Agric. Exp. Stn. Rep. No. 2054. Univ. of Rhode Island, Kingston.

Perry, M. C. 1981. Abnormal behavior of canvasbacks equipped with radio transmitters. J. Wildl. Manage. 45:786–789.

Ramakka, J. M. 1972. Effects of radio-tagging on breeding behavior of male woodcock. J. Wildl. Manage. 36:1309–1312.

SAS Institute Inc. 1985. SAS® Language Guide for Personal Computers, Version 6 Edition. SAS Institute Inc., Cary, NC. 429 pp.

Schladweiler, J. L. and J. R. Tester. 1972. Survival and behavior of hand reared mallards released in the wild. J. Wildl. Manage. 36:1118–1127.

Siegfried, W. R., P. G. H. Frost, I. J. Ball, and D. F. McKinney. 1977. Effects of radio-packages on African black ducks. S. Afr. J. Wildl. Res. 7:37–40.

Smith, H. R. 1980. Growth, reproduction and survival in *Peromyscus leucopus* carrying intraperitoneally implanted transmitters. Pages 367–374 *in* C. J. Amlaner, Jr. and D. W. Macdonald eds. A Handbook on Biotelemetry and Radio Tracking. Pergamon Press, Oxford, England.

Smith, H. R. and G. Whitney. 1977. Intraperitoneal transmitter implants—their biological feasibility for studying small mammals. Pages 109–117 *in* F. M. Long ed. Proc. 1st Int. Conf. Wildl. Biotelemetry. Univ. of Wyoming, Laramie.

Sokal, R. R. and R. J. Rohlf. 1981. Biometry, 2nd. ed. Freeman, San Francisco, CA. 859 pp.

Warner, R. E. and S. L. Etter. 1983. Reproduction and survival of radio-marked hen ring-necked pheasants in Illinois. J. Wildl. Manage. 47:369–375.

Webster, A. B. and R. J. Brooks. 1980. Effects of radiotransmitters on the meadow vole, *Microtus pennsylvanicus*. Can. J. Zool. 58:997–1001.

Wenger, C. R. and J. T. Springer. 1981. Reducing bias in predator-prey research involving telemetered young ungulates. Pages 13–19 *in* F. M. Long ed. Proc. 3rd Int. Conf. Wildl. Biotelemetry. Univ. of Wyoming, Laramie.

Wywialowski, A. P. and F. F. Knowlton. 1983. Effect of simulated radio-transmitters on captive black-tailed jack rabbits. Pages 1–11 *in* D. G. Pincock ed. Proc. 4th Int. Wildl. Biotelemetry Conf. Applied Microelectronics Institute and Technical Univ. of Nova Scotia, Halifax.

CHAPTER

4

Estimating Animal Locations

One of the most common uses of wildlife radio-tracking is estimating an animal's location. The techniques used to obtain locations of instrumented animals can be divided into two broad categories: those using triangulation and all other nontriangulation procedures. Each is discussed in the following sections.

Regardless of the procedures used, hardware is an important component of any radio-tracking study, and its performance and reliability often dictate the success or failure of the study. Therefore, the investigator should be familiar with various types of equipment and their use and limitations, in order to insure selection of the proper equipment. We do not attempt to provide guidelines or recommendations for equipment selection, antenna configurations, etc., as it is not within the scope of this book, and numerous publications dealing with radio-tracking hardware are readily available. General references that may be helpful to the uninitiated radio-tracking user include Brander and Cochran (1971), Cochran (1980), Mech (1983), and Kenward (1987). Amlaner and Macdonald (1980) provide a wide variety of papers, ranging from generalized reviews of the principles of radio-tracking and discussions of study design to technical papers on circuit construction and data analysis. Gilmer et al. (1981) provide recommendations for the design and use of receiving equipment for aerial tracking. Representatives of companies that manufacture radio-tracking equipment are also helpful. However, the most practical information is usually from other researchers who have extensive field experience with radio-tracking equipment.

Nontriangulation Location Techniques

Homing-in on the Animal

A simple, effective nontriangulation technique for obtaining positional data is to follow the transmitted signal's increasing strength until the instrumented animal is actually observed. Direct observation is particularly effective for large animals which occupy open environments (e.g., swans on lakes, lions on a savannah, antelope on grasslands, and polar bears on ice) and animals which are unwary, slow moving, or relatively sedentary (e.g., porcupines, skunks, tortoises). The only error in this technique is introduced by the quality of the maps or aerial photos and the investigator's ability to locate the position of the observed animal on the map. Variations on this general technique include circling a small area and assuming the animal is within the area, sandwiching the animal between the receiving point and a physical barrier (e.g., a mink along a lake shore), or searching a narrow belt of habitat and assuming the animal is within this habitat (e.g., a river otter along a waterway). Any point location derived from these techniques has some error because the animal is not actually seen; however, in some situations these techniques may still provide better location estimates than triangulation.

A disadvantage of all such "homing" techniques, however, is that they are usually very time intensive, thus limiting the number of animals that can be studied. Further, actually sighting the animal on the ground, or getting near it, may be disruptive. Continual harassment from tracking the animal may introduce perturbations into its normal movements and/or behavior, thus biasing the study.

Aerial Tracking

The use of aircraft for radio-tracking is effective and efficient for locating large numbers of animals quickly and is particularly valuable when working in rugged or inaccessible areas. Often, aerial tracking is the preferred method to locate animals during periods of dispersal or migration, when long distance movements occur over a short period of time. The general principle of aerial tracking is to mount one or more antennas on an aircraft and fly over the study area in a systematic way while scanning the radio frequencies of instrumented animals. Once a radio signal has been detected, the investigator uses the relative strength of the signal to direct the aircraft toward the animal. If the study is being conducted in open country and the animals are large, visual locations may be obtained from the aircraft. As we have previously discussed, inaccuracies in visual locations due to error in locating the animals' position on a map should usually be small and of little concern. For many radio-tracking studies, safety considerations preclude the low-altitude flights needed to observe animals. Visual locations from aircraft are particularly impractical when studying relatively small animals or those in forested or mountainous areas. In these situations the investigator locates the animal by either flying completely around the animal while keeping the strongest radio signal on the side of the plane facing the circle or making repeated passes near the animal from different directions (Gilmer et al. 1981, Mech 1983).

Errors in excess of 0.5 km have been reported for aerial location estimates made when the animal was not visually located (Garrott et al. 1987, Harrington et al. 1987). There is a wide variety of factors which may affect the accuracy of aerial location estimates, including altitude above ground, air speed, location procedures, prevalence of landmarks, and investigator fatigue or discomfort due to motion sickness (Gilmer et al. 1981, Mech 1983). Because of the potential for substantial inaccuracies in aerial location estimates, it may be appropriate to test accuracy, depending on the quality of location estimates needed to meet the objectives of the study.

Testing the accuracy of aerial location estimates follows the same procedures outlined for triangulation studies (see Chapter 5). In general, the investigator should place transmitters in a variety of vegetative and topographical situations throughout the study area and then attempt to locate them using the procedures that will be employed during actual data collection. Because aerial tracking directly produces the point estimate (i.e., the location the investigator marks on the map), the measure of error is the distance between the actual and estimated locations. Equations 5.1, 5.2, and 5.3 can then be used to calculate the error and precision of aerial location estimates, with the exception that replicate location estimates should not be attempted unless a different observer is used for each replication, to assure independence. The error standard deviation, obtained by accuracy testing, can then be used to construct 95% confidence areas around each location estimate by delineating a circle centered on the point estimate with a radius of 1.96 times the error standard deviation, where 1.96 is a z statistic.

In some aerial-tracking studies substantial errors may be caused in location estimates by the investigator's inability to determine exactly where he or she is located. Examples of radio-tracking studies in which this might be a problem include tracking marine mammals and pelagic birds offshore, polar bears (*Ursus maritimus*) on the pack ice, and herons in the Everglades. In these situations there are a number of commercial navigational aids which may be helpful in determining the location of the aircraft. The LORAN-C system is a widely used and relatively inexpensive navigational aid that utilizes a receiver–microprocessor in the aircraft and a series of ground-based transmitters. The on-board computer uses the timing of incoming radio signals from several transmitters to calculate the aircraft's position, which is displayed almost instantaneously as x and y coordinates (see Englert 1982, Burhans 1983, and Lert 1984 for easy-to-understand articles on LORAN-C).

LORAN stands for long range aid to navigation. Originally, LORAN-C was developed for marine navigation, but it is now commonly available for aircraft. Typical systems provide the estimated location as a readout of latitude and longitude, which is not a convenient coordinate system for most radio-tracking studies. Dodge et al. (1986) provide programs written for laptop portable computers which will convert longitude and latitude to UTM coordinates. Snyder (1982) describes algorithms for the conversion of longitude and latitude to various coordinate systems, including UTM coordinates.

The LORAN-C system uses low-frequency radio waves (90–110 kHz) that follow the earth's curvature. These frequencies are very stable (i.e., they do not skip), and are not affected by the ionosphere. Signals are sent from a chain of three to five stations, with one station the master and the remainder slaves. The master station transmits groups of pulses that are received by the slave stations. Each slave station transmits similar groups of pulses and adds a fixed time delay. The LORAN-C receiver in the aircraft receives both groups of pulses and calculates the time difference between them. To illustrate, if the time difference at the receiver is equal between received signals from a single slave transmitter and the master (taking into account the original delay added by slave stations), the receiver is on the straight line of locations equidistant from both master and slave transmitters. As the time difference deviates from equal delays, the receiver is positioned on an hyperbola defined by the time delay between the two transmitters and their positions. With a second slave station and the master station, a second hyperbola is defined. The intersection of the two hyperbolas defines the location of the receiver.

The accuracy of the position estimate depends on the distance to the stations, the number of stations, the closeness of the receiver to the lines connecting the stations, and the angle of intersection of the hyperbolas. Under optimal conditions, error of LORAN-C locations will range from under 100 m to 1 km (Patric et al. 1988). However, there are specific geographic areas (e.g., much of the central United States) that are too far from transmitters to produce reliable locations, because the LORAN-C system is primarily used in nautical navigation (maritime). The daytime range of LORAN-C is up to 1900 km, while nighttime ranges can be as great as 3200 km.

LORAN-C is the current state of the art in determining a position in the field. The future holds the promise of satellites providing location. One such system is the Navstar Global Positioning System (Underwood 1983). The U.S. Department of Defense plans to place 18 satellites in 11,000-mile-high orbits, with locations accurate to 20 m.

Satellite Tracking

Another possible solution to the problem of tracking highly mobile animals in remote areas is the use of satellites (Buechner et al. 1971). Satellite tracking of free-ranging animals was first attempted in 1970 (J. J. Craighead et al. 1971, F. C. Craighead et al. 1972). However, early equipment was too heavy (11 kg) and bulky to be useful for most animal studies. Since that time, animal-borne satellite transmitters (technically called "platform terminal transmitters") have become increasingly lighter and more compact. Three pioneering studies using these smaller transmitters in conjunction with NIMBUS satellites have demonstrated the potential utility of satellite tracking. Kolz et al. (1980) instrumented four polar bears (*Ursus maritimus*) off the coast of northern Alaska with transmitter packages that weighed 5.6 kg. Although they experienced several technical problems, a female bear was tracked for 390 days, during which she traveled approximately 1600 km. Schweinsburg and Lee (1982) also tracked polar bears in the Canadian Arctic region, with similar results, tracking one animal approximately 1000 km over 266 days. The third tracking experiment involved attaching a transmitter to the carapace of a loggerhead turtle (*Caretta caretta*) released off the coast of Mississippi in the Gulf of Mexico (Timko and Kolz 1982). The turtle was tracked over 2000 km, until the transmitter was washed ashore approximately 8 months later.

The next generation of animal-borne satellite transmitters was designed for the TIROS/ARGOS satellite system and provided increased reliability and performance in a smaller package. One such transmitter design has evolved beyond the experimental prototype stage to the point at which several investigators are routinely collecting data on relatively large numbers of caribou (*Rangifer tarandus*) and polar bears (*Ursus maritimus*) (Pank et al. 1985, Harrington et al. 1987, Fancy et al. 1988). The weight of these new transmitter packages, complete with attachment devices, ranged from 1.6 to 2.1 kg. Other satellite transmitters are being developed for birds. Fuller et al. (1984) and

Strikwerda et al. (1985, 1986) have described a solar-powered transmitter package weighing 160 g, with an expected life of at least 200 days. The transmitter has been successfully used to track a free-ranging bald eagle (*Haliaeetus leucocephalus*) and the migratory movements of swans (*Olor columbianus*) nesting in arctic Alaska.

Over the past 5–8 years there has been a proliferation of experimental studies using satellite transmitters. Marine species are the most difficult to track with conventional VHF telemetry, making them particularly attractive for studies using satellites. Species which have been instrumented include loggerhead (*Caretta caretta*), Kemp's ridley (*Lepidochelys kempi*), and leatherback (*Dermochelys coriacea*) sea turtles (Stoneburner 1982, Byles 1987); manatee (*Trichechus manatus*) (Mate et al. 1987); dugong (*Dugong dugong*) (Marsh and Rathbun 1987); basking sharks (*Cetorrhinus maximus*) (Priede 1984); walrus (*Odobenus rosmarus*) (Fancy et al. 1988); and a variety of seals and whales (Mate et al. 1986, Fancy et al. 1988). Terrestrial mammals which have been instrumented include grizzly bears (*Ursus arctos*), musk ox (*Ovibos moschatus*), Dall's sheep (*Ovis dalli*), wolves (*Canis lupus*), mule deer (*Odocoileus hemionus*), elk (*Cervus canadensis*) (Fancy et al. 1988), camels (*Camelus dromedarius*) (Grigg 1987), Iranian ibex (*Capra aegagrus*), mountain sheep (*Ovis canadensis*), moose (*Alces alces*), and African elephants (*Loxodonta africana*) (Tomkiewicz and Beaty 1987). These studies represent a substantial investment in the research and development of satellite instrumentation, which is rapidly leading the field of wildlife radio-tracking into a new era. At present, satellite transmitters are at least ten times as expensive as conventional transmitters, and many radio-tracking users feel the costs are prohibitively high. Harrington et al. (1987), however, present some convincing economic data for a study of caribou in Labrador and Newfoundland which indicate that satellite transmitters can be more cost effective than conventional VHF transmitters in some situations. These authors as well as Fancy et al. (1988) enumerate the many advantages of satellite radio-tracking and we strongly recommend that biologists and researchers contemplating major radio-tracking studies become familiar with these publications.

As with any other system which remotely estimates the location of a transmitter, errors in the location estimates are inherent. With satellite radio-tracking, the location of the transmitter is determined from the Doppler shift in the transmitter carrier frequency as a result of the movement of the satellite. There are a variety of factors which affect the quality of location estimates, including movement and altitude of the animal (location calculations assume that the transmitter is at sea level). The two most important factors, however, are oscillator stability, which is affected by temperature, and the number of Doppler measurements obtained during each satellite overpass. Fancy et al. (1988) provide the best documentation of location accuracy for satellite transmitters. These investigators used techniques similar to those we have recommended for studying the accuracy of locations obtained from aircraft. They found a mean error of 829 m, with a standard error of 26 m. Ninety percent of the estimated locations were within 1.7 km of the true location, with a maximum error of 8.8 km. These data can be used to construct confidence areas associated with location estimates.

The development of the newest-generation satellite transmitters is well advanced. By the time this book is in print transmitters for medium-sized mammals and bird transmitters weighing approximately 125 g will be operational and deployed on a variety of species (Beaty et al. 1987). These new transmitters will not only be smaller and lighter, but "smarter" as well. Sensor technology coupled with microprocessors provides the capability to monitor a wide variety of environmental characteristics and behaviors, such as air or water temperature; dive time, duration, and depth; heart rate; body position; and movement. This information may be stored or "archived" within the instrument package for later transmission to the satellite and retrieval by the biologist, providing continuous data (Tomkiewicz and Beaty 1987). The technological advances in satellite radio-tracking for animals are quickly expanding the horizons of wildlife research and management, providing exciting new opportunities for field studies of free-ranging animals.

Triangulation Location Techniques

Two Receiving Stations

Triangulation is the process of estimating the location of a transmitter by using two or more directional bearings obtained from known locations remote from the transmitter's position. In this section we present several methods for calculating point location estimates from directional bearings and the confidence area associated with these point estimates. However, the reader should be aware that new technology is appearing which uses hyperbolic triangulation, similar to that used by LORAN, rather than directional triangulation, as in the methods that we discuss (Yerbury 1980, Lemnell 1980, Lemnell et al. 1983). The animal is located by the time delay of its transmitter's response to the towers, hence the name "reverse LORAN." Although the basic method is different (i.e., timing of signals is used rather than direction), the principles of testing the accuracy of the system are the same.

Two general approaches are used to estimate an animal's location with a triangulation system, depending on the number of bearings used for determining the location estimates. The majority of investigators using radio-tracking to collect positional data have relied on two bearings to estimate an animal's location. Fixed-location towers (Fig. 4.1), vehicle-mounted receiving systems

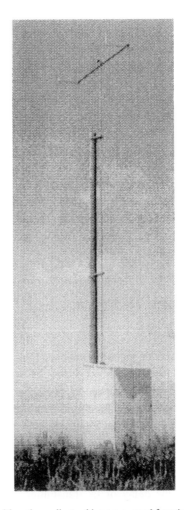

Figure 4.1 A fixed-location radio-tracking tower used for triangulation studies.

4 Estimating Animal Locations

Figure 4.2 A mobile radio-tracking tower used for triangulation studies.

(Fig. 4.2), or hand-held antenna and receivers are used to locate the animal. Bearings to the animal are taken from a compass rosette such as that pictured in Fig. 4.3 or by sighting along the antenna with a hand-held compass.

Figure 4.4 illustrates the general principles of triangulation with two bearings. Note in Fig. 4.4 that the bearings α_1 and α_2 are not angles in the mathematical sense, but rather azimuths or compass bearings from true north. To calculate the location (x_l, y_l), α_1 and α_2 must be converted to angles (in radians) by

$$\beta = (90 - \alpha) \times (\pi/180)$$

If β is less than zero, add $2 \times \pi$ to the β to make the result positive. Figure 4.5 illustrates the relationship between bearings and angles. Taking each of the tower locations as the centers of polar coordinate systems, the following equations with four unknowns (x_l, y_l, r_1, r_2) can be written to determine the animal's location:

$$x_l = r_1 \cos(\beta_1) + x_1$$
$$x_l = r_2 \cos(\beta_2) + x_2$$
$$y_l = r_1 \sin(\beta_1) + y_1$$
$$y_l = r_2 \sin(\beta_2) + y_2$$

Figure 4.3 A compass rosette and pointer mounted on a radio-tracking tower used to determine the bearing to the transmitter being tracked.

The variables r_1 and r_2 are the respective distances from towers 1 and 2 to the animal. Because the two variables of interest are x_t and y_t, the above equations can be combined by solving for r_1 and r_2 to yield two equations:

$$\frac{x_t - x_1}{\cos \beta_1} = \frac{y_t - y_1}{\sin \beta_1}$$

$$\frac{x_t - x_2}{\cos \beta_2} = \frac{y_t - y_2}{\sin \beta_2}$$

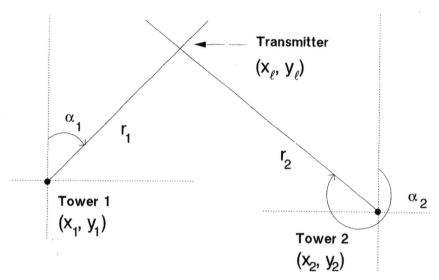

Figure 4.4 The triangulation process. The two solid lines represent the two bearings from the receiving points and intersect at the source of the signal. The distances from the transmitter to towers 1 and 2 are represented by r_1 and r_2, respectively.

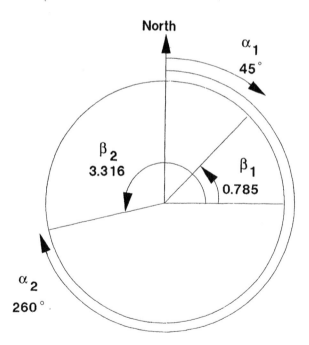

Figure 4.5 The relationship between bearings (α_1 and α_2) from a receiver to the transmitter and angles (β_1 and β_2). Angles are calculated in radians; to convert to degrees, multiply by $180°/\pi = 57.295°$.

With the trigonometric identity $\tan A = \sin A/\cos A$, and solving for y_l, the following equation is obtained:

$$y_l = \frac{(x_2 - x_1)\tan \beta_1 \tan \beta_2 - y_2 \tan \beta_1 + y_1 \tan \beta_2}{\tan \beta_2 - \tan \beta_1} \qquad (4.1)$$

By replacing y_l back into one of the original equations, the result for x_l is obtained:

$$x_l = \frac{(y_l - y_2)}{\tan \beta_2} + x_2 \qquad (4.2)$$

or, solving for x_l directly without including y_l:

$$x_l = \frac{x_1 \tan \beta_1 - x_2 \tan \beta_2 + y_2 - y_1}{\tan \beta_1 - \tan \beta_2}$$

Implementation of these relationships is somewhat more complicated in a computer program because various undefined conditions must be checked for, such as $\tan 90°$, which is ∞.

Equations 4.1 and 4.2 provide estimates of the coordinates where the animal is located. However, as is discussed later, even the best receiving systems are not perfect; that is, any bearing recorded is not the *exact* bearing from the receiving point to the transmitter, but an estimate. Thus, the recorded bearing is a random variable with some associated standard deviation. In order to develop a procedure for estimating a confidence area, the bearing error is assumed to be normally distributed with standard deviation σ and a mean of zero. Estimating the standard deviation of receiving systems and verifying that the mean of the bearing error is really zero is considered in Chapter 5.

For the above example, assume that estimates of the standard deviations for towers 1 and 2 are s_1 and s_2, respectively. The estimate of the recorded bearing for tower 1 is

$$\hat{\alpha}_1 = \alpha_1 + \varepsilon_1$$

where $\hat{\alpha}_1$ is the estimated bearing, consisting of the true bearing α_1 plus some random error, ε_1. Because ε_1 has been assumed to be normally distributed, with a mean zero and a standard deviation of σ_1 (estimated by s_1), a 95% confidence interval can be constructed for α_1 by

$$\Pr(\hat{\alpha}_1 - 1.96s_1 < \alpha_1 < \hat{\alpha}_1 + 1.96s_1) = 0.95$$

The value 1.96 is the z statistic to provide the 95% confidence interval. Thus, $\hat{\alpha}_1 \pm 1.96s_1$ provides a range or interval estimate of α. The same procedure is used for the second tower, that is, $\hat{\alpha}_2 \pm 1.96s_2$.

4 Estimating Animal Locations

Based on these two interval estimates, a confidence area for the animal's location is described by the polygon constructed from the intersection of the two bearing interval estimates (Fig. 4.6). Based on the assumptions, the probability that the animal will be within the shaded area is:

$$\Pr(\hat{\alpha}_1 - 1.96s_1 < \alpha_1 < \hat{\alpha}_1 + 1.96s_1) \times \Pr(\hat{\alpha}_2 - 1.96s_2 < \alpha_2 < \hat{\alpha}_2 + 1.96s_2)$$
$$= 0.95 \times 0.95$$
$$= 0.9025$$

That is, the interval estimates of the bearings (dashed lines of Fig. 4.6) must include the true bearing for both towers. The probability of this being the case for each of the towers is 0.95, so the probability that the shaded area in Fig. 4.6 will include the true location of the animal is $0.95^2 = 0.9025$. The polygon ABCD in Fig. 4.6 is constructed by calculating the four intersections of $\hat{\alpha}_1 \pm 1.96s_1$ and $\hat{\alpha}_2 \pm 1.96s_2$. The equations previously discussed can be used to determine these intersections. The area of this polygon is then calculated as

$$\text{area} = \frac{x_A(y_D - y_B) + x_B(y_A - y_C) + x_C(y_B - y_D) + x_D(y_C - y_A)}{2}$$

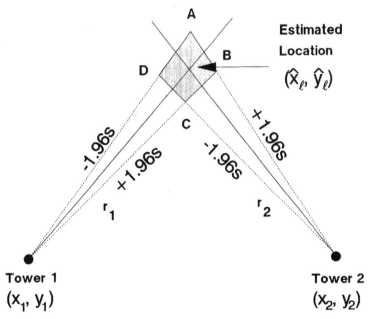

Figure 4.6 Construction of a confidence area for the animal's location determined from the bearings for two towers. The dashed lines represent the four bearings used to construct the four corners of the error polygon. The observed bearings (solid lines) have $\pm 1.96s$ added to them to obtain the dashed lines.

Note that the order of points around the polygon ABCD changes if the estimated location is below the line connecting towers 1 and 2. When this happens, the corners B and D in Fig. 4.6 switch locations; that is, the order of points around the polygon is now ADCB. The appropriate area formula is then

$$\text{area} = \frac{x_A(y_B - y_D) + x_D(y_A - y_C) + x_C(y_D - y_B) + x_B(y_C - y_A)}{2}$$

Hupp and Ratti (1983) provide a much simpler method of calculating the area of the error polygon, but their formula only applies to the special case when the distances from the location estimate to each tower are identical, that is, $r_1 = r_2$. Thus, the formula they present has little application in typical radio-tracking studies.

Because the two-tower triangulation problem is so common in radio-tracking data analysis, we have included a SAS program in Fig. 4.7 to calculate the estimate of location and the size of the associated error polygon. This program avoids determining the order of the four corners of the error polygon by calculating the area of the two triangles that make up the polygon. This program could be used to process a file of bearings to produce a file of the estimates of animal locations.

The area of the error polygon is a measure of the precision of the point estimate derived from the intersection of two bearings. For any pair of bearings, the precision of the estimates of a transmitter's location is a function of the distance from the towers taking the bearings, the precision of the bearings from the receivers to the transmitter, and the angle of intersection of the two bearings (Springer 1979). A common misconception in the wildlife literature is that the most precise estimate of location is obtained at the two points equidistant from each tower (i.e., $r_1 = r_2$), with angles of intersection of 90°. In fact, the most precise estimates of location are obtained immediately beside one of the towers where the two bearings intersect with an angle of 90°. This error polygon is smallest, because the distance across the arc created by $\alpha \pm 1.96s$ is the smallest. When all points with angles of intersection of 90° are plotted, a circle is formed centered between the two receiving points, with a diameter equal to the distance between the points (Fig. 4.8).

The shape and the size of the error polygons can be mapped, because they are a function of the estimated location. Thus, in Fig. 4.9, the size and the shape of each of the error polygons is fixed by the estimated location relative to the position of the two towers. The shape and the size of polygons are mirror images of each other on the other side of the line connecting the two towers.

4 Estimating Animal Locations

```
* ---------------------------------------------------------------- *
|    SAS program to estimate the location (xl, yl) of a            |
|       transmitter, given the locations of two towers (x1, y1,    |
|       and x2, y2) and the bearings from the towers to the        |
|       transmitter (alpha1, alpha2).                              |
* ---------------------------------------------------------------- *;
data triang2;
   array alpha{2} alpha1 alpha2;
   array beta{2} beta1 beta2;
   array poly{2,4};
   drop i beta1 beta2 pi x y ok d1 d2 d3 sp;
   pi=4*atan(1);
   s=1.5;
   input x1 y1 x2 y2 alpha1 alpha2;
   do i=1 to 2;
      beta{i}=(90-alpha{i})*(pi/180);
      end;
   link calcu;
   xl=x; yl=y;
   if ok then do;
      /* If estimate is on top of one of the towers,
         the area is zero. */
      if (abs(xl/x1-1)<0.0001 & abs(yl/y1-1)<0.0001)
       | (abs(xl/x2-1)<0.0001 & abs(yl/y2-1)<0.0001) then area=0;
      else do;
         /* Calculate area of error polygon */
         beta1=(90-(alpha1-1.96*s))*(pi/180); /* -- coordinate */
         beta2=(90-(alpha2-1.96*s))*(pi/180);
         link calcu; poly{1,1}=x; poly{2,1}=y;
         beta1=(90-(alpha1+1.96*s))*(pi/180); /* ++ coordinate */
         beta2=(90-(alpha2+1.96*s))*(pi/180);
         link calcu; poly{1,2}=x; poly{2,2}=y;
         beta1=(90-(alpha1-1.96*s))*(pi/180); /* -+ coordinate */
         link calcu; poly{1,3}=x; poly{2,3}=y;
         /* Get area of half of error polygon on -+ side */
         d1=sqrt((poly{1,1}-poly{1,2})**2+(poly{2,1}-poly{2,2})**2);
         d2=sqrt((poly{1,2}-poly{1,3})**2+(poly{2,2}-poly{2,3})**2);
         d3=sqrt((poly{1,1}-poly{1,3})**2+(poly{2,1}-poly{2,3})**2);
         sp=(d1+d2+d3)/2;
         area=sqrt(sp*(sp-d1)*(sp-d2)*(sp-d3));
```

Figure 4.7 SAS code to calculate the estimate of an animal's location and the size of the associated error polygon using bearings from two receiving points. (*Figure continues.*)

```
      beta1 = (90-(alpha1+1.96*s))*(pi/180);  /* +- coordinate */
      beta2 = (90-(alpha2-1.96*s))*(pi/180);
      link calcu; poly{1,4}=x; poly{2,4}=y;
      /* Now get area on other side of polygon */
      d2 = sqrt((poly{1,2}-poly{1,4})**2+(poly{2,2}-poly{2,4})**2);
      d3 = sqrt((poly{1,1}-poly{1,4})**2+(poly{2,1}-poly{2,4})**2);
      sp = (d1+d2+d3)/2;
      area=area+sqrt(sp*(sp-d1)*(sp-d2)*(sp-d3));
    end;
  end;
  else area= .;
  return;
calcu:
  if abs(beta1-beta2) < pi/90
    | abs(abs(beta1-beta2)-pi) < pi/90 then do;
    x= .; y= .; ok=0; end;
  else do;
    x = (x1*tan(beta1)-x2*tan(beta2)+y2-y1)/(tan(beta1)-tan(beta2));
    y = ((x2-x1)*tan(beta1)*tan(beta2)-y2*tan(beta1)+y1*tan(beta2))
        /(tan(beta2)-tan(beta1));
    ok=1;
    end;
  return;
  /* The following input consists of:
       1) X coordinate of tower 1,
       2) Y coordinate of tower 1,
       3) X coordinate of tower 2,
       4) Y coordinate of tower 2,
       5) Bearing from tower 1 to transmitter, and
       6) Bearing from tower 2 to transmitter.
     Thus, for the first record, the position of tower 1
     is (1,1), the position of tower 2 is (2,1), the bearing
     from tower 1 to the transmitter is 45 degrees, and the
     bearing from tower 2 to the transmitter is 315 degrees. */
  cards;
1 1 2 1  45 315
1 1 2 1   0 315
1 1 2 1 359 315
1 1 2 1 180 225
1 1 2 1 179 225
;
proc print; run;
```

Figure 4.7 (*Continued*)

4 Estimating Animal Locations

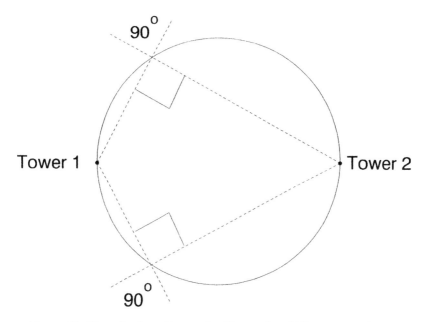

Figure 4.8 When all points with an angle of intersection of 90° are connected, a circle is formed. The most precise location is beside a tower, because the distance across the error arc is so short.

Saltz and Alkon (1985) do not feel that polygon areas are the most appropriate and useful measure of the quality of a location estimate. They argue that areas of error polygons may be independent of their linear dimensions; that is, a reduction in error polygon area may not be accompanied by corresponding changes in the length of diagonals connecting opposite corners of the polygon. Since the length of the largest diagonal always represents the maximum possible displacement of the true transmitter location, they conclude that it is the most sensitive and reliable criterion for judging quality of location estimates. For the case of a circle, the largest diagonal and the area are proportional, and therefore both measures provide the same information. However, error polygons are not circular. In fact, many error polygons are very long and narrow. Saltz and Alkon (1985) argue that because a polygon is long and narrow, the longest diagonal makes a good measure of error. However, for the same length diagonal, a fatter, more circular polygon could also be constructed. For a long, narrow polygon, much more area is encompassed by the circle formed from the longest diagonal relative to the area of the polygon than for a fatter, more circular polygon (Fig. 4.10). Thus, we believe that the area is the best measure of precision. As shown in the following discussion, the area of the error polygon is analogous to the area of a confidence ellipse developed from statistically rigorous procedures. Hence, the area of the error polygon also has some theoretical reasons associated with its use.

Three or More Receiving Stations

Use of three or more receiving stations provides major advantages in estimating the transmitter's location, mainly that erroneous bearings can be detected. Erroneous bearings may occur due to signal reflection caused by rugged terrain, mistakes in recording data, or mistakes in detecting the signal. However, the use of three or more bearings for triangulation presents a somewhat more com-

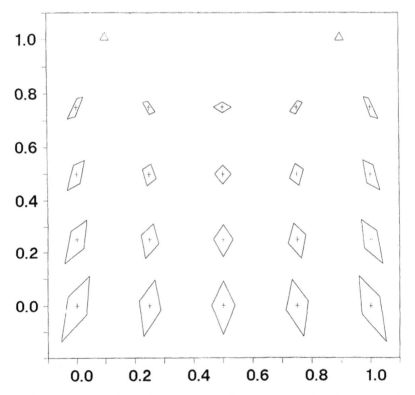

Figure 4.9 Error polygons for selected points from a two-tower triangulation system. The towers are located at the triangles at the top of the graph, with the axes providing a relative measure of distance. Only the half of the map below the two towers is shown, because the upper half is a mirror image.

4 Estimating Animal Locations

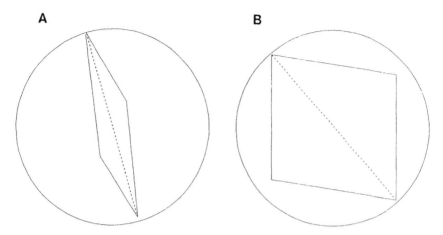

Figure 4.10 The problem with using the length of the longest diagonal of a polygon as a measure of the quality of location estimates obtained by triangulation using two receiving points. (A) The circle formed by the longest diagonal includes much more area relative to the area of the polygon than for (B).

plicated estimation problem. Although attempts have been made to extend the polygon approach to this situation (Springer 1979), we feel that more rigorous and appropriate analytical techniques are provided by the maximum likelihood, Huber, and Andrews estimators presented by Lenth (1981). By specifying a probability model for the angles $\beta_1, \beta_2, \ldots, \beta_n$, a maximum likelihood estimator of the animal's location (x_l, y_l) can be derived. In particular, a logical choice for the model is the von Mises distribution (see Batschelet 1981 for a description), having the probability density function

$$f(\beta_i; \mu_i, \kappa) = [2\pi I_0(\kappa)]^{-1} \exp[\kappa \cos(\beta_i - \mu_i)]$$

where I_0 is a modified Bessel function of the first kind and order zero and κ is the concentration parameter. We calculated parameter κ, used in all three of Lenth's (1981) estimators, as

$$A = \exp\left[\frac{-1}{2}\left(\frac{s\pi}{180}\right)^2\right] \tag{4.3}$$

$$\kappa^{-1} = 2(1 - A) + \frac{(1 - A)^2(0.48794 - 0.82905A - 1.3915A^2)}{A} \tag{4.4}$$

where s is the bearing standard deviation obtained through system testing (see Chapter 5). The parameter κ is the same for all bearings; hence, all receiving points are presumed to have the same sampling error distribution (an assumption which can be verified during system testing). Conversion of s to κ is relatively easy, as the following SAS (SAS Institute Inc. 1985) code shows.

```
*  ------------------------------------------------------------  *
|     SAS code to calculate the value of the concentration       |
|        parameter, kappa, of the von Mises distribution from    |
|        a standard deviation. Typically, the standard deviation |
|        is measured in studies estimating the precision of the  |
|        bearings, but the parameter kappa is needed for use in  |
|        Lenth's estimators of transmitter location.             |
*  ------------------------------------------------------------  *;
data stok;
   drop pi A;
   pi = 4*atan(1);
   do s=0.25 to 5 by 0.25;
      link calcu;
      output;
      end;
   do s=6 to 10 by 1;
      link calcu;
      output;
      end;
   do s=15 to 50 by 5;
      link calcu;
      output;
      end;
*  ------------------------------------------------------------  *
|     Subroutine to calculate kappa from s                       |
*  ------------------------------------------------------------  *;
calcu: A=exp(-1*(s*pi/180)**2/2);
       kappa=1/(2*(1-A)+((1-A)**2*(0.48794-0.82905*A-1.3915*A*A))/A);
return;
proc print; proc plot; plot kappa*s='*'; run;
```

This code will generate a table of s for the values specified in the three DO loops, along with the corresponding values of κ and also plot the function.

Extensive discussion of the von Mises and other distributions on the circle is given by Mardia (1972). An example of three values of s in the von Mises distribution is shown in Fig. 4.11. As the standard deviation becomes small,

4 Estimating Animal Locations

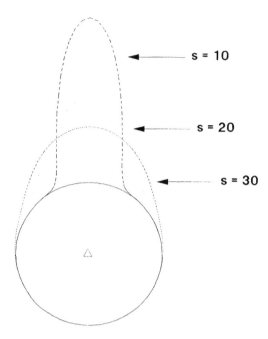

Figure 4.11 Examples of the von Mises distribution plotted around a circle for values of s of 10°, 20°, and 30°. The mean of each distribution is $\alpha = 0°$, giving a value of $\mu = 90°$. The receiving station (tower) is in the middle of the circle, with the transmitter directly north. The circle provides the zero baseline.

the distribution becomes more pointed; that is, the probability of observing bearings in the same direction increases as s gets smaller.

The full development of the maximum likelihood estimator of x_t and y_t from two or more bearings is given by Lenth (1981), and an extension of Lenth's work to incorporate a Bayesian statistical approach is described by Guttorp and Lockhart (1988). In the following section we review Lenth's development, mainly to provide a background for the statistically inclined reader with an interest in the derivation of the estimator. Readers without a quantitative background may want to skip to the next section. Intuitively, the maximum likelihood estimates of x_t and y_t are the values that maximize the probability of observing the recorded bearings. For example, in Fig. 4.12 each of the bearings from the three towers happens to be the most probable value because each of the bearings coincides with the mean of the von Mises distribution. In contrast, in Fig. 4.13 the observed bearings do not correspond exactly with the estimated location, and thus there is a need for a rigorous statistical estimation

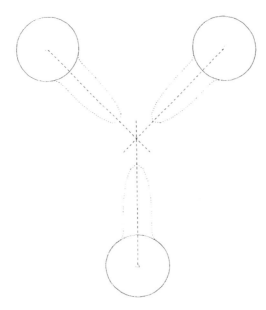

Figure 4.12 Example of a three-tower triangulation system, with the observed bearings plotted with their respective von Mises probability density functions. Each bearing happens to be the mode of its respective von Mises distribution.

procedure. The likelihood of the three observed bearings is the product of the von Mises probability functions, so the log likelihood is given by the sum of the logarithms of the von Mises distributions, that is,

$$\log L = n \log(2\pi) - n \log I_0(\kappa) + \kappa \sum_{i=1}^{n} \cos(\beta_i - \mu_i) \qquad (4.5)$$

The expected (mean) bearing from the ith tower to the transmitter (μ_i) can be expressed as $\mu_i = \tan^{-1}[(y_t - y_i)/(x_t - x_i)]$, so that Eq. 4.5 can be differentiated with respect to x_t and y_t and the resulting set of equations solved to provide maximum likelihood estimates of the transmitter's location ($\hat{x}_t$, $\hat{y}_t$). Lenth (1981) also suggests an iterative algorithm to compute ($\hat{x}_t$, $\hat{y}_t$), and the variance–covariance matrix (Q). Given the estimate of the variance–covariance matrix $\hat{Q}$, we assumed that a 95% confidence ellipse for the estimated transmitter location could be constructed based on the asymptotic normality of maximum likelihood estimators. The definition of confidence ellipse area estimate [$\hat{A}(x, y)$] for a particular location ($\hat{x}_t$, $\hat{y}_t$), given k tower coordinates (x_1, y_1), (x_2, y_2), . . . , (x_k, y_k), is

4 Estimating Animal Locations

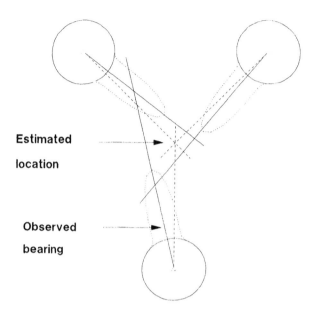

Estimated location

Observed bearing

Figure 4.13 Example of a three-tower triangulation system, with the observed bearings plotted with their respective von Mises probability density functions. The dashed lines indicate the bearings to the estimated location. The differences between the observed and estimated bearings represent the deviations or random errors (ε) in measuring the bearings.

$$\hat{A}(x_t, y_t) = \pi |\hat{Q}|^{1/2} \chi^2_{(2)(1-\alpha)} \qquad (4.6)$$

where $|\hat{Q}|$ is the determinant of the estimated variance–covariance matrix of the estimates of x_t and y_t from the maximum likelihood estimator of Lenth (1981)

$$\hat{Q} = \begin{bmatrix} \text{Var}(x_t) & \text{Cov}(x_t, y_t) \\ \text{Cov}(x_t, y_t) & \text{Var}(y_t) \end{bmatrix},$$

$|Q| = \text{Var}(x_t)\text{Var}(y_t) - \text{Cov}(x_t, y_t)^2$, π is 3.14159 . . . , and $\chi^2_{(2)(1-\alpha)}$ is the χ^2 statistic with 2 degrees of freedom at level $1 - \alpha$.

The calculation of $\hat{x}_t$, $\hat{y}_t$, and $\hat{Q}$ is performed with the FORTRAN 77 program TRIANG (available from the authors). TRIANG consists of three primary subroutines to calculate the estimates of location for the maximum likelihood, Huber, and Andrews estimators. TRIANG runs on an IBM personal computer and, to be useful, generally requires a math coprocessor because of the intensive numerical needs of the iterative process to achieve the three estimates. TRIANG originated on a Radio Shack Model 100 laptop computer for use in the field as a data recording device (White and Garrott 1984). This version only computed the maximum likelihood estimator, but did provide a graphics display to allow the user to examine the closeness of the intersection of the bearings and thus visually determine the quality of the location. To continue the precedent set in this text, a version of the Lenth (1981) maximum likelihood estimator is included as a SAS program shown in Fig. 4.14.

```
*  ---------------------------------------------------------------  *
|   SAS code to calculate the MLE of Lenth (1981).                  |
|      The purpose of this code is to calculate a maximum           |
|      likelihood estimate of an animal's location from 3           |
|      or more bearings from towers with known positions.           |
|      Input to the code consists of the date and time of           |
|      the location, the animal's identification number,            |
|      and a sequence tower identifications followed by             |
|      their respective bearings to the transmitter.                |
|      Instead of reading in the position of each of the 3          |
|      towers on each input record, the tower positions             |
|      are stored within the program and the tower id is            |
|      used to access the UTM coordinates for the tower.            |
|      The output for each input line consists of the               |
|      estimated location, area of the 95% confidence ellipse,      |
|      and a goodness-of-fit test of the estimate to the            |
|      observed bearings.                                           |
*  ---------------------------------------------------------------  *;
data trimle;
    drop pi A kappa v1 v2 v3 i j noiter converge initial
         zi di ss cs xm ym azcalc;
    label x1='MLE of x coordinate'
          y1='MLE of y coordinate'
          s='Standard deviation of bearings (degrees)'
          ellarea='Size of confidence ellipse'
          itower1='Tower number of bearing 1'
          az1='Bearing 1'
          itower2='Tower number of bearing 2'
          az2='Bearing 2'
          itower3='Tower number of bearing 3'
          az3='Bearing 3'
```

Figure 4.14 SAS program to calculate Lenth's (1981) maximum likelihood estimate of a location for three or more bearings along with the variance–covariance matrix of the estimate, the area of the 95% confidence ellipse, and a goodness-of-fit test of the estimate to the observed bearings. (*Figure continues.*)

4 Estimating Animal Locations

```
      date='Date of location'
      time='Time of location'
      id='Animal identification'
      modulat='Modulating or steady signal';
array itower{3} itower1 itower2 itower3;
array az{3} az1 az2 az3;
array beta{3};
array sinb{3};
array cosb{3};
array utm{2,8};
array amat{2,2};
array bmat{2};
pi=4*ATAN(1);
* Set Bearing Standard Deviation (in degrees) here;
s=1.5;
*Convert bearing standard deviation to kappa;
A=exp((s*pi/180)**2*(-0.5));
kappa=1/(2*(1-A)+(1-A)**2*(0.48794-0.82905*A-1.3915*A**2)/A);
* Set coordinates of towers ;
utm{1,1}=746917;  utm{2,1}=4391247;
utm{1,2}=748353;  utm{2,2}=4390596;
utm{1,3}=748212;  utm{2,3}=4392147;
utm{1,4}=749953;  utm{2,4}=4391223;
utm{1,5}=751023;  utm{2,5}=4392418;
utm{1,6}=750142;  utm{2,6}=4393673;
utm{1,7}=751674;  utm{2,7}=4393630;
utm{1,8}=750985;  utm{2,8}=4395022;
informat date yymmdd8. time time8.;
*Modify this statement for changes in number of towers;
input id $ date time modulat $
    itower1 az1 itower2 az2 itower3 az3;
format date date. time time.;
*Code to summarize bearings and plot;
do i=1 to dim(az);
    /* Convert bearing to an angle */
    beta{i}=(90-az{i})*(pi/180);
    sinb{i}=sin(beta{i});
    cosb{i}=cos(beta{i});
end;
noiter=0;
```

(Figure continues.)

```
   *Obtain initial estimate;
   x1=0; y1=0;
   do i=1 to dim(az);
      x1=x1+utm{1,itower{i}};
      y1=y1+utm{2,itower{i}};
      end;
   x1=x1/dim(az); y1=y1/dim(az);
   converge=0;
   initial=1;
iter:
   do i=1 to 2;
      bmat{i}=0;
      do j=1 to 2;
         amat{i,j}=0;
         end;
      end;
   do i=1 to dim(az);
      if az{i} ^= . then do;
         zi=sinb{i}*utm{1,itower{i}}-cosb{i}*utm{2,itower{i}};
         if initial then do;
            ss=sinb{i};
            cs=cosb{i};
            end;
         di=sqrt((x1-utm{1,itower{i}})**2+(y1-utm{2,itower{i}})**2);
         ss=(y1-utm{2,itower{i}})/(di**3);
         cs=(x1-utm{1,itower{i}})/(di**3);
         amat{1,1}=amat{1,1}+sinb{i}*ss;
         amat{2,2}=amat{2,2}+cosb{i}*cs;
         amat{1,2}=amat{1,2}-cosb{i}*ss;
         amat{2,1}=amat{2,1}-sinb{i}*cs;
         bmat{1}=bmat{1}+ss*zi;
         bmat{2}=bmat{2}-cs*zi;
         end;
      end;
   ym=(amat{1,1}*bmat{2}-amat{2,1}*bmat{1})
      /(amat{1,1}*amat{2,2}-amat{2,1}*amat{1,2});
   xm=(bmat{1}-amat{1,2}*ym)/amat{1,1};
   initial=0;
   if abs(xm-x1)/xm < 0.00001 and abs(ym-y1)/ym < 0.00001 then converge=1;
   x1=xm; y1=ym;
   noiter=noiter+1;
   if converge = 0 and noiter < 40 then goto iter;
```

Figure 4.14 *(Continues)*

```
if noiter >= 40 then do;
   put 'ERROR--Number of iterations exceed limit';
   put '    ' _N_= itower1= az1= itower2= az2= itower3= az3= ;
   end;
*Convergence achieved-get vc mat;
v1=0; v2=0; v3=0;
do i=1 to dim(az);
   if az{i} ^= . then do;
      di=sqrt((x1-utm{1,itower{i}})**2+(y1-utm{2,itower{i}})**2);
      cs=(x1-utm{1,itower{i}})/(di**3);
      ss=(y1-utm{2,itower{i}})/(di**3);
      v1=v1+sinb{i}*ss;
      v2=v2+cosb{i}*ss+sinb{i}*cs;
      v3=v3+cosb{i}*cs;
      end;
   end;
v2=v2*(-0.5);
ellarea=pi*cinv(0.95,2)*sqrt(1./((v1*v3-v2*v2)*kappa**2));
*Construct chi-square goodness of fit test;
chisq=0.;
do i=1 to dim(az);
   if az{i} ^= . then do;
      /* SAS code to calculate bearing from tower
         to estimated location */
      xm=x1-utm{1,itower{i}};
      ym=y1-utm{2,itower{i}};
      azcalc=atan(ym/xm);
      if xm < 0 & ym < 0 then azcalc=azcalc-pi;
      else if xm < 0 & ym > = 0 then azcalc=azcalc+pi;
      if azcalc < 0 then azcalc=azcalc+2*pi;
      /* Now convert angle in radians to bearing in degrees */
      azcalc=90-azcalc*(180/pi);
      if azcalc < 0 then azcalc=azcalc+360;
      if azcalc > 360 then azcalc=azcalc-360;
      /* Now determine difference between observed and
         expected bearings */
      xm=min(az{i},azcalc);
      ym=max(az{i},azcalc);
      if xm < 90 & ym > 270 then xm=xm+360;
      chisq=chisq+((xm-ym)/s)**2;
      put az{i}= azcalc= xm= ym= chisq=;
      end;
```

(*Figure continues.*)

```
    else chisq=.;
    end;
gofprob=1-probchi(chisq,dim(az)-2);

/* Each input line consists of:
    1) animal identification number,
    2) date (yy/mm/dd) of location,
    3) time (hh:mm:ss) of location,
    4) code for Steady or Modulating signal,
    5) tower number for first tower,
    6) bearing from the first tower,
    7) tower number for the second tower,
    8) bearing from the second tower,
    9) tower number for the third tower,
    10) bearing from the third tower.
    For the first observation, the bearing from
    tower 1 was 103, from tower 2 was 297, and
    from tower 3 was 229.                     */
cards;
210  84/06/06   14:38:43 S   1 103 2 297 3 229
210  84/06/06   15:04:30 S   1 101 2 298 3 233
210  84/06/06   15:34:32 S   1 102 2 298 3 229
210  84/06/06   17:07:31 M   1  99 2 300 3 229
210  84/06/06   17:40:48 M   1 165 2 281 3  .
210  84/06/06   18:07:51 S   1 150 2 284 3 221
210  84/06/06   18:38:40 S   1 146 2 284 3 222
;
proc print; run;
```

Figure 4.14 (*Continued*)

Maximum likelihood Eq. 4.5 assumes that κ is constant for each of the n towers. If for some reason the standard deviation is not the same for all the towers, then differing values of κ can be used to "weight" the individual bearings

$$\log L = n \log(2\pi) \sum_{i=1}^{n} [\kappa_i \cos(\beta_i - \mu_i) - \log I_0(\kappa_i)]$$

Modification of program TRIANG or the SAS code in Fig. 4.14 is straightforward to include this extension.

When the standard deviation of bearings from each of the towers is known from prior testing of the triangulation system (and hence a constant value of κ

4 Estimating Animal Locations

is not estimated from the observed bearings), a χ^2 test of goodness of fit can be constructed if three or more bearings are taken. This test examines the null hypothesis that all of the bearings came from the same point, that is, that none of the signals are reflections of the true signal. For each bearing, the squared difference between the observed bearing (β_i) and the bearing to the estimated location ($\hat{\mu}_i$) standardized by the standard deviation (s_i) has a χ^2 distribution with 1 degree of freedom. However, 2 degrees of freedom are lost because estimates of x_l and y_l are used to calculate μ_i. Thus, the goodness-of-fit test is

$$\chi^2 = \sum_{i=1}^{n} \left(\frac{\beta_i - \hat{\mu}_i}{s} \right)^2$$

with $n - 2$ degrees of freedom. Two warnings about this test must be offered. First, be sure that the units of β_i, μ_i, and s are the same; that is, do not find the difference between β_i and μ_i in radians and divide by s in degrees. Second, the difference between $\beta_i = 359°$ and $\mu_i = 1°$ is $2°$, not $358°$. Checks for β_i and μ_i straddling $0°$ must be incorporated into the test, as in the SAS code shown in Fig. 4.14.

Performance of Lenth's Estimators

As discussed later, one of the primary cases in which three or more bearings should be used for location estimates is when the topography of the study area results in signal reflection or bounce. Estimating a location from multiple bearings, when not all bearing intersections are in a small area, requires an analytical procedure which is robust to outliers. The Huber and Andrews estimators were developed by Lenth (1981) for this situation. Both of these estimators give approximately the same estimate of location as the maximum likelihood estimator for data without outliers. That is, the robust estimators are derived from the maximum likelihood estimator. For sets of bearings in which one or more are outliers (bounce or reflected signals), the robust estimators provide the capability to weight outliers much less than the nonreflected bearings, and thus reliable estimates of location are obtained even in the presence of some unreliable bearings. The weight of an individual bearing is based on how close intersections formed with other bearings approach the center of all possible intersections, with an iterative procedure used to determine the final weight of each bearing and hence the estimate of location.

The performance of Lenth's (1981) maximum likelihood, Huber, and Andrews estimators was tested with four 3-tower triangulation systems erected in a study area where reflected signals were common (Garrott et al. 1986). Five replicate bearings were obtained from each tower to transmitters placed at 20

TABLE 4.1
Number and Proportion of Surveyed Transmitter Locations Included within 95% Confidence Ellipses Generated Using Three Triangulation Estimators Provided by Lenth (1981)[a]

Confidence area ellipse (ha)	Maximum likelihood		Huber		Andrews	
	No. of points	Coverage (%)	No. of points	Coverage (%)	No. of points	Coverage (%)
0.0–0.1	7	57	9	44	3	100
0.1–0.2	39	51	36	56	20	100
0.2–0.3	62	57	44	68	25	92
0.3–0.4	141	29	82	51	51	80
0.4–0.5	81	42	80	49	60	60
0.5–0.6	29	41	45	27	15	80
0.6–0.7	11	36	11	36	9	33
0.7–0.8	2	100	9	22	9	33
0.8–0.9	4	0	7	0	16	0
0.9–1.0	2	0	10	0	4	0
>1.00	8	0	45	0	34	12
No estimate	14		22		154	
Total	400		400		400	

[a] Data are from Garrott et al. (1986).

surveyed locations scattered throughout the monitoring area of each three-tower triangulation system. The replicate bearings from each tower were randomly ordered. Location estimates were then calculated using each of the three Lenth (1981) estimators by combining one bearing from each of the three towers. This procedure resulted in 15 location estimates for each transmitter location, five each derived from the maximum likelihood, Huber, and Andrews estimators. The resulting data (Table 4.1) allow comparison of the three estimators when applied to identical triangulation problems.

One measure of the ability of these estimators to produce reliable location estimates is to determine how often the calculated 95% confidence ellipses cover the true location of the transmitter. Coverage for each of the estimators is presented as a function of confidence ellipse size in Table 4.1. For 95% confidence ellipses of <0.6 ha generated by the Andrews estimator, the ellipse covered (included) the actual transmitter location for 78% of the estimates, compared to 50% and 41% for the Huber and maximum likelihood estimators, respectively (Table 4.1). Although the Andrews estimator provided the best coverage, it was the least likely to generate an estimate. The algorithm failed for 154 of the 400 locations. When the Huber and maximum likelihood estimators were applied to the locations where the Andrews estimator failed, only 10% and 12%, respectively, of the respective confidence ellipses covered the true location. Thus, when the Andrews estimator failed to generate an estimate, the other two estimators were also unlikely to generate a correct estimate.

The distance between the actual and estimated locations is another measure of the performance of the three estimators. Again, the Andrews estimator was superior to the Huber and maximum likelihood estimators, with consistently smaller distances between the true location and the estimated location for confidence ellipses <0.6 ha (Fig. 4.15). Location estimates generated by all three estimators with confidence ellipses >0.6 ha, however, were unreliable, since coverage declined and the average distance between the true and estimated location increased (Table 4.1, Fig. 4.15).

From these results, we conclude that the Andrews estimator is the best estimator for calculating location estimates when reflected signals are common. Further, since all three estimators perform similarly when no bounce signals are present, the Andrews estimator appears to be the best choice when triangulating with three or more bearings, regardless of the presence or absence of reflected signals. Although the Andrews estimator will often fail to generate a

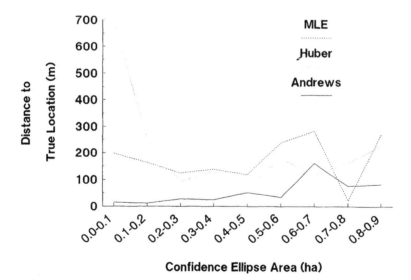

Figure 4.15 Mean distance (m) between estimated transmitter locations, generated by three estimators, from true locations as a function of confidence ellipse size (ha). The Andrews estimator has the smallest error distance for confidence ellipses <0.6 ha.

location estimate when bearings do not adequately converge, erroneous location estimates are still not uncommon (Table 4.1). Most of these "bad" locations can be eliminated by subjectively establishing a maximum confidence ellipse size for acceptable location estimates. Based on results of our tests, we decided that 0.6 ha was our maximum confidence ellipse size. However, each triangulation system is unique, requiring independent testing to establish a maximum confidence ellipse size appropriate for that system and the objectives of the study.

Data Quality Control and Censoring

An alternative technique to Lenth's estimators commonly used in triangulation studies in which precise locations are required is to obtain three or more bearings, but use ad hoc criteria to select the best pair of bearings to estimate the location. The two most prevalent criteria used are (1) the two bearings with an intersection nearest to 90° and (2) the bearings from the two receiving points providing the closest intersection to one of the receiving points (Deat et al. 1980). However, these criteria do not always provide the best estimate, as illustrated in Fig. 4.16. In this example towers 2 and 3 have accurate bearings, but the bearing from tower 1 has a 35° error. The bearing intersection from towers 1 and 3 is nearest to 90° and that from towers 1 and 2 intersect closest to one of the towers. Both are incorrect. The maximum likelihood and Huber estimators also calculated incorrect location estimates. The Andrews estimator failed to generate an estimate, indicating that the three bearings were inadequate for estimating a location, which is the appropriate result when the true location is unknown. Thus, the Andrews estimator also provides considerable improvement over subjective criteria for evaluating the quality of location estimates. Basically, the Andrews estimator provides "safety" with its estimates, providing no estimate, as opposed to inadvertent erroneous estimates.

Ideally, bearings that will produce poor location estimates should be identified in the field so that corrective measures or adjustments in the data collection procedures can be made immediately, minimizing data loss. A simple approach to field evaluation of triangulation data is to plot the bearings on a map and visually assess how close the bearings converge. Although this technique is better than no field evaluation of data, it is somewhat cumbersome, tedious, and time consuming. An alternative solution is the use of a portable battery-powered computer. There is a wide range of suitable portable computers on the market, from relatively inexpensive ($300–400) to sophisticated machines costing in excess of $1000.

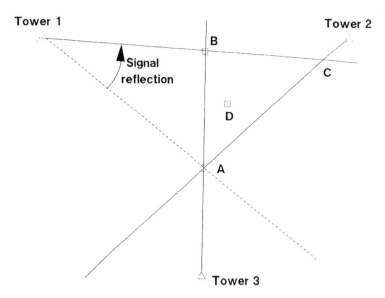

Figure 4.16 An example of how inaccurate location estimates can result in a three-tower triangulation system when signal reflection causes one bearing to be incorrect. (A) The position of the transmitter. (B) Selected by the Huber estimator and also on the basis of the bearing intersection closest to 90°. (C) Selected on the basis of closeness of the bearing intersection to one of the towers. (D) Selected by the maximum likelihood estimator. The Andrews estimator did not generate a location estimate for this example.

Two examples of triangulation programs written for use with portable computers are provided by White and Garrott (1984) and by Dodge and Steiner (1986). The programs are similar, prompting the user to enter animal identification and bearing values from various receiving locations. The x and y coordinates of the receiving stations are stored in the memory and used to perform triangulation calculations. The computer displays the data entered, a plot of the bearing intersections, coordinates of the maximum likelihood location estimate, and confidence ellipse size. These data, along with the current date and time obtained from the computer's internal clock are written on a file on the random access memory. Note that neither program uses the Andrews estimator; therefore, the location estimates and confidence ellipse size calculated by these programs may not be correct if reflected signals are present. The plot of the bearing intersections, however, will still allow the user to assess the quality of the location estimate.

The major advantage of a portable computer for field evaluation of triangulation data is the ability to use three or more bearings and complex analytical procedures to obtain immediate feedback on the quality of the data being collected. An additional advantage is that there is little chance of omitting data from location records, as the computer can prompt the telemetry user for all pertinent information. Since data files can be directly transferred by telephone or diskette to mainframe or microcomputers, errors in transcribing the data onto handwritten data forms and manually entering the data into a computer are also eliminated.

Inherent throughout this discussion of field evaluation of data is the idea of data censoring, that is, the elimination of poor-quality bearings and/or location estimates from the data. Most telemetry users probably practice some form of data censorship, as triangulation studies are usually conducted under less than ideal conditions, necessitating an assessment of the quality of estimates and eliminating poor ones. Depending on the objectives and the design of the study, criteria used to reject location estimates may be formally established by field tests. Possible criteria include confidence ellipse size less than some cutoff value, or less rigidly established criteria such as the location does not seem probable. These censoring procedures are an important part of a triangulation study, as they have direct impacts on the quality and the quantity of data. However, seldom are censoring procedures described and the proportion of data actually used in analysis reported. We recommend that reports of triangulation studies include this information in order to aid others in evaluating the study's conclusions and designing their own studies.

Summary

1. Estimating an animal's location is a statistical problem of estimating $\hat{x}_t$ and $\hat{y}_t$. Unbiased estimates are desired, as well as a confidence area about the estimate.
2. No matter what method of locating an animal and estimating its location is used, the investigator should verify the methodology by doing some investigations to check the accuracy. That is, transmitters should be placed at known locations and field researchers doing the study should estimate these locations to provide estimates of the precision of the methods being used.
3. Because the bearings used to triangulate an animal's location lap around at 360° to 0°, the normal distribution is not an appropriate model of the distribution of a bearing. A logical probability density function for this process is the von Mises distribution.

4. To detect erroneous bearings and to guard against erroneous location estimates, at least three bearings should be taken to estimate the transmitter location. The estimators developed by Lenth (1981) are pertinent to this estimation problem.
5. When signal bounce is expected, Lenth's (1981) Andrews estimator is preferred because the erroneous signals are detected and given less weight in determining the estimate of the transmitter location. The performance of the Andrews estimator is nearly identical to the maximum likelihood estimator when no signal bounce is present, so we suggest that it be used routinely.

References

Amlaner, C. J., Jr. and D. W. Macdonald (eds.). 1980. A Handbook on Biotelemetry and Radio Tracking. Pergamon Press, Oxford, England. 804 pp.

Batschelet, E. 1981. Circular Statistics in Biology. Academic Press, New York. 371 pp.

Beaty, D. W., S. M. Tomkiewicz, Jr., and J. Carter. 1987. Accessory equipment supplementing ARGOS data collection and processing. Pages 181–190 *in* Proc. Argos Users Conf., Service Argos, Greenbelt, MD.

Brander, R. B. and W. W. Cochran. 1971. Radio-location telemetry. Pages 95–105 *in* R. J. Giles ed. Wildlife Management Techniques, 3rd ed. The Wildlife Society, Washington, D.C.

Buechner, H. K., F. C. Craighead, Jr., J. J. Craighead, and C. E. Cote. 1971. Satellites for research on free-roaming animals. BioScience 21:1201–1205.

Burhans, R. W. 1983. Using LORAN-C and time and frequency calibration. Radio-Electronics 83(July):63–65.

Byles, R. 1987. Development of a sea turtle satellite biotelemetry system. Pages 199–210 *in* Proc. Argos Users Conf., Service Argos, Greenbelt, MD.

Cochran, W. W. 1980. Wildlife telemetry. Pages 507–520 *in* S.D. Schemnitz ed. Wildlife Management Techniques, 4th ed. The Wildlife Society, Washington, D.C.

Craighead, F. C., Jr., J. J. Craighead, C. E. Cote, and H. K. Buechner. 1972. Satellite and ground radio tracking of elk. Pages 99–111 *in* S. Galler et al. Animal Orientation and Navigation. Natl. Aeronaut. Space Admin., Washington, D.C.

Craighead, J. J., F. C. Craighead, Jr., J. R. Varney, and C. E. Cote. 1971. Satellite monitoring of black bears. BioScience 21:1206–1212.

Deat, A., C. Mauget, R. Mauget, D. Maurel, and A. Sempere. 1980. The automatic, continuous and fixed radio tracking system of the Chize Forest: theoretical and practical analysis. Pages 439–451 *in* C. J. Amlaner, Jr. and D. W. Macdonald eds. A Handbook on Biotelemetry and Radio Tracking. Pergamon Press, Oxford, England.

Dodge, W. E. and A. J. Steiner. 1986. XYLOG: A computer program for field processing locations of radio-tagged wildlife, Tech. Rep. 4. U.S. Fish and Wildl. Serv., Washington, D.C. 22 pp.

Dodge, W. E., D. S. Wilkie, and A. J. Steiner. 1986. UTMTEL: a laptop computer program for location of telemetry "finds" using LORAN C. Massachusetts Cooperative Wildlife Research Unit, Amherst. 21 pp.

Englert, K. 1982. LORAN-C A marine long range navigational tool. Popular Electronics 82(August):40–47.

Fancy, S. G., L. F. Pank, D. C. Douglas, C. H. Curby, G. W. Garner, S. C. Amstrup, and W. L. Regelin. 1988. Satellite telemetry: a new tool for wildlife research and management. Resour. Publ. No. 172. U.S. Fish and Wildl. Serv., Washington, D.C. 54 pp.

Fuller, M. R., N. Levanon, T. W. Strikwerda, W. S. Seegar, J. Wall, H. D. Black, F. P. Ward, P. W. Howey, J. Partelow, and P. Lert. 1984. Feasibility of a bird-borne transmitter for tracking via satellite. Pages 1–6 *in* Proc. 8th Int. Symp. on Biotelemetry, Dubrovnik, Yugoslavia.

Garrott, R. A., G. C. White, R. M. Bartmann, and D. M. Weybright. 1986. Reflected signal bias in biotelemetry triangulation systems. J. Wildl. Manage. 50:747–752.

Garrott, R. A., G. C. White, R. M. Bartmann, L. H. Carpenter, and A. W. Alldredge. 1987. Movements of female mule deer in northwest Colorado. J. Wildl. Manage. 51:634–643.

Gilmer, D. S., L. M. Cowardin, R. L. Duval, L. M. Mechlin, C. W. Shaiffer, and V. B. Kuechle. 1981. Procedures for the use of aircraft in wildlife biotelemetry studies. Resour. Rep. No. 140. U.S. Fish and Wildl. Serv., Jamestown, ND. 19 pp.

Grigg, G. 1987. Tracking camels in central Australia. Argos Newsletter 29:1–3.

Guttorp, P. and R. A. Lockhart. 1988. Finding the location of a signal: a Bayesian analysis. J. Am. Stat. Assoc. 83:322–330.

Harrington, F. H., A. N. Veitch, and S. N. Luttich. 1987. Tracking barren-ground and woodland caribou by satellite: the more the need for PTTs, the better they work. Pages 221–242 *in* Proc. Argos Users Conf., Service Argos, Greenbelt, MD.

Hupp, J. W. and J. T. Ratti. 1983. A test of radio telemetry triangulation accuracy in heterogeneous environments. Pages 31–46 *in* D. G. Pincock ed. Proc. 4th Int. Wildl. Biotelemetry Conf. Applied Microelectronics Institute and Technical Univ. of Nova Scotia, Halifax.

Kenward, R. E. 1987. Wildlife Radio Tagging. Academic Press, San Diego, CA. 222 pp.

Kolz, A. L., J. W. Lentfer, and H. G. Fallek. 1980. Satellite radio tracking of polar bears instrumented in Alaska. Pages 743–752 *in* C. J. Amlaner, Jr. and D. W. Macdonald eds. A Handbook on Biotelemetry and Radio Tracking. Pergamon Press, Oxford, England.

Lemnell, P. A. 1980. An automatic telemetry system for tracking and physiology. Pages

453–456 *in* C. J. Amlaner, Jr. and D. W. Macdonald, eds. A Handbook on Biotelemetry and Radio Tracking. Pergamon Press, Oxford, England.

Lemnell, P. A., G. Johnsson, H. Helmersson, O. Holmstrand, and L. Norling. 1983. An automatic radio-telemetry system for position determination and data acquisition. Pages 76–93 *in* D. G. Pincock ed. Proc. 4th Int. Wildl. Biotelemetry Conf. Applied Microelectronics Institute and Technical Univ. of Nova Scotia, Halifax.

Lenth, R. V. 1981. On finding the source of a signal. Technometrics 23:149–154.

Lert, P. 1984. Loran-C for lightplanes. Air Progress December:11–20.

Mardia, K. V. 1972. Statistics of Directional Data. Academic Press, New York. 357 pp.

Marsh, H. and G. B. Rathbun. 1987. Tracking dugongs. Argos Newsletter 29:9.

Mate, B., G. Rathburn, and J. Reed. 1986. An Argos-monitored radio tag for tracking manatees. Argos Newsletter 26:3–7.

Mate, B. R., J. P. Reid, and M. Winsor. 1987. Long-term tracking of manatees through the Argos satellite system. Pages 211–220 *in* Proc. Argos Users Conf., Service Argos, Greenbelt, MD.

Mech, L. D. 1983. Handbook of Animal Radio-Tracking. Univ. of Minn. Press, Minneapolis. 107 pp.

Pank, L. F., W. L. Regelin, D. Beaty, and J. A. Curatolo. 1985. Performance of a prototype satellite tracking system for caribou. Pages 97–118 *in* R. W. Week and F. M. Long eds. Proc. 5th Int. Conf. Wildl. Biotelemetry, Univ. of Wyoming, Laramie.

Patric, E. F., T. P. Husband, G. G. McKiel, and W. M. Sullivan. 1988. Potential of LORAN-C for wildlife research along coastal landscapes. J. Wildl. Manage. 52:162–164.

Priede, I. G. 1984. Argos tracks a shark. Argos Newsletter 19:6–7.

SAS Institute Inc. 1985. SAS® Language Guide for Personal Computers, Version 6 Edition. SAS Institute Inc., Cary, NC. 429 pp.

Saltz, D. and P.U. Alkon. 1985. A simple computer-aided method for estimating radiolocation error. J. Wildl. Manage. 49:664–668.

Schweinsburg, R. E. and L. J. Lee. 1982. Movement of four satellite-monitored polar bears in Lancaster Sound, Northwest Territories. Arctic 35:504–511.

Snyder, J. P. 1982. Map projections used by the U.S. Geological Survey, Bull. 1532. U.S. Geological Survey, Washington, D.C. 313 pp.

Springer, J. T. 1979. Some sources of bias and sampling error in radio triangulation. J. Wildl. Manage. 43:926–935.

Stoneburner, D. C. 1982. Sea turtle (*Caretta caretta*) migration and movements in the South Atlantic Ocean, NASA SP-457. Natl. Aeronaut. Space Admin., Washington, D.C. 74 pp.

Strikwerda, T. E., H. D. Black, N. Levanon, and P. W. Howey. 1985. The bird-borne transmitter. Johns Hopkins APL Tech. Dig. 6:60–67.

Strikwerda, T. E., M. R. Fuller, W. S. Seegar, P. W. Howey, and H. D. Black. 1986.

Bird-borne satellite transmitter and location program. Johns Hopkins APL Tech. Dig. 7:203–208.

Timko, R. E. and A. L. Kolz. 1982. Satellite sea turtle tracking. Mar. Fish. Rev. 44:19–24.

Tomkiewicz, S. M., Jr. and D. W. Beaty. 1987. Wildlife satellite telemetry: a progress report—1987. Pages 191–198 *in* Proc. Argos Users Conf., Service Argos, Greenbelt, MD.

Underwood, D. 1983. Navstar:supernav or overkill? Motor Boating & Sailing 82(September):88–92.

White, G. C. and R. A. Garrott. 1984. Portable computer system for field processing biotelemetry triangulation data. Colo. Div. Wildl. Game Inf. Leafl. 110:1–4.

Yerbury, M. J. 1980. Long range tracking of *Crocodylus porosus* in Arnhem Land, Northern Australia. Pages 765–776 *in* C. J. Amlaner, Jr. and D. W. Macdonald eds. A Handbook on Biotelemetry and Radio Tracking. Pergamon Press, Oxford, England.

CHAPTER

5

Designing and Testing Triangulation Systems

Triangulation systems vary in complexity from an investigator using a hand-held antenna to obtain directional bearings from two locations, to vehicle mounted receiving systems, to automated systems which simultaneously determine bearings from numerous fixed radio-tracking towers. Regardless of the sophistication of the triangulation system, bearings are only an estimate of the direction from the receiving point to the transmitter. Bearing errors may be caused by many factors, such as receiving and transmitting locations, terrain, equipment, weather, observers, power lines, and vegetation (Slade et al. 1965, Tester 1971, Cederlund et al. 1979, Springer 1979, Hupp and Ratti 1983, Lee et al. 1985, Garrott et al. 1986). An important concept to understand when using triangulation procedures is that because bearings are not measured exactly (i.e., observed bearings are estimates of the true but unknown bearings), only an estimate of an animal's location is obtained. Thus, in addition to an estimate of an animal's location (point estimate), an estimate of the probable area (area estimate) where the animal is likely to be found is also desirable (see Chapter 4).

Despite the wide use of triangulation techniques, few investigators perform rigorous tests of the accuracy of their triangulation systems. Several studies have documented bearing errors in excess of $10°$ (Hupp and Ratti 1983, Lee et al. 1985, Garrott et al. 1986), which would result in location estimates far from the true location of the transmitters. The amount of error which is acceptable for a given study generally depends on the objectives of the study. A triangulation system which produces point estimate errors of 0.5–1.0 km may be adequate for studies of the seasonal movements of a migratory elk herd that travels hundreds of kilometers. The same triangulation system, however, probably would be unacceptable if the objective of the study is to determine the proportion of time that cow elk occupy various vegetative communities during the calving season. Hence, when an investigator decides that triangulation techniques must be used to meet the objectives of his or her study, there are a multitude of decisions which must be made with respect to designing and testing the triangulation system, data collection procedures, and analytical tools employed to obtain point and area location estimates. This chapter is devoted to discussing these topics.

Measuring Accuracy of Directional Bearings

No matter how much time, effort, and thought an investigator devotes to designing a triangulation system, the quality of location estimates produced by the system is unknown until it has been tested in the field. The importance of testing the triangulation system cannot be overemphasized. All too often, investigators fail to test their triangulation systems and are content to consider point estimates derived from bearing intersections as exact locations, with little or no consideration of accuracy or precision. Regardless of the sophistication of the system, it must be tested (1) to determine the standard deviation of directional bearings which is needed to estimate precision (see Chapter 4), (2) to determine whether the system is capable of producing location estimates of adequate accuracy to meet the objectives of the study, and (3) to establish the field protocol to be used during data collection. Generally, the complexity and intensity of the testing procedure employed depend on the anticipated or desired precision of location estimates, which, in turn, is dictated by study objectives. Measuring bearing accuracy may only involve a day or two of fieldwork for studies requiring only approximate location estimates, such as seasonal movements. In contrast, studies of habitat use by animals occupying complex vegetative mosaics require very precise locations and, hence, considerable time and effort must be devoted to testing and, if necessary, improving the triangulation system.

The accuracy of directional bearings has two components, bias and precision, which are measured by studying bearing errors. Error (ε) is the difference between the true bearing (α) and the bearing estimated by the receiving system ($\hat{\alpha}$), where the circumflex indicates an estimate of the parameter. Figure 5.1

5 Designing and Testing Triangulation Systems

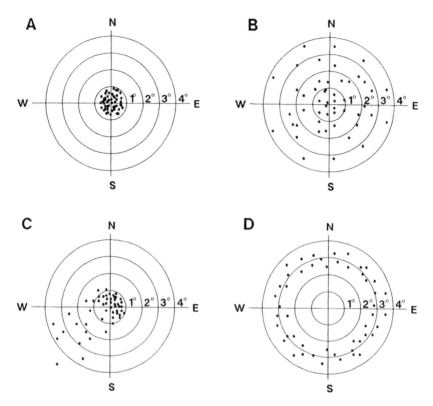

Figure 5.1 Plots of bearing error illustrating the concepts of bias and precision for receiving systems used in radio-tracking studies. Error distribution would be expected with well-designed and operated tower/vehicle systems (A) and hand-held antenna (B). (C) A directional problem with signal interference. (D) A biased receiving system.

graphically illustrates the concepts of bias and precision of bearing errors. If the study objectives require accurate locations, the receiving system should be capable of determining bearings to transmitters which are both unbiased and very precise, as illustrated in Fig. 5.1A. One could expect this quality of performance from a well-designed fixed tower or vehicle-mounted system which employs stacked Yagi antennae and a "null" combiner. If coarser location estimates are acceptable, a less precise receiving system may be adequate, but bearing errors should still be unbiased, as depicted in Fig. 5.1B. The proper operation of a hand-held two- or three-element Yagi antenna and compass would be expected to provide this level of performance.

Figure 5.1C and D illustrate typical problems with receiving systems. In Fig. 5.1C the receiving system is unbiased, but precision in one direction is much poorer than in other quadrants of the compass. The directionality of the problem indicates that something in the direction of the imprecise bearings is interfering with the reception of the transmitter's signals. Such problems can be caused by electrical interference or "noise" from power lines, objects too close to the antenna, or signal reflection and/or absorption due to cliffs, buildings, dense vegetation, etc. The error distribution in Fig. 5.1D depicts a precise receiving system that is biased in all directions. This problem is usually caused by a misaligned compass rosette or pointer on a fixed tower or vehicle-mounted system, or, if a hand-held compass is used, an improper magnetic declination setting.

In order to measure bearing accuracy, multiple measurements of error are needed so that

$$\varepsilon_{ij} = \alpha_i - \hat{\alpha}_{ij} \tag{5.1}$$

for each location i and replicate j (Lee et al. 1985). Calculation of a mean error ($\bar{\varepsilon}$) is then possible with

$$\bar{\varepsilon} = \frac{\sum_{i=1}^{n} \sum_{j=1}^{r} \varepsilon_{ij}}{nr} \tag{5.2}$$

where n is the number of locations and r is the number of replicates. Some sources of error, such as the inherent directionality of an antenna, are random, and the mean error of a large number of bearings should be near zero. A mean error significantly different from zero is bias. Bias problems should be investigated and corrected before further testing or data collection. Precision is a measure of the variability of the estimated bearings and may be defined by the standard deviation (SD)

$$SD = \left[\frac{\sum_{i=1}^{n} \sum_{j=1}^{r} (\varepsilon_{ij} - \bar{\varepsilon})^2}{(nr - 1)} \right]^{1/2} \tag{5.3}$$

This estimate of SD is the basis for calculating all area estimates, regardless of the analytical procedure chosen. The above formulas assume that the ε_{ij}'s are sampled from identical populations for each location. This assumption is inherent in the use of the triangulation system, because the investigator is attempting to estimate locations, and hence cannot know the variance structure of the errors a priori.

From the above definitions, it should be obvious that the only way to quantify the bias and precision of bearings is to place transmitters in a variety of

5 Designing and Testing Triangulation Systems

known locations throughout the study area and obtain multiple bearing estimates on each transmitter location from the receiving points. The testing scheme should be designed to provide a balance between the subjective decisions characteristic of radio-tracking and an objective systematic approach. For example, if a vehicle-mounted receiving system will be used, the test should not be performed from random receiving points along the road system. Such a test may place the receiving system in areas of poor signal reception, such as depressions in topography, or in stands of dense vegetation, where tracking would not normally be attempted. On the other hand, testing should not be designed to maximize performance of the receiving system, such as placing all test transmitters at elevated points within line of sight of the receiving antenna. Instead, the investigator should strive to design a testing procedure that simulates actual data collection as closely as possible.

To illustrate testing procedures for quantifying the accuracy of directional bearings, we will use an example from a radio-tracking study conducted in northwestern Colorado (Garrott et al. 1986). The objectives of the study required relatively precise locations, so a series of permanent receiving towers was erected on ridge tops in triangular patterns with animals to be simultaneously tracked from three towers. Twenty test locations were used on each tracking area, which was delineated by a three-tower system. The locations were subjectively selected to represent a variety of topographical and vegetative situations, from open ridge tops to steep forested slopes (Fig. 5.2). If logistics are not prohibitive, an alternative to subjective placement of test locations would be to evenly space test locations in a grid pattern throughout the study area. Transmitter locations on a grid would allow the investigator to produce isopleth maps of the estimated precision of location estimates similar to those in Figs. 5.7–5.11 (Cederlund et al. 1979, Lemnell et al. 1983). Such maps would help delineate areas that would produce poor-quality location estimates and aid in attempts to enhance the design of the triangulation system (e.g., adding another receiving point in the problem area).

Test transmitter and tower locations were surveyed to obtain the precise location of each in UTM coordinates. An alternative to surveying would have been to estimate the locations using maps or aerial photos. Such locations, however, are less accurate, due to distortion of photos, map inaccuracies, and observer limitations. Although visually determined locations may be adequate for some studies, fixed towers were expected to provide very good precision. This expected precision justified the additional effort to survey in all locations, as any inaccuracies in test locations introduce additional error into the test results.

Analysis of Wildlife Radio-Tracking Data

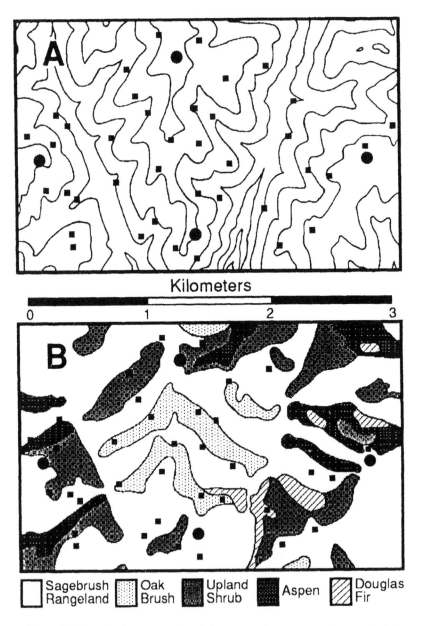

Figure 5.2 Map of a four-tower triangulation system (three towers used at one time) illustrating the position of the receiving towers (circles) and the test transmitter locations (squares) with respect to topography (A) and major plant communities (B). Topographical lines in (A) represent 61-m (200-foot) contour intervals.

5 Designing and Testing Triangulation Systems

Once the x and y coordinates of the receiving towers and test transmitter locations were determined, the true bearing (α) from each receiving point to each test transmitter location was calculated. This was done using basic trigonometric formulas for solving angles and sides of a right triangle. Figure 5.3 provides an example of how these calculations were performed for two test transmitters located in opposite directions from the receiving point. Sides a of right triangles 1 and 2 were calculated by subtracting the x coordinate of the receiving location from the x coordinates of transmitter locations 1 and 2, respectively. Side b is calculated similarly, using the y coordinates. Thus, side $1a$ equals 428 m (737312 − 737740), side $1b$ equals 708 m (4408928 − 4409636), side $2a$ equals 582 m (737312 − 736730), and side $2b$ equals 697 m (4408928 − 4408231). Note that negative values were ignored, as we were only interested in the distance between locations and not the position of locations with respect to each other.

Since we now know two sides of the right triangles, angles θ_1 and θ_2 can be solved as

$$\text{tan of angle} = \frac{\text{length of opposite side}}{\text{length of adjacent side}} = \frac{a}{b}$$

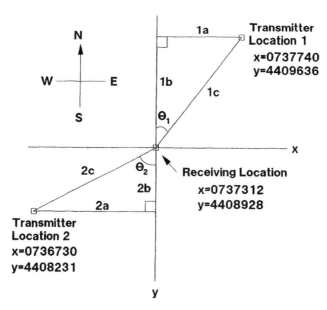

Figure 5.3 Examples of trigonometric principles used to calculate the true bearings from a receiving point to test transmitter locations.

The hypotenuse of each triangle (side c), which is the distance between the receiving and transmitting locations, can also be solved as:

$$c^2 = a^2 + b^2$$

giving

$$c = \sqrt{a^2 + b^2}$$

Thus, angle θ_1 is 31° (tan θ_1 = 428/708 = 0.6045), and angle θ_2 is 40° (tan θ_2 = 582/697 = 0.8350). The hypotenuse of triangle 1 equals 827 m ($\sqrt{428^2 + 708^2}$) and the hypotenuse of triangle 2 equals 908 m ($\sqrt{582^2 + 697^2}$). The true bearing (α_1) from the receiving location to transmitter location 1, located to the northeast, is θ_1 or 31°; however, since transmitter location 2 is to the southwest of the receiving location, the true bearing is θ_2 + 180°, or 220°. To obtain true bearings for transmitter locations in the southeast quadrant, the calculated angle of the right triangle would be subtracted from 180°, and for transmitter locations in the northwest quadrant the calculated angle would be subtracted from 360°.

In actual practice the SAS code (Fig. 5.4) could be used to calculate the bearing from the receiving location to the transmitter.

This piece of code can be included in a DATA step program and called by

```
atan2:   /* SAS code to calculate bearing from
            point (x1, y1) to point (x2, y2) */
         pi = atan(1.)*4;
         xp = x2-x1; yp = y2-y1;
         if xp > 0 then theta = atan(yp/xp);
         else if xp < 0 then theta = pi+atan(yp/xp);
         else if xp = 0 then do;
            if yp < 0 then theta = pi*3/2;
            else if yp > 0 then theta = pi/2;
            else theta = .; end;
         if theta < 0 then theta = theta+2*pi;
         /* Now convert angle in radians to bearing in degrees */
         alpha = 90-theta*(180/pi);
         if alpha < 0 then alpha = alpha+360;
   return;
```

Figure 5.4 SAS code to calculate bearings from receiving locations to known transmitter locations.

5 Designing and Testing Triangulation Systems

the user with the LINK command. Note that in the computer code, the checking for quadrant (NE, SW, etc.) is taken care of with the checks on the sign of xp and yp.

After calculating true bearings from each receiving point to the test locations, the transmitters were deployed. The most commonly used technique is to affix the transmitters to wooden stakes placed at the test locations (Cederlund et al. 1979, Springer 1979, Hupp and Ratti 1983, Lee et al. 1985, Garrott et al. 1986). This "fence post" technique is attractive because it is simple and convenient, but several authors have suggested that fence post tests do not accurately duplicate field conditions during actual data collection (Cederlund et al. 1979, Cochran 1980, Lee et al. 1985). Two specific characteristics not mimicked by fence post studies are signal absorption due to the proximity of the animal's body to the transmitter antenna (Cochran 1980, Hupp and Ratti 1983) and signal attenuation or modulation due to the movement of animal-carried transmitters (Cederlund et al. 1979, Lee et al. 1985). Techniques that have been used in an attempt to simulate these conditions include attaching the test transmitters to bottles of saline solution (Hupp and Ratti 1983) or to tethered dogs (Cederlund et al. 1979) and having a person move the transmitter in a standardized manner at each test location (Lee et al. 1985). Ideally, the test procedure should simulate actual data collection as closely as possible, although practical considerations may restrict the investigator to fence post tests. In this situation the investigator should recognize that signals from test transmitters are optimal and that, with transmitters carried by animals, the precision of bearing errors may be somewhat poorer than indicated by the test results.

Before testing the precision of directional bearings, tests were conducted to determine whether the compass rosette pointers at the base of the towers were oriented correctly with the antenna. Such tests are required for all receiving systems for which some sort of directional indicator is aligned with the antenna receiving pattern (usually fixed towers and some vehicle-mounted systems). Transmitters were placed at five of the test locations within the line of sight of each tower and in a variety of directions from the towers, with five bearings obtained for each transmitter. It was important that each replicate bearing be an independent observation, that is, previously determined bearings could have no influence on the value of the current bearing. In this study we achieved independence by randomly ordering the sequence in which bearings were obtained and assuring anonymity of transmitter frequency–location pairings. In addition, a dummy rosette was placed over the actual compass rosette to prevent placing the pointer exactly where the previous reading had been.

TABLE 5.1
Example of Data from a Test
of Pointer–Antenna Orientation[a]

Transmitter frequency	True bearing	Estimated bearing	Error true-estimated
150	50	53	−3
		56	−6
		54	−4
		55	−5
		54	−4
190	78	82	−4
		84	−6
		86	−8
		83	−5
		80	−2
230	133	137	−4
		137	−4
		139	−6
		138	−5
		140	−7
270	169	174	−5
		172	−3
		174	−5
		174	−5
		175	−6
310	206	210	−4
		209	−3
		213	−7
		212	−6
		208	−2
		Mean	−4.76

[a] Five replicate bearings were obtained for each of five test transmitter locations. The estimated bearings are always greater than the true bearings. The consistently negative errors indicate misalignment of the pointer, and the wide range of errors within each transmitter location indicates relatively poor precision.

5 Designing and Testing Triangulation Systems

After a bearing was determined, the pointer and the antenna mast were held stationary, while the dummy rosette was removed to record the actual bearing. The dummy rosette was then put back in place and rotated to a new position for determining the next bearing.

An example of antenna–pointer orientation data is provided in Table 5.1. A calculated mean error of $-4.76°$ indicates that the antenna and the pointer were not aligned. Plotting histograms of the bearing errors (Fig. 5.5) showed that the errors were approximately normally distributed and skewed to the negative side of zero, confirming the need to readjust the pointer by approximately 5°. After the pointer was moved, the test was repeated to confirm proper alignment. In this example all bearing errors were similar. If, however, most bearing errors cluster around zero, but bearings from one or two test transmitters are incorrect, the problem is not with antenna–pointer alignment. In this case the investigator should look for other problems in the direction of the problem transmitters, such as electrical lines, objects close to the receiving antenna, and reflected signals.

With all towers properly oriented, bearings were taken on all test transmitter locations on the study area (Garrott et al. 1986). Independence of replicates was assured as described for orientation procedures. These data were then used

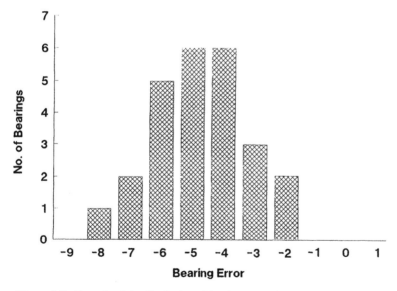

Figure 5.5 Example of the distribution of bearing errors from a test of antenna–pointer orientation, with the pointer out of alignment by approximately 5°.

in Eqs. 5.1–5.3 to calculate the mean error and precision of each receiving system. If the precision is not adequate to meet the objectives of the study, the investigator must attempt to improve the performance of the triangulation system. As discussed in the previous section, some possible solutions are improving equipment, rearranging or adding additional receiving points, decreasing size of the area to be monitored, and increasing the number of bearings used to determine each location. Failure to adequately improve the performance of the triangulation system may necessitate abandonment of the technique or redefinition of the objectives of the study.

Special Considerations for Mobile Triangulation Systems

In our discussion of testing triangulation systems, we used an example involving receiving towers at fixed locations. Triangulation systems based on permanent towers are a good choice when study objectives require precise location estimates, and the duration and intensity of the radio-tracking schedule justifies the additional expense of erecting multiple receiving systems. Two disadvantages of fixed receiving locations, however, are that the investigator is limited to working intensively in a relatively small area and study animals must be present in high enough densities to provide sufficient numbers of instrumented animals for meaningful sample sizes. In the example presented in the previous section, neither of these restrictions was a problem, because the areas occupied by individual deer were relatively small (<1.3 km^2) compared to the area being monitored (6.6 km^2) and many more deer were present than we could possibly monitor. Many studies require the tracking of animals that are in low densities over large areas or involve the tracking of highly mobile animals. In these situations hand-held and vehicle-mounted receiving systems provide the flexibility to move over a large area and remain close to the animals being monitored. There are, however, a number of special problems that the investigator must consider when using a mobile triangulation system.

Present throughout the discussion in this chapter is the assumption that the location of the receiving unit is known and, hence, has no variance. This condition can be met with hand-held and vehicle-mounted antenna systems simply by establishing a series of locations throughout the study area with known coordinates that will be used for all tracking. Occasionally, an investigator will be forced to obtain bearings from unknown locations. A common solution to this problem is to obtain bearings from the unknown receiving point to trans-

5 Designing and Testing Triangulation Systems

mitters placed at known locations (beacons) on the study area. The estimators provided by Lenth (1981) can then be used to estimate the position of the receiving unit by reversing the bearings, that is, converting the bearings from the receiving unit to the beacons to bearings from the beacons to the receiving unit and treating the beacons as towers. In this fashion the position of the mobile receiver is located relative to the beacons. The question that arises from this procedure is how much error is introduced into the fixes taken on instrumented animals from the uncertainties of the receiver positioning process? Because the Lenth (1981) estimators assume that the receiving locations are known, enough readings on beacons would be required to estimate the location of the mobile receiver with precision much greater than the precision of the animal locations. In reality, this may prove impractical. Unless study objectives can be met with relatively "coarse" location estimates, triangulation from unknown locations is not recommended.

Because mobile units are moved from one location to the next, a problem arises in determining the bearing direction relative to true north. With fixed towers, a compass rosette and a pointer are used, with the antenna receiving path and pointer oriented by using beacons at known locations. Since the receiving system is fixed once the orientation process has been completed, the investigator can directly read bearings for instrumented animals from the compass rosette. Vehicle-mounted receiving systems may also utilize a compass rosette and an orientation procedure similar to that used for fixed towers. However, because the vehicle is constantly moved from one location to the next, the position of the rosette relative to north will change with every move, necessitating reorientation. The number of beacons and replicate bearings needed to adequately orient the system is dependent on the desired quality of the location estimates.

To calculate the number of beacons and replicate bearings needed, assume a receiving precision of s; that is, a bearing is read from the mobile receiver with a standard deviation of s or variance s^2. To position the compass rosette correctly with respect to true north, a reading θ_1 is taken, and thus the variance of θ_1 is s^2. Because the location of the mobile unit and the location of the beacon are both known, the true bearing from the mobile unit location to the beacon is known (α). Now a bearing is taken to the animal (θ_2), again with variance s^2. Then the azimuth (θ) from the mobile receiver to the animal calculated from true north is

$$\theta = \theta_2 + (\theta_1 - \alpha)$$

Because α is known (not a random variable), and because the variance of a sum is the sum of the variances, the variance of θ is

$$\begin{aligned}\text{Var}(\theta) &= \text{Var}(\theta_2) + \text{Var}(\theta_1) \\ &= s^2 + s^2 \\ &= 2s^2\end{aligned} \quad (5.4)$$

Thus, the confidence ellipse calculated for the location of the animal should be calculated with an SD of $\sqrt{2}s$.

The variance of θ can be reduced by taking readings on more than one beacon, producing a mean for θ_1 rather than a single bearing. Thus, if n different beacons are used, the variance of $\bar{\theta}_1$ is s^2/n, and this value would be substituted into Eq. 5.4 to produce a variance of θ as

$$\begin{aligned}\text{Var}(\theta) &= \text{Var}(\theta_2) + \text{Var}(\bar{\theta}_1) \\ &= s^2 + \frac{s^2}{n}\end{aligned}$$

Thus,

$$\text{SD} = \sqrt{\frac{n+1}{n}}\, s$$

Rather than take bearings on different beacons to orient a tower, repeated readings on a smaller number of beacons could be used for the same purposes. The assumption made in taking repeated readings from the same beacon is that the replicates are independent, that is, the first bearing taken does not influence the observer with respect to the result of the second reading of the same beacon. Also, the beacons should be located in the same interval of the compass rosette in which animal locations will eventually be made. Again, this procedure may prove cumbersome and impractical if only a few animals will be tracked before the mobile unit is moved to a new position. An alternative technique to avoid the problem of reorienting the mobile receiving system at each new location is the use of special compass devices which are not affected by proximity to the metal vehicle (Cederlund and Lemnell 1980).

When hand-held antennas are used to obtain bearings (and for some mobile receiving systems), a compass rosette and a pointer are not used. Instead, bearings are obtained by sighting down the antenna mast with a compass, a quick and simple technique. This procedure has proven adequate for many general movement studies, but it is approximate, at best, because there is no way to be sure one is sighting directly down the receiving path of the antenna. It is also

5 Designing and Testing Triangulation Systems

difficult to accurately align the compass needle with magnetic north. Such techniques, therefore, will significantly decrease the precision of the bearings and should be avoided when study objectives require precise location estimates.

A final problem involving mobile triangulation systems is the timing of directional bearings for each location estimate. In all of our previous discussions of triangulation in Chapter 4 and in this chapter, we assumed that all bearings used to calculate a location estimate were taken simultaneously. Mobile triangulation systems are often operated by a single person, with a certain amount of time elapsing between each bearing as the receiving system is moved to each new location. If the animal is moving during the triangulation process, it will be at a new location for each bearing (Fig. 5.6). The result is that all the bearings are "averaged" to obtain a single location estimate. It should be obvious that the movement of the animal will produce relatively imprecise location estimates, if an estimate can be calculated at all. The magnitude of the problem is directly related to the distance the animal moves between the readings of the first and last bearings. The error introduced into the location estimates cannot be addressed. We can only assume that the error is variable and can be substantial. Hence, it is impossible to assess the precision of the location

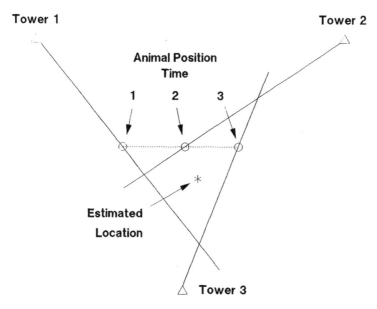

Figure 5.6 When bearings are not taken simultaneously on a moving animal, large errors in the estimated location may result.

estimates. In situations in which the investigator can assume that the animal is sedentary, such as birds on a roost or nest, the location estimators described in Chapter 4 would still be applicable with nonsimultaneous bearings.

Designing the Triangulation System

Two major factors which influence the precision of location estimates derived from triangulation are the number of bearings used for each estimate and the location of the receiving points with respect to the transmitter. In order to better understand the impacts of these factors on triangulation accuracy, we present the results of computer simulations performed by White (1985) to determine the optimal locations of various numbers of receiving points to maximize the precision of location estimates. Although these simulations were conducted to aid in the placement of fixed antenna towers, the results apply to any situation in which instrumented animals within an area must be monitored, whether hand-held, vehicle-mounted, or fixed antennas are used.

In these simulations the maximum likelihood estimator developed by Lenth (1981) (see Chapter 4) was used to estimate locations, and the variance–covariance matrix of the estimated locations was used to estimate the precision. Because no reflected signals are simulated, the results are the same as if the Huber or Andrews estimators developed by Lenth (1981) had been used. Precision was assessed at locations by the areas of the confidence ellipses. Two measures of precision for the study area were considered: the average confidence ellipse area over all locations and the maximum confidence ellipse area of a location on the study area. The average confidence ellipse was obtained by integrating the confidence ellipse area function over the study area, given a set of tower coordinates. The maximum confidence ellipse area was obtained by locating the position on the study site where the least accurate location estimate was obtained.

The shape of the study site is important in locating towers to maximize the precision of location estimates for an area. In these simulations a square and a rectangle with the length twice the width were examined. Extensions to study sites of other shapes become possible once the principles operating to optimize tower placement on square and rectangular areas are developed.

The definition of the confidence ellipse area $[A(x, y)]$ for a particular location (x_l, y_l), given k tower coordinates $(x_1, y_1), (x_2, y_2), \ldots, (x_k, y_k)$ is

$$A(x_l, y_l) = \pi |Q|^{1/2} \chi^2_{(2)(1-\alpha)} \tag{5.5}$$

5 Designing and Testing Triangulation Systems

where $|Q|$ is the determinant of the variance–covariance matrix of the estimates of x_l and y_l from the maximum likelihood estimator of Lenth (1981), π is 3.14159..., and $\chi^2_{(2)(1-\alpha)}$ is the χ^2 statistic with 2 degrees of freedom at level $1 - \alpha$. Thus, the average confidence ellipse area (C) for the entire study site was

$$C = \frac{\int_{x_{min}}^{x_{max}} \int_{y_{min}}^{y_{max}} A(x_l, y_l) dx_l \, dy_l}{(x_{max} - x_{min})(y_{max} - y_{min})} \tag{5.6}$$

that is, the integral of the confidence ellipse area function over the study site divided by the total area of the study site. The maximum confidence ellipse area was defined as the

$$A_{max} = \max[A(x, y) \in R(X, Y)] \tag{5.7}$$

that is, the location with the maximum confidence ellipse area (least precise estimate) of all possible (X, Y) locations on the study site. Additional details of the methodology are provided by White (1985).

No constraints were put on tower locations for cases with three or more towers. That is, towers could be located off the study site if such positions would provide more precise estimates of location than would locating the towers within the study site. For the two-tower case, the towers had to be located off the study site so that all locations within the boundaries could be estimated with bearings from two towers. More exactly, the line connecting the two towers (baseline) could not traverse the study site.

Spatial dimensions were expressed in terms of unit squares. That is, the square study site was defined as the square with lower left coordinates of (0, 0) and upper right coordinates of (1, 1). Similarly, the rectangle was defined with lower left coordinates of (0, 0) and upper right coordinates of (2, 1). Units for area were thus expressed as a proportion of this unit square. For example, if the study site was 4 km², then an area of 0.1 units² would be 10% of 4 km², or 0.4 km².

Coordinates of tower locations which minimize the average confidence ellipse area on a square and a rectangle are shown in Tables 5.2 and 5.3, respectively. Contour plots of the area of the confidence ellipse for two to six towers on a square are shown in Figs. 5.7–5.11. Similar contour shapes occurred on a rectangular study site, but larger average confidence ellipses were observed because of the greater distance between towers and the larger study site (cf. Tables 5.2 and 5.3).

TABLE 5.2

Location of Two to Six Towers on a Square Study Site Which Minimize the Average Confidence Ellipse Area for Estimates of Triangulated Locations[a]

Number of towers	Tower coordinates	Average ellipse area
2	(0.101, 1.007), (0.899, 1.007)	4.01E-3
3	(0.500, 0.175), (0.817, 0.671), (0.183, 0.671)	7.27E-4
4	(0.156, 0.500), (0.500, 0.844), (0.844, 0.500), (0.500, 0.156)	4.68E-4
5	(0.156, 0.454), (0.336, 0.839), (0.834, 0.319), (0.432, 0.154), (0.781, 0.765)	3.26E-4
6	(0.166, 0.500), (0.287, 0.841), (0.712, 0.841), (0.834, 0.500), (0.712, 0.159), (0.287, 0.159)	2.50E-4

[a]The average confidence ellipse area was generated with a bearing standard deviation of 1°. (From White 1985.)

TABLE 5.3

Location of Two to Six Towers on a Rectangular Study Site Which Minimize the Average Confidence Ellipse Area for Estimates of Triangulated Locations[a]

Number of towers	Tower coordinates	Average ellipse area
2	(0.349, 1.036), (1.651, 1.036)	6.28E-3
3	(0.247, 0.382), (1.000, 0.673), (1.753, 0.382)	1.66E-3
4	(0.205, 0.500), (1.000, 0.761), (1.795, 0.500), (1.000, 0.239)	9.82E-4
5	(0.138, 0.550), (1.000, 0.692), (1.862, 0.551), (1.412, 0.350), (0.588, 0.350)	6.68E-4
6	(0.142, 0.500), (0.656, 0.743), (1.344, 0.743), (1.858, 0.500), (1.343, 0.256), (0.656, 0.256)	5.17E-4

[a]The average confidence ellipse area was generated with a bearing standard deviation of 1°. (From White 1985.)

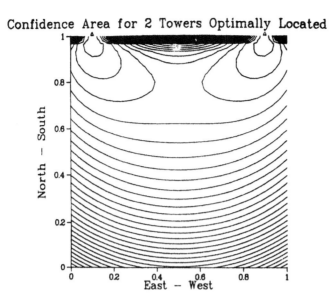

Figure 5.7 Contour plot of confidence ellipse area as a function of location on a square study site for two towers (triangles) optimally located to minimize the average confidence ellipse area (from White 1985). The isopleth nearest a tower is 0.0005 units^2, with 0.0005 increments for successively larger isopleths.

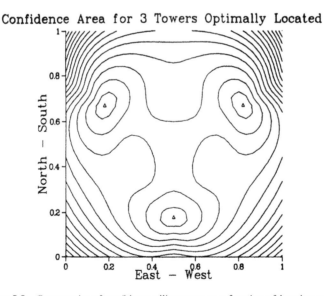

Figure 5.8 Contour plot of confidence ellipse area as a function of location on a square study site for three towers (triangles) optimally located to minimize the average confidence ellipse area (from White 1985). The isopleth nearest a tower is 0.0008 units^2, with 0.0001 increments for successively larger isopleths.

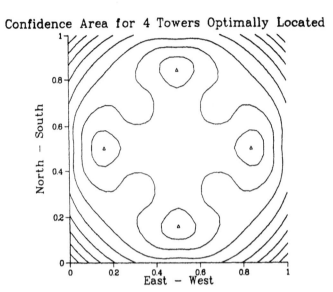

Figure 5.9 Contour plot of confidence ellipse area as a function of location on a square study site for four towers (triangles) optimally located to minimize the average confidence ellipse area (from White 1985). The isopleth nearest a tower is 0.0005 units^2, with 0.0001 increments for successively larger isopleths.

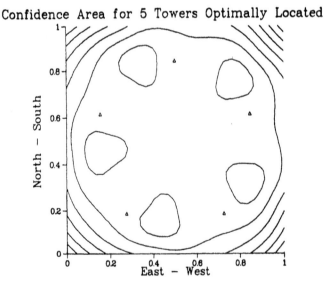

Figure 5.10 Contour plot of confidence ellipse area as a function of location on a square study site for five towers (triangles) optimally located to minimize the average confidence ellipse area (from White 1985). The isopleth nearest a tower is 0.0004 units^2, with 0.0001 increments for successively larger isopleths.

5 Designing and Testing Triangulation Systems

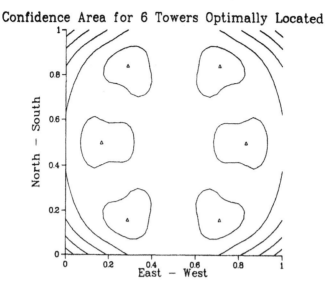

Figure 5.11 Contour plot of confidence ellipse area as a function of location on a square study site for six towers (triangles) optimally located to minimize the average confidence ellipse area (from White 1985). The isopleth nearest a tower is 0.0003 units^2, with 0.0001 increments for successively larger isopleths.

Coordinates of tower locations which minimize the maximum confidence ellipse area on a square or rectangle are shown in Tables 5.4 and 5.5, respectively. Contour plots of the confidence ellipse area for these tower locations were similar to those in Figs. 5.7–5.11, but the towers were positioned nearer the edge of the study site. The tower coordinates in Tables 5.4 and 5.5 tended to give a more uniform coverage of the study site, but at the expense of a larger average confidence ellipse area than the positions given in Tables 5.2 and 5.3.

White (1985) considered a unit square and a rectangle for the study area, but it would be easy to extrapolate these results to other shapes of study sites. For example, a circular study site could be overlaid on any of the contour plots in Figs. 5.7–5.11. Both the average and maximum confidence ellipse areas would be smaller than the values listed in Tables 5.2 or 5.4, because the maximum values occur in the corners of the square and rectangular areas (Figs. 5.7–5.11). Based on the patterns shown in Figs. 5.7–5.11, a circle would be the optimal shape of the study area to maximize the triangulation precision.

From Tables 5.2–5.5 and Figs. 5.7–5.11, two general principles are suggested to guide a researcher in positioning towers on areas of other shapes.

TABLE 5.4
Location of Two to Six Towers on a Square Study Site Which Minimize the Maximum Confidence Ellipse Area for Estimates of Triangulated Locations[a]

Number of towers	Tower coordinates	Average ellipse area
2	(−0.107, 1.055), (1.107, 1.055)	1.18E-2
3	(0.500, −0.016), (1.005, 0.686), (−0.005, 0.686)	1.78E-3
4	(0.057, 0.221), (0.221, 0.943), (0.943, 0.779), (0.779, 0.057)	9.71E-4
5	(0.336, 0.510), (0.211, 0.994), (0.929, 0.145), (0.204, 0.012), (0.941, 0.838)	6.72E-4
6	(0.233, 0.500), (0.184, 1.003), (0.817, 1.002), (0.765, 0.500), (0.817, −0.002), (0.184, −0.003)	4.85E-4

[a] The maximum confidence ellipse area was generated with a bearing standard deviation of 1°. (From White 1985.)

TABLE 5.5
Location of Two to Six Towers on a Rectangular Study Site Which Minimize the Maximum Confidence Ellipse Area for Estimates of Triangulated Locations[a]

Number of towers	Tower coordinates	Average ellipse area
2	(0.123, 1.114), (1.877, 1.114)	1.76E-2
3	(1.000, 0.001), (1.856, 0.700), (0.144, 0.700)	3.81E-3
4	(−0.013, 0.404), (0.616, 0.767), (2.013, 0.596), (1.384, 0.233)	2.12E-3
5	(1.000, 0.500), (0.231, 0.961), (1.769, 0.039), (0.231, 0.039), (1.769, 0.961)	1.33E-3
6	(0.057, 0.225), (0.253, 0.963), (1.134, 0.706), (1.943, 0.775), (1.747, 0.037), (0.866, 0.294)	1.03E-3

[a] The maximum confidence ellipse area was generated with a bearing standard deviation of 1°. (From White 1985.)

5 Designing and Testing Triangulation Systems

Towers should not be placed in the corners of an area, but rather along the sides or in the interior to provide the best precision. In addition, towers should be equally spaced over the study site in basic geometric forms, such as a equilateral triangle for three towers and a diamond for four towers.

Before designing a triangulation system, the investigator should decide on the range of precision of location estimates needed to meet the objectives of the study. This step in the planning process is often overlooked, but it is of primary importance to the success of any study requiring triangulation techniques. In studies in which determining the general location of the animal is all that is required, the precision level of location estimates may be easily attained. Other study objectives, however, may require relatively good precision, dictating a more thorough approach to designing the triangulation system.

For example, consider habitat studies in which triangulation methods are often used. The precision of the locations should be adequate to locate an animal within a habitat type or to minimize the proportion of locations where the confidence ellipse includes the boundary between two habitat types. Suppose that the animals are exploiting resources along the ecotone between two habitat types. A reasonable definition of this edge habitat might be 100 m on either side of the habitat boundary. In addition, measurement of the habitat map shows that less than 10% of the area is within 100 m of a habitat boundary. Hence, if instrumented animals select locations at random, less than 10% of the locations should include a boundary in the confidence ellipse. Locations are desired with the precision necessary to classify a tagged animal's location to a habitat, if it is located more than 100 m from the edge of this habitat. An animal within 100 m of the edge of a habitat would be classified as on the boundary. Thus, the study requires the precision necessary to provide ellipses with maximum dimensions of less than 200 m. To approximate this situation, a circle of radius <100 m is necessary. Hence, the maximum confidence ellipse cannot exceed $\pi 100^2 m^2$, with this maximum still giving portions of the study site with lower precision than would be required because of the use of a circle to approximate the confidence ellipse.

Given this maximum allowable confidence ellipse size, the number of towers necessary to provide this precision for a given bearing standard deviation can be selected from Table 5.6. To illustrate, suppose the study site is a 2×2 km^2 area and a maximum confidence ellipse of $\pi 100^2 m^2$ is required. To translate these conditions to the unit square of Table 5.6, a confidence ellipse is needed with area

$$\pi 100^2 m^2 \times \frac{10^{-6} \text{ km}^2}{1 \text{ m}^2} = 0.0314 \text{ km}^2$$

TABLE 5.6
Average Confidence Ellipse Area as a Function of the Tower Bearing
Standard Deviation and the Number of Towers for a Square Study Site[a]

Standard deviation (degrees)	Number of towers				
	2	3	4	5	6
5.00	.999E-1	.181E-1	.111E-1	.810E-2	.621E-2
4.75	.902E-1	.163E-1	.100E-1	.732E-2	.561E-2
4.50	.810E-1	.147E-1	.901E-2	.657E-2	.503E-2
4.25	.723E-1	.131E-1	.804E-2	.586E-2	.449E-2
4.00	.640E-1	.116E-1	.713E-2	.520E-2	.398E-2
3.75	.563E-1	.102E-1	.627E-2	.457E-2	.350E-2
3.50	.490E-1	.888E-2	.546E-2	.398E-2	.305E-2
3.25	.423E-1	.766E-2	.471E-2	.343E-2	.263E-2
3.00	.360E-1	.653E-2	.402E-2	.293E-2	.224E-2
2.75	.303E-1	.549E-2	.337E-2	.246E-2	.189E-2
2.50	.250E-1	.454E-2	.279E-2	.203E-2	.156E-2
2.25	.203E-1	.368E-2	.226E-2	.165E-2	.126E-2
2.00	.160E-1	.290E-2	.179E-2	.130E-2	.998E-3
1.75	.123E-1	.222E-2	.137E-2	.997E-3	.764E-3
1.50	.902E-2	.163E-2	.101E-2	.734E-3	.561E-3
1.25	.626E-2	.114E-2	.732E-3	.510E-3	.390E-3
1.00	.401E-2	.727E-3	.468E-3	.326E-3	.250E-3
0.75	.226E-2	.409E-3	.263E-3	.183E-3	.140E-3
0.50	.100E-2	.183E-3	.117E-3	.813E-4	.622E-4
0.25	.248E-3	.457E-4	.294E-4	.203E-4	.156E-4
0.00	.000E-0	.000E-0	.000E-0	.000E-0	.000E-0

[a] From White (1986).

The proportion of the total study site represented by this average confidence ellipse is

$$\frac{0.0314 \text{ km}^2 \text{ maximum confidence ellipse}}{2 \times 2 \text{ km}^2 \text{ study site}} = 0.00785$$

This confidence ellipse size corresponds to approximately three towers with a bearing precision of 3.25°, or two towers with a bearing precision of 1.4° (Table 5.6). Unfortunately, one is never sure what the bearing precision will actually be until the triangulation system has been field tested (see "Measuring Accuracy of Directional Bearings"). Thus, the investigator must rely on the range of bearing precision reported for other studies using similar equipment in comparable terrain.

5 Designing and Testing Triangulation Systems

The above example can be reversed to determine the size of an area that can be monitored from a fixed number of towers with a given bearing SD. Usually, the research budget determines the number of towers and the number of people available for monitoring, and the bearing SD is fixed by the equipment available.

Suppose that equipment for three towers is available and, from previous work, the bearing SD is expected to be approximately 1.5°. For this example, assume that locations are required to be accurate to 50 m, so that the average confidence ellipse translated to a circle is $\pi 50^2 = 0.0078$ km². From Table 5.6, the average confidence ellipse is 0.00163 units² for three towers with a bearing SD of 1.5°. Thus, the following proportion to solve for a (size of study area) is set up:

$$\frac{0.0078 \text{ km}^2}{a \text{ km}^2} = \frac{0.00163 \text{ units}^2}{1 \text{ units}^2}$$

The size of a square study site which can be monitored with the required precision with the three towers is thus $a = 4.7$ km². Therefore, the dimensions of the study area would be the square root of a or approximately 2.2 km on one side.

In general, when study objectives require relatively precise location estimates, the investigator should consider using bearings from at least three receiving points. Figure 5.12 provides a graphic presentation of selected data

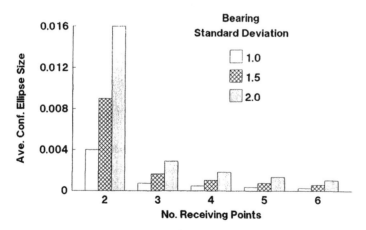

Figure 5.12 The effects of bearing precision (standard deviation) and the number of receiving points on the average confidence ellipse size of location estimates.

from Table 5.6 to illustrate the dramatic difference (approximately six-fold) in precision between location estimates derived from two versus three bearings. Although precision continues to increase with each additional receiving point, the magnitude of the increase gained diminishes substantially. Also illustrated in Fig. 5.12 is the increase in the precision of location estimates made possible by increasing bearing precision. In general, bearing precision is determined by the design of the receiving antenna system. There is some latitude in design of antenna systems, but improvements are limited. Hence, once the antenna system has been designed and is in place, the only practical way to improve the precision of location estimates is by increasing the number of receiving points used to derive the estimates.

The results of these simulations do not allow for study sites with elevational gradients that may prevent signal reception or cause signal reflection (bounce) (Tester 1971, Macdonald and Amlaner 1980, Brewer 1983, Hupp and Ratti 1983, Lee et al. 1985). Bearing errors from reflected radio signals may be large ($>5-10°$), but although such errors are easily detected when testing a receiving system, they are often impossible to identify when the transmitter location and, hence, the true bearing, is unknown. Such erroneous bearings would result in location estimates far from the animal's true location. It is important, therefore, when designing a triangulation system that is intended to be used in an area with significant topography, to minimize as much as possible the reception of reflected signals.

The only quantitative data on this problem is provided by Garrott et al. (1986). In this study all transmitter locations that were within line of sight of the receiving antenna produced bearings with small mean errors and SDs. However, 52% of the transmitter locations that were not within line of sight of the receiving antenna produced bearings with large mean errors and/or large SDs that were attributed to signal reflection. These data suggest that the best approach to minimizing problems with reflected signals is to locate receiving stations within view of the area to be monitored. If the study area is large, the best approach to handle this problem may be to divide the study site into smaller regions and to position receiving points within these subdivisions, to provide optimal coverage within this restricted area.

Placing receiving stations at the highest elevations will usually provide the best line-of-sight positions and reduce bounce signals, but this may require deviating from the optimal arrangement of receiving stations shown in Figs. 5.7–5.11. Figure 5.13 provides an idea of the sensitivity of the average confidence ellipse area function to the position of one of the three towers in a

5 Designing and Testing Triangulation Systems

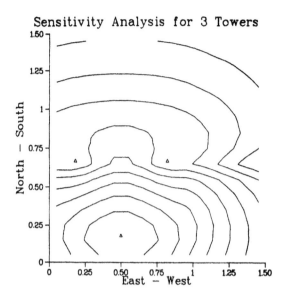

Figure 5.13 Contour plot of the change in the average confidence ellipse area as a function of the location of the bottom tower (triangle) (from White 1985). The isopleth around the lower tower is 0.0008 units^2, with successive isopleths in increments of 0.0001 units^2. Thus, for the first isopleth around the lower tower, the average confidence ellipse area for the study area is 0.0008 units^2 when the tower is positioned on this isopleth. If the tower is moved even farther from the optimal position to the 0.0009 units^2 isopleth, the average confidence ellipse area for the study area becomes 0.0009 units^2.

triangle configuration. For example, if the third tower cannot be located exactly in the position indicated in Table 5.2, how much does the average confidence ellipse size increase? From Fig. 5.13, the isopleth enclosing this tower is the 0.0008 contour, indicating that this tower can be positioned over a fairly large area without significant increase in the average confidence ellipse area from the optimum of 0.000727 given in Table 5.2. The sensitivity of the average confidence ellipse area function to tower position decreases steadily as the number of towers increases, so that the exact positioning of a tower becomes less important as the number of towers increases.

In reality, more than one of the towers may be placed in a suboptimal position due to topography or other logistic constraints. The only procedure to estimate exactly this loss of precision is to analyze the specific situation with the procedures developed by White (1985). A geographic information system (GIS) such as MOSS (Data Systems Support Group 1979) could be used in conjunction with digitized topographic maps to optimize the number and position of receiving stations for a given study area. MOSS and many other GIS systems are capable of producing a line-of-sight map (MOSS function VISTA), that is, a map of locations visible from a particular receiver location (even 15 m in the air!). Given the position of towers that can see a set of locations, a contour map of expected confidence ellipse size can be constructed using a combination of program TRIANG and MOSS, or other contour mapping algorithms, such as PROC PLOT of SAS (SAS Institute Inc. 1985). Thus, receiving points could be located to maximize the area from which line-of-sight radio signals could be received and thus minimize the expected confidence ellipse size.

In addition to placing receiving points where they provide the maximum line-of-sight coverage of the study area, more receiving points can be added to help overcome problems with reflected signals. In the previous chapter we presented an analytical procedure (the Andrews estimator) for location estimators that is robust to outliers (reflected signals). With three bearings the estimator allows the detection of inaccurate location estimates, but this cannot provide an accurate estimate if one or more of the signals is reflected. If only one of four bearings is reflected and the other three intersect within a small area, accurate location estimates can be obtained. By field testing the triangulation system, the proportion of erroneous bearings due to signal reflection obtained during actual data collection can be estimated (see "Measuring Accuracy of Directional Bearings"). This information then allows calculation of the probability of obtaining three accurate bearings, the minimum needed for accurate location estimates, while tracking free-ranging instrumented animals.

For example, Garrott et al. (1986) found by field testing that 33% of the test locations on their study area produced reflected signals at the receiving towers. Therefore, they estimated that only 30% of the locations for free-ranging deer would be usable with only three towers, that is, $(1.0 - 0.33)^3 = 0.30$ is the probability of obtaining three accurate bearings. With four towers the estimated proportion of usable locations increased to 60%. Defining p as the probability of obtaining a reflected signal and n as the number of receiving towers, the fraction of usable locations (u), assuming at least three bearings must converge, is

$$u = \sum_{i=3}^{n} \binom{n}{i} (1 - p)^i p^{(n-i)} \tag{5.8}$$

Thus, if field testing revealed a problem with reflected signals, the investigator can calculate the number of receiving points needed to obtain an acceptable

5 Designing and Testing Triangulation Systems

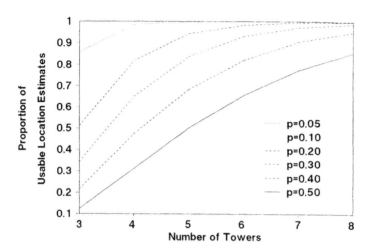

Figure 5.14 Plot of expected number of nonreflected bearings as a function of the number of towers and different probabilities of reflected signals (p).

proportion of accurate location estimates and redesign the triangulation system before data collection begins. These calculations assume that the proportion of reflected signals for instrumented animals will be the same as for the test locations. Equation 5.8 is the binomial probability density function summed for values of three or more. The term $\binom{n}{i}$ is a binomial coefficient equal to $n!/i!(n-i)!$. The exclamation points mean factorial, that is, $6! = 6 \times 5 \times 4 \times 3 \times 2 \times 1$, with $0! = 1$. The remainder of Eq. 5.8 is the probability of getting i "good" signals, that is, $(1-p)^i$, times the probability of getting $n - i$ "bad" signals, that is, $p^{(n-i)}$. The binomial coefficient $\binom{n}{i}$ provides the total number of possibilities of getting i "good" signals and $n - i$ "bad" signals and thus, when multiplied by the probability of a single case of i "good" signals and $n - i$ "bad" signals, gives the total probability of this situation. The summation from 3 to n provides the sum of the probabilities for all the cases with three or more "good" signals. A plot of the number of usable bearings as a function of the number of towers and several values of the probability of a reflected signal (p) is shown in Fig. 5.14, and the SAS code (SAS Institute Inc. 1985) to generate the data for this figure is shown in Fig. 5.15.

Another consideration in the design of a triangulation system may be whether the confidence ellipse area is the best measure of the "quality" of a location estimate, or whether the linear error distance might be a better measure

```
title 'Probability of obtaining 3 non-reflected bearings';
title2 'given n towers and p=Pr(reflected signal)';
data;
    label n='Number of towers'
          i='Number of non-reflected signals'
          p='Probability of reflected signal'
          u='Number of usable signals';
    * Open file for output values for input to Lotus,
      where a PIC file will be created for Freelance;
    file 'lotus.dat';
    keep p psym n u;
    do p=0.05,0.1,0.2,0.3,0.4,0.5;
        psym=p*10;
        do n=3 to 8;
            u=0;
            do i=3 to n;
                /* gamma(x+1) = x! */
                u=u+(gamma(n+1)/(gamma(i+1)*gamma(n-i+1)))
                    *(1-p)**i*p**(n-i);
            end;
            output;
            put p u n;
        end;
    end;
proc plot; plot u*n=psym; run;
```

Figure 5.15 SAS code to generate the expected proportion of usable (nonreflected) signals, given the probability of a reflected signal and the number of towers.

for the expected use of the data. Saltz and Alkon (1985) suggest that the maximum length across an error polygon may be the best measure of "quality." Saltz and White (1990) have shown that the length of the major axis of the confidence ellipse correlates well with error distance. In Figs. 5.16–5.18 the shape and size of confidence ellipses are shown for two-, three-, and four-tower triangulation systems. The length of the major axis of the confidence ellipse may vary substantially even though the ellipse area remains the same. Note that Fig. 5.16 corresponds to Fig. 4.9, in which the error polygons are mapped instead of the confidence ellipses derived from the Lenth estimator. Saltz and White (1990) recommend the length of the major axis of the confidence ellipse as the best linear measure of error.

5 Designing and Testing Triangulation Systems

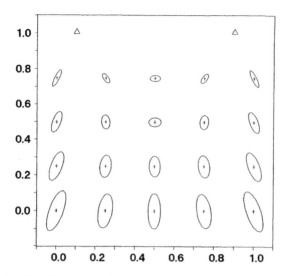

Figure 5.16 Map of confidence ellipse shape and size for a two-tower triangulation system. The axes reference the unit square, and the two towers are shown as triangles positioned at the top of the square.

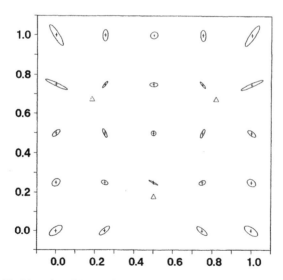

Figure 5.17 Map of confidence ellipse shape and size for a three-tower triangulation system. The axes reference the unit square, and the three towers are shown as triangles positioned as an equilateral triangle in the middle of the square.

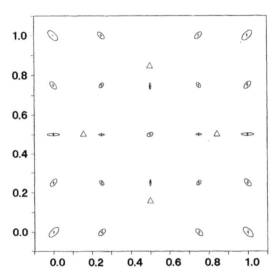

Figure 5.18 Map of confidence ellipse shape and size for a four-tower triangulation system. The axes reference the unit square, and the four towers are shown as triangles positioned as a diamond in the middle of the square.

Summary

1. The accuracy of a triangulation system is determined by the bias and precision of bearings from the receiving stations to the transmitters. Field studies are required to document the precision and bias of a system. Specifically, the bearing standard error must be estimated from field studies with transmitters at known locations.
2. Before a triangulation study can be designed, the range of precision of the location estimates must be appraised to meet the objectives of the study. Once the required precision is determined, the combination of study area size, number of towers, and bearing SD can be determined to provide the needed precision.
3. The occurrence of reflected, or "bounced," signals is common in rugged terrain. The Andrews estimator of Lenth (1981) will provide an adequate estimation procedure in the presence of reflected signals, but the precision of the triangulation system will not be as good as for a system with no reflected signals. If the proportion of reflected signals is estimated from a field evaluation of the triangulation system, the

5 Designing and Testing Triangulation Systems

proportion of usable location estimates can be determined based on the binomial distribution.
4. The precision of a radio-tracking system for a particular study area can be improved by positioning the receiving stations at optimal locations determined from computer analysis. Geographic information systems can assist in this process by providing line-of-sight maps, that is, maps of the terrain that can be seen from the top of a receiving tower.
5. Mobile receiving stations require additional information to determine the orientation of the compass rose and position. Usually, the position is determined a priori, but often radio beacons are used to orient the compass rose. Errors associated with this process must be incorporated into the final estimate of the precision of the estimated location.

References

Brewer, L. W. 1983. Radio-tracking the northern spotted owl in Washington state: a discussion of equipment and technique. Pages 12–21 *in* D. G. Pincock ed. Proc. 4th Int. Wildl. Biotelemetry Conf. Applied Microelectronics Institute and Technical Univ. of Nova Scotia, Halifax.
Cederlund, G. and P. A. Lemnell. 1980. A simplified technique for mobile radio tracking. Pages 319–322 *in* C. J. Amlaner, Jr. and D. W. Macdonald eds. A Handbook on Biotelemetry and Radio Tracking. Pergamon Press, Oxford, England.
Cederlund, G., T. Dreyfert, and P. A. Lemnell. 1979. Radiotracking techniques and the reliability of systems used for larger birds and mammals. Swedish Environ. Protection Board, Solna, pm 1136. 102 pp.
Cochran, W. W. 1980. Wildlife telemetry. Pages 507–520 *in* S. D. Schemnitz, ed. Wildlife Management Techniques, 4th ed. The Wildlife Society, Washington, D.C.
Data Systems Support Group. 1979. MOSS user's manual version II, WELUT-79/07. U.S. Fish and Wildl. Serv., Fort Collins, CO. Not consecutively paginated pp.
Garrott, R. A., G. C. White, R. M. Bartmann, and D. M. Weybright. 1986. Reflected signal bias in biotelemetry triangulation systems. J. Wildl. Manage. 50:747–752.
Hupp, J. W. and J. T. Ratti. 1983. A test of radio telemetry triangulation accuracy in heterogeneous environments. Pages 31–46 *in* D. G. Pincock ed. Proc. 4th Int. Wildl. Biotelemetry Conf. Applied Microelectronics Institute and Technical Univ. of Nova Scotia, Halifax.
Lee, J. E., G. C. White, R. A. Garrott, R. M. Bartmann, A. W. Allredge, and G. C. White. 1985. Assessing the accuracy of a radiotelemetry system for estimating mule deer locations. J. Wildl. Manage. 49:658–663.
Lemnell, P. A., G. Johnsson, H. Helmersson, O. Holmstrand, and L. Norling. 1983. An automatic radio-telemetry system for position determination and data acquisi-

tion. Pages 76–93 *in* D. G. Pincock ed. Proc. 4th Int. Wildl. Biotelemetry Conf. Applied Microelectronics Institute and Technical Univ. of Nova Scotia, Halifax.

Lenth, R. V. 1981. On finding the source of a signal. Technometrics 23:149–154.

Macdonald, D. W. and C. J. Amlaner, Jr. 1980. A practical guide to radio tracking. Pages 143–159 *in* C. J. Amlaner, Jr. and D. W. Macdonald eds. A Handbook on Biotelemetry and Radio Tracking. Pergamon Press, Oxford, England.

SAS Institute Inc. 1985. SAS® Language Guide for Personal Computers, Version 6 Edition. SAS Institute Inc., Cary, NC. 429 pp.

Saltz, D. and P. U. Alkon. 1985. A simple computer-aided method for estimating radio-location error. J. Wildl. Manage. 49:664–668.

Saltz, D. and G. C. White. 1990. Comparing different measures of the error involved in radio-telemetry location. J. Wildl. Manage. 54:169–174.

Slade, N. A., J. J. Cebula, and R. J. Robel. 1965. Accuracy and reliability of biotelemetric instruments used in animal movement studies in prairie grasslands of Kansas. Trans. Kans. Acad. Sci. 68:173–179.

Springer, J. T. 1979. Some sources of bias and sampling error in radio triangulation. J. Wildl. Manage. 43:926–935.

Tester, J. R. 1971. Interpretation of ecological and behavioral data on wild animals obtained by telemetry with special reference to error and uncertainties. Pages 383–408 *in* Proc. Symp. Biotelemetry. Univ. of Pretoria, South Africa.

White, G. C. 1985. Optimal locations of towers for triangulation studies using biotelemetry. J. Wildl. Manage. 49:190–196.

CHAPTER

6

Simple Movements

Since the first transmitter was attached to an animal nearly three decades ago, the study of animal movements has been a primary goal of radio-tracking studies. In recent years a gradual shift has taken place from descriptive movement studies to quantitative investigations aimed at studying activity patterns, habitat use, and survival. Although the primary goal of these studies may not be to collect data on animal movements per se, ancillary data on the movements of instrumented animals are nearly always available. This chapter deals with a variety of subjects related to the analysis of movement data. We have arranged the topics in the order in which an investigator may conduct his or her analysis, starting with identifying and correcting errors in data files and graphic presentations of the data, then moving to quantitative techniques for analyzing characteristics of long-distance movements, fidelity to specific areas, and associations between animals. Home range and habitat utilization are relatively complex and are presented in separate chapters.

Identifying and Correcting Data Errors

The first step in the analysis of movement data collected with radio-tracking techniques should always be detection of errors in the data. No matter how carefully data are recorded, transcribed, and entered into the computer, mistakes are inevitable. Two programs are available from the authors for routine checking of radio-tracking data: BIOCHECK and BIOPLOT. BIOCHECK provides three methods for detecting errors. First, locations can be checked to see that they are correctly ordered through time. That is, dates and times should increase chronologically for each animal. Second, the distances between consecutive locations are checked against a value supplied by the user to see that the distance is a reasonable value. For instance, an entry error may have been made that results in a mountain goat (*Oreamnos americanus*) moving 60 km between two locations. If this movement is during a nonmigratory period, the user would suspect an error. The distance between location i (x_i, y_i) and the next location (x_{i+1}, y_{i+1}) is determined from the formula

$$d_i = \sqrt{(x_{i+1} - x_i)^2 + (y_{i+1} - y_i)^2}$$

Obviously, the distance moved between two locations is a function of the time between the two locations, so an additional check can be performed to verify that the rate of movement (speed) is reasonable:

$$\text{speed} = \frac{d_i}{t_{i+1} - t_i}$$

Again, the user must supply the value that is used to check the data. The two values for distance and rate ("speed limit") should be selected to identify a few of the more extreme locations that are correct, so that most of the incorrect locations are identified. The third test performed by BIOCHECK determines which of the locations are outside a rectangle specified by the user. This check is done to be sure that all locations are inside of a reasonable area. However, the test is limited to a rectangle, and often can be performed more efficiently with the BIOPLOT program, which presents the data graphically. BIOCHECK is available both as a FORTRAN code (see Appendix 3) and as an interactive SAS procedure.

BIOPLOT is used to plot the data on the screen of an IBM personal computer or an IBM-compatible model and allows the interactive removal of locations from the data file, or flags them, so that the user can verify their validity. The user can move the cursor to a particular location and request that it be deleted or just flagged. In the process, the coordinates and time of the location are identified. BIOPLOT is useful for visually checking for location errors. The code is available as a BASIC program (see Appendix 4).

A useful commercial database package that can perform nearly the same plotting function is Reflex. Reflex can split the computer screen into two "views" of the data, one a listing of the (x, y) coordinates, the second a plot

6 Simple Movements

of the locations. As the cursor is moved down the list of locations, a second cursor moves around the plot of data to identify the plotted location. Thus, outliers can be identified in the list view and corrected or deleted as necessary.

Still Graphics

Graphic presentation of movement data is generally in the form of two-dimensional plots. Thus, one dimension of the data is necessarily not used because the data consist of the three-dimensional vector (x, y, and t). For example, if the time dimension is ignored, a map of an animal's movements is obtained. Often, a map is constructed with various symbols representing the animal's locations. We term this type of presentation a "symbol plot." No attempt is made to connect the symbols to illustrate the time dimension. Symbol plots are particularly useful for demonstrating areas of high use, such as centers of activity. Habitat use is also conveniently displayed when the symbol plot is constructed on a habitat map.

One approach to maintaining the order of the locations and indicating the differing amounts of time involved for each movement is to connect consecutive locations with distinguishing lines. We term this type of display a "vector plot." In Fig. 6.1 elk (*Cervus elaphus*) movements with less than 10 days between locations are shown with solid lines; 10–30 days, with dashed lines; and >30 days, with dotted lines. Another possibility for identifying line types is to use different colors. However, colors are not generally suitable for publication because of the increased cost involved.

The development of relatively inexpensive microcomputers and associated software has made the graphic presentation of radio-tracking data readily available to most biologists. The preparation of professional quality graphics is not an easy task, but the preparation of quick graphics is simple. Such plots are important in that a researcher may view his own data, catching mistakes in recording or entry, and facilitating his interpretation of the data.

Sophisticated graphics capabilities are available through commercial software. As an example, spread sheet packages such as Lotus 1-2-3 or Quattro have graphics capabilities to plot a set of (x, y) data, with the consecutive points connected by a solid line, or only the points can be plotted in a choice of symbols (i.e., a symbol plot). Numerous other packages also provide the same capabilities. One advantage of Lotus 1-2-3 is that the graphics image can be saved to a file and later loaded into Freelance Plus, a graphics package for the interactive manipulation of images on the screen. Using Freelance Plus, the

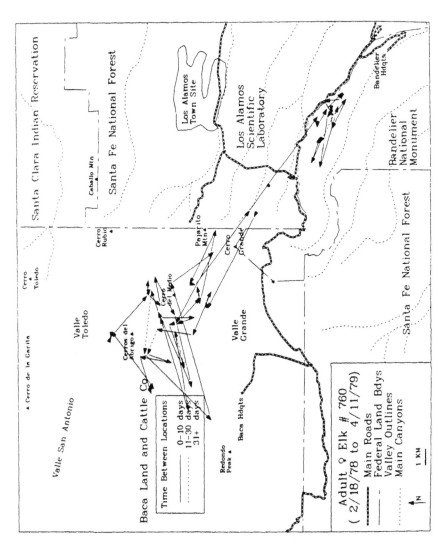

Figure 6.1 Computer-generated plot of the movements of an adult female elk (*Cervus elaphus*). Dotted, dashed, and solid lines indicate the amount of time between consecutive locations and trace the movements of the animal chronologically through time.

6   Simple Movements

Lotus image can be refined and improved as desired. Furthermore, Freelance Plus provides the capability to enter a map on which to display the radio-tracking locations, thus putting them in perspective of terrain, habitat, etc., as desired. Plots such as Fig. 6.1 can be generated with the Lotus 1-2-3–Freelance Plus combination.

A second advantage of using the Lotus 1-2-3–Freelance Plus combination is that Freelance Plus interfaces to numerous output devices. One such device is the Polaroid Palette, a relatively inexpensive film recorder for use with IBM-compatible microcomputers. The Palette can be used to produce presentation quality slides of radio-tracking data. Freelance Plus produces output on pen plotters for the production of publication-quality plots, laser printers for desktop publishing capabilities, or dot matrix printers for draft copies.

SAS GRAPH also provides the user with good-quality graphics capabilities for presentation of movement data. In addition, numerous output devices are supported, just as with Freelance Plus. The main drawback we see with the use of SAS GRAPH is the complexity in developing a map of the area where the animal is tracked. The interactive graphics editing capability of Freelance Plus makes map development fairly straightforward compared to the more tedious process with SAS GRAPH. The batch-oriented nature of SAS jobs would allow production of many maps of animal movement, for example, maps for each animal during each season or month. Another possibility for innovation is the use of the three-dimensional graphics capabilities of SAS GRAPH.

For most applications a single map will be adequate for plotting movements of instrumented animals; however, investigations involving migrating waterfowl, marine mammals, pelagic birds, and other animals which move on a global scale may require many specialized maps to depict different types of movements and/or different areas of the world. For example, an investigator studying pelagic birds nesting on the Antarctic Peninsula may discover that movements of birds instrumented in one colony are restricted to the Peninsula and the South Atlantic Ocean. Birds from another colony wander extensively throughout the Pacific Ocean, traveling as far north as the Aleutian Islands of Alaska, while birds from a third colony spend much of the winter in the Indian Ocean. In this situation it would clearly be helpful to be able to depict these movements on a variety of maps. Anyone who has digitized and refined a base map of their study area can appreciate how formidable a task it would be for this investigator to digitize and prepare a series of maps to illustrate these data.

An alternative to manually producing the maps is the use of programs containing geographic databases that are designed to produce customized maps based on a series of user-defined variables such as WORLD (available from Philip Voxland, Social Sciences Research Facilities, University of Minnesota, Minneapolis, Minnesota 55455). This program is written largely in FORTRAN 77 and is capable of creating maps of the world or any part of the globe the user delineates. The program is extremely easy to learn and provides the user with the flexibility of producing maps using over 100 projections. Four databases are available for defining geographic areas. The smallest and most generalized database uses only 700 points to depict all global features, while an intermediate database uses 8000 points. The program can also use the two databases developed by the U. S. Central Intelligence Agency, World Data Bank I and II, which contain 100,000 and 5.5 million points. There is a variety of commands which allow the user to add titles, boxes, grids, tick marks, symbols, lines, borders, and shading. Movements of animals are best illustrated using the GREAT command which plots great circles between each consecutive set of locations. Final maps can then be produced on a pen plotter, using a variety of colors and pen widths. Figures 6.2 and 6.3 are examples of maps produced by the WORLD program (see Fig. 1.1 for another example).

Although we have had experience only with the software mentioned on IBM-compatible computers, many other packages are available to fulfill this

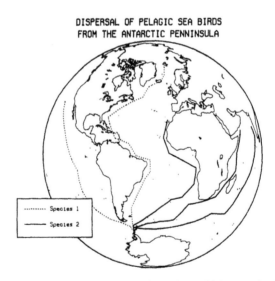

Figure 6.2 Map of fictitious movements of pelagic sea birds created using the WORLD program. This map depicts the globe using a polar azimuthal spherical projection and an 8000-point geographic database. Twelve different commands were used to produce the map and plot 60 locations in two different line styles.

6 Simple Movements

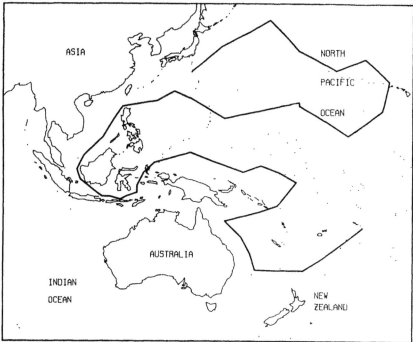

Figure 6.3 Map of fictitious movements of a sea turtle instrumented off the coast of Japan using the WORLD program. This map depicts a portion of the world centered on the equator near Indonesia and was created using the same database and a similar projection as used in Fig. 6.2. Ten different commands were used to define the area to be mapped, produce labels, and plot the 43 turtle locations.

need. Also, the graphics capabilities of other types of microcomputers, such as the Apple Macintosh, may further enhance the biologist's abilities to present radio-tracking data graphically.

Animated Graphics

Another method of graphically displaying all three dimensions of radio-tracking data is animation. Animation adds impact to the presentation of radio-tracking data. White (1979) has described the use of computer-generated movies for portrayal of radio-tracking locations through time. The computer plots the location of each animal on a base map, one frame at a time. Thus, the symbols for each animal are shown moving around the map through time. The color intensity of the symbol is enhanced when the location of the animal is based on an actual radio location; whereas, movement of the symbol at normal color intensity represents linear interpolation between actual radio fixes. The current time is also displayed on the map, either in the form of a bar that lengthens with time or in a digital clock format.

Data from multiple animals displayed simultaneously provide the means to qualitatively examine interactions between individual animals or sex or age classes as functions of season and habitat. Individuals or groups of animals (stratified by age or sex) can be distinguished by the shape or color of the symbols. Changes in the map through time can also be displayed. For example, the response of elk (*Cervus elaphus*) to a logging operation can be illustrated by showing the logging operation as shaded areas appearing and leaving through time. The impact on a caribou (*Rangifer tarandus*) herd from a new pipeline can be shown by the pipeline appearing on the map as it was constructed. Another interesting application of this technique is the addition of physiological data to the presentation. For example, a film was produced for the Colorado Division of Wildlife from data on a study of the behavioral and physiological responses of mule deer (*Odocoileus hemionus*) to snowmobiles. In the film the symbols for each deer were color-coded to reflect the behavior of the individual. For example, a resting animal was green, a moderately disturbed animal was yellow, and an animal fleeing from the disturbance was red. Hence, the film condensed the complex relationships among time, spatial distribution, movement, and behavior into one easily understood graphic presentation.

The application of this technique has been extremely limited, because few researchers have had access to mainframe computers interfaced with 16-mm film recorders. To our knowledge, the only radio-tracking films produced to date were generated on the Los Alamos National Laboratory computer network in New Mexico. Equipment to produce radio-tracking films may, however, be widely available in the near future due to the continuing expansion of the capabilities of microcomputers. The faster IBM-compatible machines can easily produce the 30 frames per second required for smooth image movement and personal computer-based animation systems are currently capable of producing professional-quality videos. Kunkel and Luchak (1986) reviewed three commercially available systems and predict that many others will be on the market in the near future. The $30,000–$60,000 purchase price of personal computer-based animation systems is generally prohibitively expensive for individual researchers, but is not unrealistic for institutions such as universities and government agencies. Undoubtedly, the price will decline as the technique becomes more popular. Video animation will provide substantial rewards from the presentation impact.

Migration and Dispersal

"Migration" and "dispersal" are terms which refer to relatively large-scale movements of animals. There appear to be no universally accepted definitions to differentiate these terms, as various authors use the same terms to refer to different types of movement. A good example of conflicting definitions used by different authors is provided by comparing the terminology used by Baker (1978) and the various authors in the book edited by Swingland and Greenwood (1983). In order to avoid confusion, we define migration as a regular, round-trip movement of individuals between two or more areas or seasonal ranges. In contrast, dispersal is a one-way movement of individuals from their natal site or an area that has been occupied for a period of time.

Regardless of the type of long-distance movement studied, there are generally four basic questions to be addressed: (1) Which animals or groups of animals move? (2) When do the movements occur? (3) Why do the animals move? (4) Where do the animals go? Determining which animals perform long-distance movements is simply a matter of instrumenting and monitoring enough animals to assure that if long-distance movements are reasonably common, at least part of the instrumented population will be involved.

One common difficulty in studying long-distance movements is the wide variation in the movement patterns of individuals. To illustrate this problem, suppose ten animals from a population are instrumented and located weekly for 1 year. Analysis of the data at the end of the study reveals that four animals remained relatively sedentary and three animals occupied specific areas for various lengths of time before moving a short distance (<2 km) and establishing new home ranges. The remaining three animals made what appeared to be long-distance dispersal movements, two traveling 10–25 km in an easterly direction, with the other animal traveling >80 km to the north.

Results such as these may allow the investigator to discuss, in detail, the movements of the study animals, but what generalizations can be made about the movement patterns of the population? An obvious solution to the problem is to instrument a larger sample of the population; however, most studies have logistic and financial constraints that limit the number of animals studied. Selectively instrumenting those animals most prone to perform the movements of interest may also help maximize the amount of pertinent data collected. Storm et al. (1976) provide an example of such a sampling strategy. In this study the investigators were particularly interested in the dispersal movements of red foxes (*Vulpes fulva*) and expected most long-distance movements to be performed by juvenile foxes as they left their natal denning area. Hence, they concentrated primarily on locating denning sites and instrumenting pups just prior to the fall, when dispersal movements were expected. This strategy resulted in a successful study of red fox dispersal.

Determination of the timing of specific types of movements is often an essential study objective. In order to study migratory and dispersal movements, the investigator must define these movements relative to the data collected. Criteria used to establish endpoints of a long-distance movement must be well defined. All too often, radio-tracking studies reporting long-distance movements present data on the timing and duration of migratory and dispersal movements without reporting the criteria used to segregate these types of movements from other, more localized movements. The lack of an explicit definition not only leaves the reader wondering how the distinctions were made, but also makes comparison between studies difficult, because of the subjectivity involved in determining when an animal initiated or completed the movement of interest. Precisely defining and reporting how the endpoints of a movement were determined assure that the data can be consistently evaluated for each study animal and that other investigators can make comparisons with their results.

An example of one possible definition for the onset and completion of seasonal migrations is presented by Garrott et al. (1987) for a mule deer (*Odocoileus hemionus*) population in northwest Colorado.

> A deer was considered to have initiated spring migration when it was 1st located outside the winter range during April or May. Boundaries of the winter range for each study area were subjectively determined from the collective winter movements of all instrumented deer. Spring migration was considered complete when an animal was first located ≤ 2 km from its mean summer range location for that year. Conversely, initiation of fall migration was defined as the first time an animal was located ≥ 2 km of its mean summer range location during mid-September to mid-November. Migration was considered complete when an animal was 1st located within the collective winter range of the radio-collared population.

Subjective judgments had to be made in formulating these definitions, the most obvious being the 2-km criterion. This example also illustrates the necessity for movement data outside the migration period in order to define the "collective winter range" and each animal's "mean summer range location," which are essential for establishing the onset and cessation of migratory movements in this example.

Many authors have attempted to correlate the timing of long-distance movements with various natural and man-caused phenomena (McCullough 1964, Hoskinson and Mech 1976, Hoskinson and Tester 1980, Nelson and Mech 1981, Garrott et al. 1987). Perhaps the most common correlations are drawn between the timing of seasonal migrations and environmental variables such as temperature, precipitation, and snow cover. Such correlations, however, are generally not very convincing in developing cause-and-effect relationships, because the environmental variables follow seasonal patterns as well. Hence, the timing of spring migration is correlated with increasing temperatures and the timing of fall migrations is correlated with decreasing temperatures. In fact, there are many environmental variables (vegetation desiccation, leaf fall, soil moisture, photoperiod, etc.) that would also probably show some correlation with the timing of seasonal movements, but are not necessarily part of cause-and-effect relationships.

In order to strengthen the inference of a cause-and-effect relationship between environmental variables and the timing of migrations, both variables must be studied for many years. Long-term studies provide a better opportunity to study variations in the timing of migrations. To illustrate this point, consider a hypothetical study of the timing of the fall migration of a herd of elk (*Cervus elaphus*). The investigator suspects that snow depth on the summer range is the primary impetus for fall migration. Animals are instrumented and studied for 2 years. During both years several major snow storms occurred in November and the elk migrated to winter range soon thereafter, supporting the investigators' hypothesis. However, an anti-hunting group points out that November is the general big-game season and insists that it is the harassment of hunters that caused the elk to migrate, not snow. Which interpretation of the data is correct? The confounding of hunting and snowfall, that is, the lack of a control for either factor, precludes conclusive results, as discussed in Chapter 2. The study is continued for another 5 years, during which time the November hunting season is continued; however, there are two abnormally dry winters and 1 year when heavy snows occur in early October. Elk migrate to winter range in November during most years; however, during the dry winters no elk leave summer range until late December. The elk also migrate in October during the year of abnormally early snowfall. A cause-and-effect relationship is still not demonstrated, but the evidence now clearly favors the hypothesis that snow, not hunter harassment, stimulates elk to migrate in the fall.

A pitfall that investigators should avoid when correlating environmental variables with the timing of long-distance movements is the idea that the sampling units in the study are the instrumented animals. This is not the case, as all animals are subjected to the same changes in the environmental variables, and hence are not independent observations. Therefore, the sampling unit is the distribution of the migratory movements for each year that the phenomenon was studied (Fig. 6.4). A 2-year study involving 25 animals each year provides the investigator with a sample size of 2, not 50. The only time that the individual animal is the sampling unit is when the factors can be measured independently for each animal, and each animal behaves independently of the others. A good example of such a study is one in which the timing of dispersal is correlated with the age or physical condition of the instrumented animals.

The time which elapses between relocations of instrumented animals (sampling interval) has a direct effect on the precision with which the timing of migratory or dispersal movements can be determined. What constitutes "adequate" precision will vary depending on study species and research objectives.

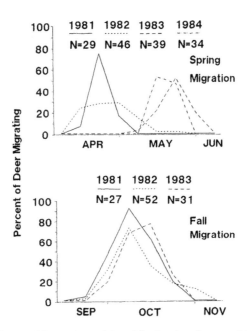

Figure 6.4 Timing of four spring and three fall migrations for a population of instrumented female mule deer (*Odocoileous hemionus*) in northwestern Colorado (from Garrott et al. 1987).

6 Simple Movements

If one is attempting to correlate the onset of long-distance movements to a specific short-term phenomenon, such as a 2-day hunting season or an abnormally heavy snowfall, intensive sampling, perhaps on a daily basis, may be required. In contrast, if movements are to be correlated with phenomena which change gradually, such as physical maturation, photoperiod, or desiccation of forage, then locating animals on a weekly or biweekly basis may be adequate.

If animals are located intensively enough during long-distance movements, one can measure, describe, and test hypotheses about the directionality of the movements. Normal statistical procedures cannot be applied to measures of directionality because the measurement (i.e., compass bearings) is circular; that is, there is no true zero, and high and low values are arbitrary. To illustrate the problem encountered when standard statistical procedures are applied to circular distributions, consider the simple problem of determining the mean direction of an animal's migratory path. Six locations are obtained for the animal during its migration, with the direction of movement for each location measured as a compass bearing from the location at time $t - 1$ to the location at time t. The five bearings are 5°, 356°, 15°, 12°, and 350°, indicating that the animal consistently traveled to the north. The arithmetic mean, however, is $(5 + 356 + 15 + 12 + 350)/5 = 148°$, which is obviously incorrect.

Because of the nature of circular distributions, specialized statistical procedures must be used. Zar (1974) presents a clear and concise overview of the most common statistical procedures used for circular distributions in the biological sciences, and the following discussion draws heavily from this reference. For those who would like a more complete discussion of the topic, we suggest the studies by Mardia (1972) and by Batschelet (1981). Another summary of the common analyses is presented by Mimmack et al. (1980), along with a description of a computer program they offer for the analysis of two-dimensional directional data.

To illustrate the various statistical procedures that can be used in analyzing the directionality of movements, we will use a fictitious sample of data collected over a 5-week period on an instrumented grizzly bear (*Ursus arctos*) in Yellowstone National Park (Table 6.1). To refresh your memory, remember that the distance (d_i) between two locations is

$$d_i = \sqrt{(x_{i+1} - x_i)^2 + (y_{i+1} - y_i)^2}$$

Thus, the value $d_1 = 468.48$ is generated from

$$d_1 = \sqrt{(279434 - 279377)^2 + (4428525 - 4428990)^2}$$

To calculate the direction of travel from one location (x_i, y_i) to a second location (x_{i+1}, y_{i+1}), first determine the relative distances traveled in the x and y directions:

$$X_i = x_{i+1} - x_i$$

and

$$Y_i = y_{i+1} - y_i$$

Then the angle in degrees (a_i) is calculated as

$$a_i = \begin{cases} \arctan(Y_i/X_i)(180°/\pi) & \text{if } X_i > 0 \\ 180° + \arctan(Y_i/X_i)(180°/\pi) & \text{if } X_i < 0 \\ 90° & \text{if } X_i = 0 \text{ and } Y_i > 0 \\ 270° & \text{if } X_i = 0 \text{ and } Y_i < 0 \end{cases}$$

where the arctan function computes an angle calculated in radians that is converted to degrees. This angle (a_i) is then converted to a bearing from true north (b_i) with the equation

$$b_i = 90 - a_i$$

If b_i is negative, that is, the direction is measured to the left of true north rather than to the right, just add 360° to the result to make the value positive. To remember the difference between angles and bearings, refer back to the section in Chapter 4 on using two towers for triangulation, where these concepts are discussed. The SAS (SAS Institute Inc. 1985) function to calculate this equation is presented in Chapter 5 in the section on measuring the accuracy of directional bearings. To illustrate these calculations, the angle between the first and second locations in Table 6.1 is calculated as

$$X_1 = 279434 - 279377$$
$$= 57$$

and

$$Y_1 = 4428525 - 4428990$$
$$= -465$$

Thus,

$$a_i = \arctan(-465/57)(180°/\pi)$$
$$= -83.012°$$

Then, the angle is converted to a bearing:

$$b_i = 90° - (-83.012)$$
$$= 173.012°$$

that is, the animal is traveling in a direction a little east of due south.

We now have a sample of angles $\{a_1, a_2, \ldots, a_n\}$ (or equivalently, bear-

6 Simple Movements

TABLE 6.1

Simulated Locations and Direction of Movement of an Instrumented Grizzly Bear (*Ursus arctos*) during the 5-Week Period Immediately After It Was Translocated into a Remote Area of Yellowstone National Park

Observation	x_i	y_i	d_i	a_i	b_i
1	279377	4428990	468.48	−83.012	173.012
2	279434	4428525	1012.83	−51.897	141.897
3	280059	4427728	550.01	−74.498	164.498
4	280206	4427198	149.64	−79.216	169.216
5	280234	4427051	845.15	−31.057	121.057
6	280958	4426615	563.39	−87.864	177.864
7	280979	4426052	765.51	−83.776	173.776
8	281062	4425291	797.86	−42.206	132.206
9	281653	4424755	324.85	−30.117	120.117
10	281934	4424592	754.91	2.809	87.191
11	282688	4424629	753.35	−84.592	174.592
12	282759	4423879	819.05	−55.747	145.747
13	283220	4423202	201.38	−34.479	124.479
14	283386	4423088	283.95	−111.052	201.052
15	283284	4422823	783.67	−13.504	103.504
16	284046	4422640	583.46	9.769	80.231
17	284621	4422739	228.11	−210.275	300.275
18	284424	4422854	683.67	−78.953	168.953
19	284555	4422183	721.01	−0.238	90.238
20	285276	4422180	30.89	60.945	29.055
21	285291	4422207	165.70	−51.124	141.124
22	285395	4422078	909.03	−61.123	151.123
23	285834	4421282	758.28	−86.673	176.673
24	285878	4420525	844.79	−26.353	116.353
25	286635	4420150	—	—	—

ings $\{b_1, b_2, \ldots, b_n\}$) that represent the bear's direction of movement from one location to the next. Two techniques for graphically depicting the data are presented in Fig. 6.5. In order to describe the sample of angles, a mean angle and its variance should be calculated. To compute the mean angle $\bar{a}$, $\bar{X}$ and $\bar{Y}$ must be determined, where

$$\bar{X} = \frac{1}{n} \sum_{i=1}^{n} \cos a_i$$

and

$$\bar{Y} = \frac{1}{n} \sum_{i=1}^{n} \sin a_i$$

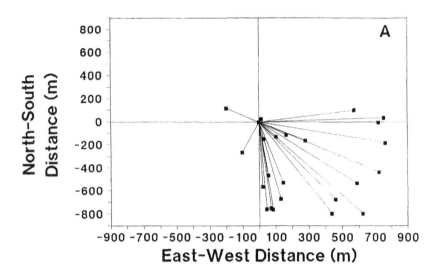

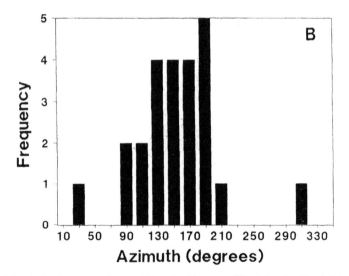

Figure 6.5 A circular scatter diagram (A) and a histogram (B) of the directional data in Table 6.1. In general, scatter diagrams are used when data are sparse and histograms are used when the data set is too large to be easily depicted in a scatter diagram.

Then r, which is a measure of the concentration of the sample of angles $\{a_1, a_2, \ldots, a_n\}$, can be calculated as

$$r = \sqrt{\bar{X}^2 + \bar{Y}^2}$$

If $r = 0$, then the mean angle is undefined, and we conclude there is no mean direction. If $r \neq 0$, then $\bar{a}$ can be determined using the relationships

$$\bar{a} = \begin{cases} \arctan(\bar{X}/\bar{Y})(180°/\pi) & \text{if } \bar{X} > 0 \\ 180° + \arctan(\bar{Y}/\bar{X})(180°/\pi) & \text{if } \bar{X} < 0 \\ 90° & \text{if } \bar{X} = 0 \text{ and } \bar{Y} > 0 \\ 270° & \text{if } \bar{X} = 0 \text{ and } \bar{Y} < 0 \end{cases}$$

which is equivalent to the arctan function shown above for determining direction from the (x_i, y_i) coordinates. The results can be checked with the formulas

$$\cos \bar{a} = \frac{\bar{X}}{r}$$

and

$$\sin \bar{a} = \frac{\bar{Y}}{r}$$

The mean angle ($\bar{a}$) can be converted to a mean bearing ($\bar{b}$) as

$$\bar{b} = 90° - \bar{a}$$

exactly the same as an individual angle is converted to a bearing. The SAS code (SAS Institute Inc. 1985) to perform the calculations for the grizzly bear data is presented in Fig. 6.6. Notice that the SAS code computes the mean direction as an angle and then converts it to a mean bearing.

Since r is a measure of concentration, $1 - r$ is a measure of dispersion, or "circular variance" (Mardia 1972). The value r may vary from 0, when the mean angle is undefined, to 1, when all angles are identical. Hence, when $1 - r = 0$, there is no dispersion, with maximum dispersion occurring when $1 - r = 1$. To obtain a statistic with a theoretical range from 0 to ∞, which is analogous to the standard deviation of linear data, the angular deviation, s, must be determined:

$$s = \frac{180°}{\pi} \sqrt{-4.60517 \log r}$$

or

$$s = \frac{180°}{\pi} \sqrt{-2 \ln r}$$

```
* ---------------------------------------------------------- *
| SAS code to compute the mean distance between consecutive  |
| locations, and the mean angle and mean bearing for direction |
| between consecutive locations. The input data file should  |
| have the variables x and y sorted in chronological order.  |
| PROC CHART is used to generate histograms of the mean      |
| direction in 2 different formats for both angles and bearings. |
* ---------------------------------------------------------- *;
data direct;
   keep i x y d a b;
   retain x1 y1 bigxsum bigysum dsum 0;
   set bear end=lastobs;
   x2=x; y2=y;
   if _n_ > 1 then do;
      d=sqrt((x2-x1)**2+(y2-y1)**2);
      link atan2;
      x=x1; y=y1; i=i-1;
      a=alpha; if a<0 then a=a+360; /* Direction of travel as angle */
   b=90-a; if b<0 then b=b+360;  /* Direction of travel as bearing */
output;
      bigxsum=bigxsum+cos(theta);   /* Cumulate sums for means */
      bigysum=bigysum+sin(theta);
      dsum=dsum+d;
   end;
   /* Save current location for next observation */
   x1=x2; y1=y2;
   if lastobs then do;
      x=x2; y=y2; i=i+1; d=.; a=.; b=.;
      output;
      dbar=dsum/_n_;                /* Average distance */
      bigxbar=bigxsum/_n_;
      bigybar=bigysum/_n_;
      x2=bigxbar; y2=bigybar;
      x1=0; y1=0; link atan2;
      abar=alpha;                   /* Average direction as angle */
   bbar=90-abar; if bbar<0 then bbar=bbar+360;    /* as bearing */
r=sqrt(bigxbar**2+bigybar**2);
      s=sqrt(-2*log(r))*180/pi;
      z=_n_*r**2;
      mu=-pi/4;  /* Set mu to bearing of 135 degrees */
      u=sqrt(2*_n_)*r*cos(theta-mu);
      put 'Mean distance between locations: ' dbar=;
      put '                        Mean X: ' bigxbar=;
```

Figure 6.6 SAS code used to calculate the mean angle for the data in Table 6.1 and to construct histogram and circular frequency plots for both angles and bearings. Selected output for the bear data in Table 6.1 is shown.

```
        put '                    Mean Y: ' bigybar=;
        put '          Mean direction (angle): ' abar=;
        put '          Mean direction (bearing): ' bbar=;
        put '              Concentration (0-1): ' r=;
        put '               Standard Deviation: ' s=;
        put '              Raleigh's z statistic: ' z=;
        put '      Expected direction of travel: ' mu=;
        put '          Test statistic for V test: ' u=;
     end;
return;
atan2:   /* SAS code to calculate angle from point (x1, y1)
            to point (x2, y2) */    pi=atan(1.)*4;
         xp=x2-x1; yp=y2-y1;
         if xp > 0 then theta=atan(yp/xp);
         else if xp < 0 then theta=pi+atan(yp/xp);
         else if xp = 0 then do;
            if yp < 0 then theta=pi*3/2;
            else if yp > 0 then theta=pi/2;
            else theta=.; end;
         if theta < 0 then theta=theta+2*pi;
         /* Now convert angle in radians to angle in degrees */
         alpha=theta*180/pi);
    return;
proc print;
proc chart;
    vbar a / type=freq midpoints=5 to 355 by 20;
    star a / type=freq midpoints=5 to 355 by 20;
    title 'Angles';
proc chart;
    vbar b / type=freq midpoints=5 to 355 by 20;
    star b / type=freq midpoints=5 to 355 by 20;
    title 'Bearings';
run;
```

```
Mean distance between locations: DBAR=559.95882883
                         Mean X: BIGXBAR=0.4303351569
                         Mean Y: BIGYBAR=-0.537932046
             Mean direction (angle): ABAR=308.65912762
           Mean direction (bearing): BBAR=141.34087238
               Concentration (0-1): R=0.6888825977
                 Standard Deviation: S=49.466165665
              Raleigh's z statistic: Z=11.863980837
      Expected direction of travel: MU=-785398163
          Test statistic for V test: U=4.8413360149
```

Figure 6.6 (*Continued*)

depending on whether base 10 or natural logarithms are used. For the r value of 0.689, calculated from the grizzly bear data, $s = 49.5°$. The values of r, and hence s, are the same whether the direction is expressed as an angle or as a bearing. In both cases the dispersion around the mean is the same, because both angles and bearings are measured in degrees.

Calculating a mean angle, $\bar{a}$, and angular deviation, s, provides descriptive statistics, but we have not yet tested whether the bear was consistently moving in one direction. That is, whether the direction of the bear's movements was random or oriented. This is tested by determining whether the distribution of the sample population $\{a_1, a_2, \ldots, a_n\}$ departs significantly from a uniform circular distribution. Critical values for $r_{(\alpha, n)}$ can be determined from tables given by Zar (1974) and by Batschelet (1981). For large values of n (sample size) not covered in these tables, Rayleigh's z can be calculated as

$$z = nr^2$$

and compared to critical values in the more general tables given by Zar (1974) and by Batschelet (1981). If the calculated value of r or z is greater than or equal to the critical value and the sample distribution is unimodal, then we may reject the null hypothesis that the direction of movement was random, in favor of the alternative hypothesis that the movements were oriented in some direction.

The tests outlined above are limited to testing the null hypothesis of no mean directionality, or randomness, versus the alternative hypothesis that there is some directionality in the movements of the animals. In many situations, however, the investigator may expect that the animal will move in a specific direction. In our grizzly bear example one may suspect that the translocated bear shows some homing tendency and travels toward its original area of occupancy. In this situation a V test (Zar 1974) may be used to test the null hypothesis of no directionality in movements, versus the alternative hypothesis that the direction of the animal's movements are clustered about some mean angle. The test statistic (u) is

$$u = (2n)^{1/2} r \cos(\bar{a} - \mu_a)$$

where μ_a is the mean angle predicted. Again, the critical values for $u_{(\alpha, n)}$ are available in tables given by Batschelet (1981). For the grizzly bear example μ_a is 135°, and from previous calculations we know $n = 24$, $r = 0.689$, and $\bar{a} = 141.3°$; then $u = 6.92 \cos(141.3° - 135°)$, which is 4.84 (Fig. 6.6). The critical value of u for $\alpha = 0.001$ and a sample size of 24 is 3.0359. Since the calculated u exceeds the critical value, we can reject the null hypothesis ($P < 0.001$) in favor of the alternative hypothesis that the direction of the bear's movement was oriented toward the area it was removed from; that is, the bear was exhibiting homing behavior.

Measuring Fidelity

Fidelity is the tendency of an animal either to return to an area previously occupied or to remain within the same area for an extended period of time. Few animals are totally nomadic; hence, for most species that have been studied, individual animals show fidelity to at least some part of the area they occupy. The degree of fidelity displayed by various species can differ widely. For example, nonmigratory animals such as the black-capped chickadee (*Parus atricapillus*) (Smith 1976), and the muskrat (*Ondatra zibethicus*) (Boutin and Birkenholz 1987) may spend their entire lives within an area no larger than several km². Other animals migrate between widely separate winter and summering areas, with some species, such as the bighorn sheep (*Ovis canadensis*) (Geist 1971), and the redshank (*Tringa totanus*) (Hale and Ashcroft 1982), returning to very specific sites, while other species, such as the barren ground caribou (*Rangifer tarandus*) (Kelsall 1968, Bergerud 1974) and the wildebeest (*Connochaetes taurinus*) (Talbot and Talbot 1963), show fidelity to seasonal ranges but not to specific sites. In contrast, pelagic species such as the Pacific salmon (*Oncorhynchus* spp.) (McKeown 1984:55), Laysan albatross (*Diomedea immutabilis*) (Fisher 1976), and green sea turtle (*Chelonia mydas*) (Burnett-Herkes 1974) spend years wandering the oceans. However, even these species return periodically to traditional sites to reproduce.

There are many questions and competing theories concerning the function and evolution of site fidelity in the animal kingdom (Shields 1983), which makes this subject an interesting topic for ecological study. From a more pragmatic perspective, the site fidelity of a species to a particular area also provides basic information which can aid biologists in understanding the dynamics of animal movements and social systems. This knowledge, in turn, aids in developing management and conservation plans for a species. Site fidelity studies are also useful in detecting the impacts of human activities, such as logging, petroleum exploration, contaminant release, and off-road vehicle use on animal movements (Davis 1970, Garrott and White 1984, Edge et al. 1985, Kuck et al. 1985).

In order to measure fidelity, a series of locations must be obtained over a period of time for individually identifiable animals. Telemetry provides an ideal tool for studies of fidelity, because each animal can be uniquely identified and tracked for long periods (years with large species). Since locations can usually be obtained whenever desired, an additional benefit of using radio-tracking is that the researcher can design sampling schemes to assure adequate data. In some studies fidelity may be measured as simply the percentage of instrumented animals returning to a specific seasonal range or, in the case of non-migratory animals, the percentage of instrumented animals remaining within a given area for a particular length of time. Often, however, the biologist will want to compare the fidelity of individual animals or groups of animals, which requires a more quantitative approach.

To test for fidelity changes quantitatively, a quantitative measure (and hence a quantitative definition) of fidelity must be developed. As a first definition, suppose that fidelity is defined as a shift in the location of the area of use of the animal. A test of whether the means of the x and y coordinates change through time would be a test of fidelity. The mean location for each animal from all locations obtained during the season of interest (Schoen and Kirchhoff 1985, Garrott et al. 1987) is calculated. The distance between an individual's mean locations in consecutive years can then be used as a measure of fidelity, and tested with multivariate methods, for example, Hotelling's T^2 test. Distance data can be tabulated and presented in a figure for visual comparison between groups (Fig. 6.7). A Mann–Whitney test on the x and y values can also be used on distance data to test for differences in fidelity between groups of animals (Schoen and Kirchhoff 1985, Garrott et al. 1987).

The definition of fidelity as a shift in the mean location from one time period to the next would not allow for an expansion of the area of use, while still maintaining the same core area. To detect such an expansion, a test for a change in the variances and covariances of the x and y coordinates would be required. An even further refined definition of the quantitative measure of fidelity would be a shift in the time spent in the various portions of the area used. For example, consider a female ring-necked pheasant (*Phasianus colchicus*) that spends its time entirely within 1 km^2. During the first year of a study, most of the time was spent near a 5-ha area of hay ground, where its nest was located. During the second year of the study, the bird still used the same square kilometer, but spent most of its time near a different 5-ha section of hay ground. Although the general area of use was still the same, and the variance–covariance matrix of the x and y coordinates might remain the same, the animal has

6 Simple Movements

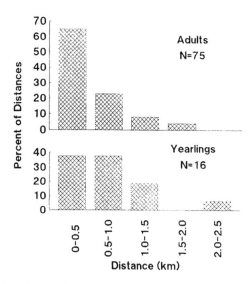

Figure 6.7 The distances between means of summer locations of instrumented mule deer (*Odocoileus hemionus*) in consecutive years.

changed its use of the area it inhabits. To detect such subtle changes, a statistical test designed to detect differences in the distribution of the locations must be used.

Such a test is the MRPP test of Mielke and Berry (1982) (also see Zimmerman et al. 1985, Biondini et al. 1988). Changes in an animal's area of utilization, such as required for testing for fidelity, is provided. The method tests whether two sets of locations come from a common probability density distribution. The approach is nonparametric, making no assumptions about the shape of the underlying distribution. The null hypothesis of the test is that the two (or more) utilization distributions are the same. The MRPP statistic is based on the within-group average of pairwise distance measures between locations compared to the average distances between locations when groups are ignored. Sample sizes for each sample need not be the same, and in fact might be quite dissimilar for the two data sets. The program for the method is shown by Biondini et al. (1988).

As an example, consider the data plotted in Fig. 6.8. Although not quantitative, the figure adequately illustrates the fidelity of this deer to a specific summering site. However, the influence of a few locations in 1985 greatly affects the estimate of overlap, hence fidelity. For the data shown in Fig. 6.8,

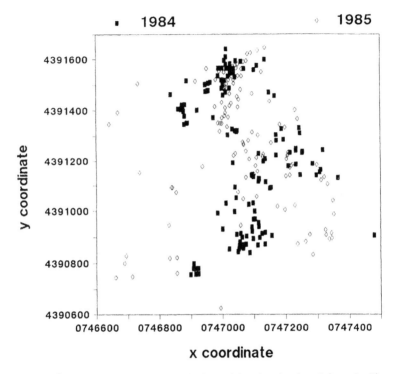

Figure 6.8 Locations of a female mule deer (*Odocoileus hemionus*) from the Piceance Basin of western Colorado that was tracked for two consecutive summers.

the MRPP statistic is highly significant ($P < 0.001$), suggesting a change in home range use as a function of year.

This level of a shift in fidelity may be more than the investigator bargained for. Such small changes might be considered a nonsignificant change in the use of an area. Thus, the researcher must decide a priori what kinds of changes in the distribution of locations from one time period to the next should be considered a change in the use of an area. An approach that we propose to eliminate this problem would be to designate specific areas (possibly based on the observations from the first period) as cells, and perform a χ^2 test of independence between the time periods. For example, we may want to define fidelity for the mule deer shown in Fig. 6.8 as a shift in use from one drainage to another drainage. Thus, we are willing to allow the deer to use the drainage where it is located in the first year differently the next year, and claim that fidelity has been demonstrated. However, if a second watershed is used extensively the second year, but used very little the first year, we would claim that the deer has not shown strong fidelity to the original drainage. This type of χ^2 test can be constructed when the area where the animal is tracked can be divided up objectively into subareas. However, the choice of how to divide up the area into subareas is subjective in most situations. Thus, the results of the test are dependent on the subareas, and so must be considered in light of these assumed subareas.

A final technique used for detecting fidelity is to collect enough locations on each instrumented animal to develop a home range estimate with one of the models described in Chapter 7. Continued data collection over several time periods will then provide a series of home range maps that can be compared for similarity. Because the procedure depends heavily on the assumptions of the home range estimation procedure, this approach to measuring fidelity is covered in detail in Chapter 7. However, we suggest that the procedures described above, which are based on the actual data, are preferable to the home range procedures, for which inferences are based on a model developed from the data. See Chapter 7 for further discussion of the problems associated with using home range estimates.

Animal Association

The question most frequently addressed, in one form or another, by biologists using radio-tracking techniques is: How do animals move through the environment? An equally interesting question, but one rarely pursued, is: How do individuals move with respect to one another? While the question of where an animal is in relation to its environment is usually directed at topics such as distribution, migration, and use of resources, the question of how animals move with respect to conspecifics is directed at a species' social behavior. Particularly productive topics of empirical research would be in fields with large bodies of theoretical literature, such as competition and population genetics.

One of the difficulties in studying the social behavior of animals with radio-tracking techniques is a lack of quantitative tools to objectively determine whether one animal's movements or behaviors are correlated with another animal's movements or behaviors. Dunn and Gipson (1977), Dunn (1979), and Dunn and Brisbin (1982) discuss two types of measurements which can be applied to radio-tracking data to explore possible interactions between the movements of instrumented animals. The first, static territorial interaction, deals with overlap of home ranges. If A_1 and A_2 represent the home range areas of animals 1 and 2, respectively, and $A_{1,2}$ is the area of overlap, then static interaction (S) is

$$S_{1,2} = \frac{A_{1,2}}{A_1}$$

and

$$S_{2,1} = \frac{A_{1,2}}{A_2}$$

where $S_{1,2}$ is the proportion of animal 1's home range overlapped by animal 2 and $S_{2,1}$ is the proportion of animal 2's home range overlapped by animal 1. Note that $S_{1,2}$ does not equal $S_{2,1}$ unless $A_1 = A_2$ (Macdonald et al. 1980). The interpretation of static interaction in the study of animal movements is limited, as the statistic does not imply any mutual awareness among the respective animals (Dunn 1979). Another limitation of the statistic is that radio-tracking location data for each animal must be forced into a shape with one of the many home range models. As we discuss in Chapter 7, these areas may not accurately represent the area the animal occupies, which could lead to erroneous home range overlaps. An example of the problems of calculating and interpreting static interaction using red fox (*Vulpes vulpes*) movement data is provided by Macdonald et al. (1980) and by Jeselnik (1981) for gray foxes (*Urocyon cinereoargenteus*).

The second statistic developed by Dunn (1979), dynamic territorial interaction, is designed to improve on the shortcomings of static interaction by measuring the degree of correlation between the movements of instrumented animals. In this technique the time series of the data, which is ignored in most home range models, is modeled by using the multivariate Ornstein–Uhlenbeck stochastic process to calculate home range estimates (see Chapter 7). Dunn (1979) presents a statistic to test the hypothesis that two animals move independently of each other. The test requires that animals are monitored simultaneously, with a constant sampling interval within each data collection period (or burst). An additional assumption is that animal movements are bivariate normally distributed (i.e., random movement about a central point).

Although this technique has been available for nearly a decade, it has not been widely utilized. This is probably due to the difficulty biologists have in interpreting the statistical presentation of the method and the rigid requirements of the data collection procedure, that is, constant time intervals between observations within a burst, and bursts far enough apart in time that the observations can be considered independent. Macdonald et al. (1980) present a less complex test for dynamic territorial interaction, although the test is also based on the bivariate normal home range model.

6 Simple Movements

In some studies the transmitters are used to lead the investigator to the animals, and direct visual contact is made. If a sufficiently large proportion of animals is individually marked, then it is possible to record the identities of the animals observed with instrumented animals. With these data the coefficient of association developed by Cole (1949) can be used as a measure of animal association. The coefficient (CA) is

$$CA = \frac{2AB}{A + B}$$

where A is the number of times animal 1 was observed during a specific time period, B is the number of times animal 2 was observed during the same time period, and AB is the number of times animals 1 and 2 were observed together throughout the time period.

In most situations, however, animals are remotely located and never observed. Under these conditions a measure of animal association can be obtained by calculating the distance between all possible pairs of instrumented animals and using cluster analysis to identify animals that are constantly near each other (Morgan et al. 1974). The input vector consists of (x_i, y_i) for each animal, with locations being taken at the same time for both animals. The distance separating pairs of animals is used to cluster them into groups.

Similar to Dunn's dynamic territorial interaction (Dunn 1979), locations of different animals taken at different times cannot be used in cluster analysis, because there is no reasonable method to determine the relationship in three-dimensional space of time to spatial distances. That is, cluster analysis techniques require a measure of distance between the two animals. If the locations of both animals are not taken at the same time, then somehow time must be taken into account to compute the distance between the members of the pair. Just because each of them was at a particular location at different times does not mean they were associating. Thus, time of locations is crucial.

To illustrate this problem, consider locations of two animals—one at (x_1, y_1, t_1) and the second animal at (x_2, y_2, t_2) with $t_1 < t_2$. Cluster analysis methods require that all the input variables be on a common scale of measurement. However, time is obviously different than the two spatial coordinates. Thus, the distance between (x_1, y_1, t_1) cannot be related to (x_2, y_2, t_2) unless time is translated into distance, that is, the same scale of measurement as x and y. Possibly $(t_2 - t_1)$ could be replaced with the average (or maximum) distance the study animal could travel during this time period, but such an approach is arbitrary.

Therefore, the conclusion is that when using cluster analysis to determine the positive or negative association between two animals, locations of all animals in the study must be taken at the same time. For some studies this may be impossible; however, this condition may be easily met for studies that rely primarily on aircraft or satellites for locating instrumented animals, as a large number of animals can usually be located in a short time period. An advantage of cluster analysis over dynamic interaction is that there are no assumptions made about the shape of the home range, the percentage of locations that adequately describe the area of occupancy, or the underlying distribution of an animal's movements.

Sample Size

No firm guidelines can be given regarding sample size requirements for a basic movements study if specific hypotheses are not being tested. The relationship between a plot of the data and what is an adequate sample size is not clear. The usual problem of whether to track several animals intensively or to tag more animals and track each individual less intensively recurs. We can only suggest that other data sets be examined to determine the trade-offs between more animals versus more intensive tracking. Naturally, the questions being asked in the specific experiment to be performed must be considered in developing this trade-off. Also, simple movement data are usually collected just to obtain the basic information needed to design an experiment, and hence are part of a descriptive phase in the research.

For the procedures that estimate a parameter or construct a test of a hypothesis, the usual approaches to sample size calculation can be performed to determine the number of observations needed to achieve the desired confidence interval length or power of the test. For example, the power of the MRPP test for differences between the locations of an animal in two time periods can be calculated to provide a sample size necessary to detect some level of difference. Dunn (1978) has developed criteria to determine how many bursts of locations are needed to test hypotheses about interaction between animals.

Several efforts to model the movements of animals with stochastic models have been presented in the literature. Siniff and Tester (1965) first discussed possible models of animal movement from radio-tracking data. Kareiva and Shigesada (1983) modeled insect movement paths as a sequence of straight line moves and turning angles, both chosen from probability distributions. McCulloch and Cain (1989) used a correlated random walk model to predict an organism's rate of spread. As this area of quantitative research matures, movement models may become useful in helping to determine sample sizes to test specific hypotheses.

Summary

1. Three simple checks for the validity of radio-tracking data are that the location is reasonable, the location is within a feasible distance of the previous location, and the rate of travel between the previous location and the current location is logical.
2. Graphic presentations of radio-tracking data require that three-dimensional data (x, y, and t) be collapsed into two dimensions, unless animation is used to present the time dimension. Rapid developments with microcomputer animation will provide inexpensive methods to present all three dimensions of radio-tracking data in the near future.
3. Migration and dispersal can be studied easily with radio-tracking techniques. However, careful definitions of migration and dispersal must be developed to distinguish these movements in the data.
4. Correlation of environmental variables with long-distance movements will not generate cause-and-effect relationships, because most of the environmental variables are also correlated.
5. Studies of the direction of movements must incorporate statistics specific to angles, as typical linear estimators such as the mean and variance will not perform correctly for directional data where the smallest (0°) and largest (360°) measurements are identical.
6. Fidelity of an individual animal to an area is determined by measurements through time, with the use of two or more home range estimators being a logical approach. The difficulty with home range estimators is the assumptions necessary to generate reasonable estimates.
7. The problem of measuring animal association is similar to measuring an individual's fidelity to an area. Animals' associations can be dynamic as well as static. Again, comparison of home range estimates is a logical approach, but one fraught with difficulties.

References

Baker, R. R. 1978. The Evolutionary Ecology of Animal Migration. Holmes & Meier, New York. 1012 pp.
Batschelet, E. 1981. Circular Statistics in Biology. Academic Press, New York. 371 pp.

Bergerud, A. T. 1974. The role of the environment in the aggregation, movement and disturbance behavior of caribou. The behavior of ungulates and its relation to management, Vol. 2 IUCN New Series Publ. 24. Morges, Switzerland. 584 pp.

Biondini, M. E., P. W. Mielke, Jr., and E. F. Redente. 1988. Permutation techniques based on Euclidean analysis spaces: a new and powerful statistical method for ecological research. COENOSES 3:155-174.

Boutin, S. and D. E. Birkenholz. 1987. Muskrat and round-tailed muskrat. Pages 315-324 in M. Novak, J. A. Baker, M. E. Obbard, and B. Malloch Wild Furbearer Management and Conservation in North America. Ontario Ministry of Natural Resources, Toronto.

Burnett-Herkes, J. 1974. Returns of green sea turtles (*Chelonia mydas* Linnaeus) tagged at Bermuda. Biol. Conserv. 6:307-308.

Cole, L. C. 1949. The measurement of interspecific association. Ecology 30:411-424.

Davis, J. L., 1970. Elk use of spring and calving range during and after controlled logging. M.S. Thesis, Univ. of Idaho, Moscow. 51 pp.

Dunn, J. E. 1978. Optimal sampling in radio telemetry studies of home range. Pages 53-70 in H. H. Shugart, Jr. ed. Time Series and Ecological Processes. SIAM, Philadelphia, PA.

Dunn, J. E. 1979. A complete test of dynamic territorial interaction. Pages 159-169 in F. M. Long ed. Proc. 2nd Int. Conf. Wildl. Biotelemetry. Univ. of Wyoming, Laramie.

Dunn, J. E. and I. L. Brisbin, Jr. 1982. Characterizations of the multivariate Ornstein-Uhlenbeck diffusion process in the context of home range analysis, Stat. Lab. Tech. Rep. No. 16. Univ. of Arkansas, Fayetteville. 72 pp.

Dunn, J. E. and P. S. Gipson. 1977. Analysis of radio telemetry data in studies of home range. Biometrics 33:85-101.

Edge, W. D., C. L. Marcum, and S. L. Olson. 1985. Effects of logging activities on home-range fidelity of elk. J. Wildl. Manage. 49:741-744.

Fisher, H. I. 1976. Some dynamics of a breeding colony of Laysan albatrosses. Wilson Bull. 88:121-142.

Garrott, R. A. and G. C. White. 1984. Methodology for assessing the impacts of oil shale development on the Piceance Basin mule deer herd. Pages 228-231 in R. D. Comer et al. eds. Issues and Technology in the Management of Impacted Western Wildlife. Thorne Ecological Institute, Boulder, CO.

Garrott, R. A., G. C. White, R. M. Bartmann, L. H. Carpenter, and A. W. Alldredge. 1987. Movements of female mule deer in northwest Colorado. J. Wildl. Manage. 51:634-643.

Geist, V. 1971. Mountain Sheep—A Study in Behavior and Evolution. Univ. of Chicago Press, Chicago, IL. 383 pp.

Hale, W. G. and R. P. Ashcroft. 1982. Pair formation and pair maintenance in the redshank, *Tringa totanus*. Ibis 124:471-490.

Hoskinson, R. L. and L. D. Mech. 1976. White-tailed deer migration and its role in wolf predation. J. Wildl. Manage. 40:429–441.

Hoskinson, R. L. and J. R. Tester. 1980. Migration behavior of pronghorn in southeastern Idaho. J. Wildl. Manage. 44:132–144.

Jeselnik, D. L. 1981. Comparative analysis of movement, behavior and habitat utilization of free-ranging gray foxes. M.S. Thesis, Univ. of Georgia, Athens. 104 pp.

Kareiva, P. M. and N. Shigesada. 1983. Analyzing insect movement as a correlated random walk. Oecologia 56:234–238.

Kelsall, J. P. 1968. The Migratory Barren-ground Caribou in Canada. Dept. of Indian Affairs and Northern Development, Can. Wildl. Serv., Ottawa. 339 pp.

Kuck, L., G. L. Hompland, and E. H. Merrill. 1985. Elk calf response to simulated mine disturbance in southeast Idaho. J. Wildl. Manage. 49:751–757.

Kunkel, G. and G. H. Luchak. 1986. Animating corporate presentations. PC Magazine 5:220–236.

Macdonald, D. W., F. G. Ball, and N. G. Hough. 1980. The evaluation of home range size and configuration using radio tracking data. Pages 405–424 in C. J. Amlaner Jr., and D. W. Macdonald eds. A Handbook on Biotelemetry and Radio Tracking. Pergamon Press, Oxford, England.

Mardia, K. V. 1972. Statistics of Directional Data. Academic Press, New York. 357 pp.

McCulloch, C. E. and M. L. Cain. 1989. Analyzing discrete movement data as a correlated random walk. Ecology 70:383–388.

McCullough, D. R. 1964. Relationship of weather to migratory movements of black-tailed deer. Ecology 45:249–256.

McKeown, B. A. 1984. Fish Migration. Timber Press, Portland, OR. 224 pp.

Mielke, P. W., Jr. and K. J. Berry. 1982. An extended class of permutation techniques for matched pairs. Commun. Stat. 11:1197–1207.

Mimmack, G. M., P. D. Morant, T. M. Crowe, and A. McKenzie. 1980. Biological applications of a computer programme package (DIRECT) for statistical analysis of two-directional data. S. Afr. J. Wildl. Res. 10:29–37.

Morgan, B. J. T., M. J. A. Simpson, J. P. Hanby, and J. Hall-Craggs. 1974. Visualizing interaction and sequential data in animal behaviour: theory and application of cluster-analysis methods. Behaviour 56:1–43.

Nelson, M. E. and L. D. Mech. 1981. Deer social organization and wolf predation in northeastern Minnesota. Wildl. Monogr. 77:1–53.

SAS Institute Inc. 1985. SAS® Language Guide for Personal Computers, Version 6 Edition. SAS Institute Inc., Cary, NC. 429 pp.

Schoen, J. W. and M. D. Kirchhoff. 1985. Seasonal distribution and home-range patterns of Sitka black-tailed deer on Admiralty Island, southeast Alaska. J. Wildl. Manage. 49:96–103.

Shields, W. M. 1983. Optimal inbreeding and the evolution of philopatry. Pages 132–159 in I. R. Swingland and P. J. Greenwood The Ecology of Animal Movement. Clarendon Press, Oxford.

Siniff, D. B. and J. R. Tester. 1965. Computer analysis of animal movement data obtained by telemetry. BioScience 15:104–108.

Smith, S. M. 1976. Ecological aspects of dominance hierarchies in black-capped chickadees. Auk 93:95–107.

Storm, G. L., R. D. Andrews, R. L. Phillips, R. A. Bishop, D. B. Siniff, and J. R. Tester. 1976. Morphology, reproduction, dispersal, and mortality of midwestern red fox populations. Wildl. Monogr. 49:1–82.

Swingland, I. R. and P. J. Greenwood (eds.). 1983. The Ecology of Animal Movement. Clarendon Press, Oxford. 311 pp.

Talbot, L. M. and M. H. Talbot. 1963. The wildebeest in western Masailand, East Africa. Wildl. Monogr. 12:1–88.

White, G. C. 1979. Computer generated movies to display biotelemetry data. Pages 210–214 *in* F. J. Long ed. Proc. 2nd Int. Conf. Wildl. Biotelemetry. Univ. of Wyoming, Laramie.

Zar, J. H. 1974. Biostatistical Analysis. Prentice-Hall, Englewood Cliffs, NJ. 620 pp.

Zimmerman, G. M., H. Goetz, and P. W. Mielke, Jr. 1985. Use of an improved statistical method for group comparison to study effects of prairie fire. Ecology 66:606–611.

CHAPTER

7

Home Range Estimation

Home range is defined as "that area traversed by the individual in its normal activities of food gathering, mating and caring for young" (Burt 1943:351). Thus, the units of home range are areal, for example, m^2. In practice, the term "home range" is used to describe two aspects of animal movement. First, the basic map of locations produced from tracking an animal is often referred to as the animal's home range. The second use of the term "home range" refers to the numerical estimate of the area used by the animal. Most of this chapter explores the mechanics of the numerical estimation process, but the importance of the map of locations should be emphasized. All too often, the assumptions of a numerical estimate are so seriously violated that the number becomes worthless, but the map of locations is not distorted. The practical information gained from the tracking data is accommodated with the map, but may be lost with the numerical estimate.

The key word in the above definition is "normal." Home range is not *all* the area that an animal traverses during its lifetime, but rather the area where it *normally* moves. Excursions to the area outside its normal area should not be considered part of the home range (Burt 1943:351). Therefore, objective criteria with a biological basis are needed to select movements that are "normal." In the literature two criteria have been used. The first is the subjective evaluation of the researcher, and thus does not provide a repeatable, objective approach for other researchers. The second is to use a probability level; for example, the estimated home range includes the animal's locations 95% of the time. This leads to a more precise probabilistic definition of home range: the probability of finding an animal at a particular location on a plane. This density function has been called the "utilization distribution" (Jennrich and Turner 1969, VanWinkle 1975, Dunn and Gipson 1977, Ford and Krumme 1979, Anderson 1982). The home range estimate is calculated by drawing equal height contours around the utilization distribution (Anderson 1982). The home range is specified by the contour such that 95% (or some other percentage) of the animal's locations are within the contour. This criterion is certainly objective, and thus repeatable, but is arbitrary. Why should 5% or any other percentage of an animal's movements not be considered normal (cf. Anderson 1982)? However, for purposes of comparison among studies, an objective repeatable method is needed, and a 95% area meets this criterion. Other definitions have been used to exclude excursions from what is considered the "normal" home range. Examples include deleting linear, short time frame movements from the data, but these definitions also include arbitrary criteria.

A second inadequacy of Burt's (1943) definition of home range is that a time frame is not specified. Some time period must be defined over which the home range is estimated to implement a home range estimator. This time frame could be as long as the lifetime of the animal, but generally shorter periods are of a more practical value.

The primary considerations in estimating home range must be (1) the purpose of the effort and (2) whether the methods used to collect the data satisfy this purpose. Sampling considerations must be taken into account. If the purpose of the study is to estimate the average home range for the population of animals for the month of January, then an adequate sample of animals must be radio-marked and an adequate number of observations on each animal must be taken during January to satisfactorily estimate each home range. A two-stage sampling effort is required. First, a sample of the animal population must be captured and radio-marked, with the requisite assumption that this sample of animals be representative of the entire population. Next, each of the radio-marked animals is sampled via telemetry to determine its locations during the month of January. Samples must be taken throughout the month at all hours of the day in order to be representative of an animal's movements for the month. Generally, a representative sample means a random sample, with each animal equally likely to be radio-marked, and each time of day equally likely to be chosen as the time to determine the location of the transmitter.

In this chapter we discuss the assumptions required for various home range

7 Home Range Estimation

estimators. The strengths and inadequacies of each estimator are addressed. A review structured much like this presentation is provided by Macdonald et al. (1980).

Independence of Observations

Tracking data are three dimensional, as is pointed out throughout this book. Thus, the closer in time two locations are taken, the less likely they are to be statistically independent. Stated differently, given the animal's location at time t, the expected change in location would be small for a small increase in time, $t + \Delta t$. As the difference between the two times becomes greater, the probability that the second location can be known, given the first, becomes less likely. A general rule of thumb to determine whether two locations taken at times t_1 and t_2 can be considered statistically independent is whether sufficient time has elapsed for the animal to move from one end of its home range to the other. However, more sophisticated approaches have been developed by Dunn (1978a) and by Dunn and Brisbin (1982).

All but one of the estimators discussed in this chapter require that the input data (i.e., locations) be statistically independent. Each location contributes as much information as every other location. If two locations are not independent, the sum of the information contributed by the two data points is not 2 units, but less than 2 units, because one of the locations can be used to make a reasonable guess of the other. To further illustrate this point, 100 locations taken 5 seconds apart on a radio-marked elk would not contribute nearly as much information on the elk's home range for a 30-day period as only ten locations each taken 3 days apart. With locations taken 5 seconds apart, the estimate of home range would be very small and might even be zero if the animal had been resting during the 500-second interval. Only one of the locations (the first in the sequence) really contributes any information on the animal's home range. The other 99 locations inflate the sample size, but contribute little or no additional information.

The same process of sample size inflation occurs for the home range estimators discussed in the following sections, when two or more locations are not independent. The sample size used to determine the estimate is much smaller than it appears. Only the estimator by Dunn and Gipson (1977) takes into account the time dimension of the data, and thus does not generate erroneous answers when there is a lack of statistical independence in the input data.

The above example is not a practical case, because the 100 locations taken

5 seconds apart in time are not a representative sample of the 30-day interval. If the objective of the study is to estimate the home range over the 500-second interval, the 100 locations taken at 5-second intervals would be a representative sample, and the argument could be made that the home range is properly estimated. The 100 observations would not be statistically independent on the time scale of 30 days, but they would be independent on the time scale of 500 seconds.

The bottom line on the question of statistical independence of consecutive observations is: "Has the time interval been properly sampled for which the home range estimate is to apply?" If a representative sample has been taken, then the observations can be assumed to be independent relative to the time frame of the sample. Generally, a random sample of time must be taken to provide a representative sample. Each instant in time throughout the entire time interval must have an equal chance of being sampled. In practice a systematic sample of time might be taken that would be a representative sample.

Minimum Convex Polygon

The oldest and most common method of estimating home range is the minimum area polygon (Mohr 1947). The minimum area polygon is constructed by connecting the outer locations to form a convex polygon (Fig. 7.1), and then calculating the area of this polygon. Eddy (1977) provides an efficient FORTRAN algorithm for determining the coordinates of the polygon. Given these coordinates clockwise around the polygon, the area (A) is calculated as

$$\hat{A} = \frac{\left[x_1(y_n - y_2) + \sum_{i=2}^{n-1} x_i(y_{i-1} - y_{i+1}) + x_n(y_{n-1} - y_1) \right]}{2}$$

The advantages of the convex polygon are (1) simplicity, (2) flexibility of shape, and (3) ease of calculation. However, the disadvantages are major. The size of the home range estimate increases indefinitely as the number of locations increases (Jennrich and Turner 1969), because the minimum area polygon is estimating total area utilized, not the area utilized in normal movements. As more locations are taken, the chances that a location is taken that is outside the existing polygon continues to be a finite probability. This phenomenon is true even if the animal is in a restricted area, for example, if the animal is in a square pen. The probability of taking a location when the animal is exactly in a corner would be zero. Yet, the pen describes the true home range. Theoretically, a finite sample would never generate an estimate exactly equal to the area

7 Home Range Estimation

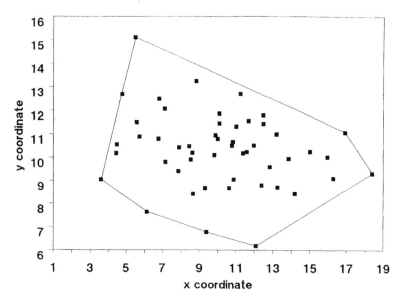

Figure 7.1 Construction of a minimum area polygon by connecting the outer locations to form a convex polygon. The estimate of home range is 0.007668 ha for these hypothetical data generated from a bivariate normal distribution. The data values are given in Table 7.1 so that the reader may verify this estimate.

TABLE 7.1
Simulated Data from a Bivariate Normal Distribution with
$\mu_x = \mu_y = 10$, $\sigma_x = 3$, $\sigma_y = 1.5$, and $\text{cov}(x, y) = 0$

Observation no.	t	x(m)	y(m)
1	24OCT86:14:06:28	10.6284	8.7061
2	25OCT86:21:32:18	11.5821	10.2494
3	26OCT86:11:13:00	15.9756	10.0359
4	27OCT86:02:33:30	10.0038	10.8169
5	01NOV86:08:02:56	11.3874	10.1993
6	02NOV86:18:09:32	11.2546	12.7176
7	05NOV86:04:12:04	16.2976	9.1149
8[a]	07NOV86:08:13:17	18.3951	9.3318
9	10NOV86:09:07:12	12.3938	8.8212
10	20NOV86:02:02:32	8.6500	8.4404
11[a]	20NOV86:03:19:29	12.0992	6.1831
12	23NOV86:15:56:39	5.7292	10.9079
13[a]	24NOV86:09:51:48	5.4973	15.1300

(*continues*)

TABLE 7.1
(*Continued*)

Observation no.	t	x(m)	y(m)
14	25NOV86:14:52:56	7.8972	10.4456
15	26NOV86:05:40:39	12.4883	11.8111
16	02DEC86:04:50:57	10.0896	11.4690
17	03DEC86:00:35:40	8.4350	10.4925
18	03DEC86:12:19:43	13.2552	8.7246
19	07DEC86:00:20:25	13.8514	9.9629
20	07DEC86:10:05:10	10.8396	10.6994
21	07DEC86:17:00:30	7.8637	9.4293
22	08DEC86:17:18:20	6.8118	12.4956
23	13DEC86:01:06:41	11.6917	11.5600
24[a]	13DEC86:05:38:09	3.5964	9.0637
25	15DEC86:18:49:15	10.7846	10.5355
26[a]	16DEC86:16:42:14	16.9375	11.0807
27	17DEC86:07:05:19	9.8753	10.9715
28	17DEC86:21:29:58	13.2040	11.0077
29[a]	18DEC86:15:01:57	6.1340	7.6522
30	20DEC86:13:03:37	7.1120	12.0681
31	21DEC86:20:49:01	8.8229	13.2519
32	21DEC86:22:26:11	4.7925	12.6987
33	22DEC86:18:06:00	15.0032	10.2604
34	23DEC86:08:32:42	11.9726	10.5340
35	27DEC86:09:51:28	9.8157	10.1214
36	29DEC86:12:20:03	6.7730	10.8152
37	30DEC86:01:21:24	11.0163	11.3384
38	31DEC86:22:43:14	9.2915	8.6962
39	04JAN87:01:18:52	4.4533	10.1955
40	04JAN87:11:20:56	14.1811	8.4525
41	07JAN87:22:10:49	8.5240	9.9342
42[a]	08JAN87:12:02:33	9.3765	6.7882
43	11JAN87:12:58:54	10.8769	9.0810
44	13JAN87:05:58:34	12.4894	11.4518
45	14JAN87:02:22:31	8.6165	10.2106
46	14JAN87:19:36:24	7.1520	9.8179
47	15JAN87:05:59:05	5.5695	11.5134
48	20JAN87:23:37:29	12.8300	9.6083
49	27JAN87:04:17:46	4.4900	10.5646
50	30JAN87:13:40:10	10.0929	11.8786

[a]Locations on the boundary of the convex polygon. Note that one of the squares designating locations in Fig. 7.1 touches the boundary, but is actually inside the polygon, and so is not part of the set of seven locations that define the polygon.

of the square pen. However, as more locations are taken, the probability of getting points farther into each of the four corners increases, so the estimate of home range size would be expected to continue to increase as more data are collected. Thus, the home range estimate is a function of the number of locations utilized to generate the estimate. Two estimates are not comparable if one is based on 100 data points, while the second is based on 500 data points, because the second estimate is expected to be somewhat larger. The additional 400 locations may provide further outlying locations, which lead to a larger polygon.

One approach to correct the problem of increasing home range size with increasing sample size is to eliminate the "outliers" before the home range polygon is calculated. We have calculated 95% polygons based on an ordering criterion for the locations. Consider a list of numbers. This list can be ordered (ranked) in descending order, with the largest number last. Nonparametric confidence intervals can be placed on the 95th quantile, based on the ranking of values in the list (Conover 1971:110–115). However, ordering a one-dimensional list of numbers is straightforward, whereas ordering a two-dimensional list (i.e., x and y coordinates of location) is not obvious, because ordering by only the x or y coordinate is not reasonable. We assumed that the list could be ordered based on the amount of area each location contributes to the minimum area polygon. Thus, to select the first location (the one contributing the most area to the polygon), each of the points on the polygon boundary is deleted one at a time, with the area attributed to that point being the difference in area between the original polygon and a new polygon constructed with the one data point eliminated. The location on the original polygon that contributes the most area is selected as the highest ranking location. The entire procedure is then repeated for the polygon formed when the highest ranking point is eliminated, and continues in this fashion until enough points have been ordered to form the 95% polygon, with 5% of the points ranked.

The inadequacy of the above procedure is demonstrated when two locations are closely spaced, but are far from the majority of locations (Fig. 7.2). The area contributed to the polygon by each of these two outlying locations is small when the other point is still included in the polygon boundary. That is, exclusion of one of these two points, but including the other, eliminates only a small sliver of area from the polygon. To effectively eliminate the two points from the analysis, they have to be simultaneously deleted. Therefore, to construct a 95% polygon, all possible combinations of 5% of the points in the data set must be considered. For example, to construct a 95% polygon for 100 data points, all possible sets of five points must be eliminated, one group at a time,

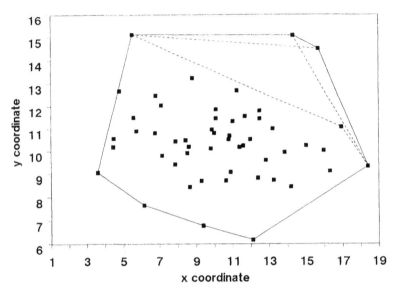

Figure 7.2 The problems encountered when only one location at a time is deleted from the minimum area polygon. Both of the outliers must be deleted to remove the area inside the large dashed-line triangle.

and a polygon area calculated. Note that if none of the five points is on the polygon boundary, then the polygon area is the same as the original, so the area doesn't actually have to be calculated.

Kenward (1987) provides three programs to compute "probability polygons" following strategies similar to that described above. He uses simple rules to exclude the farthest locations from a center of activity (such as a nest), or from the median x and y coordinates, or from the arithmetic mean or harmonic mean of the coordinates. Note that the median or arithmetic mean or harmonic mean don't necessarily have to be in the home range! Envision an animal with its area of use along the inside of a river curve—a boomerang-shaped home range. Any of Kenward's algorithms can be "tricked" into giving spurious estimates, because his simple rules are not always going to be biologically reasonable. Hartigan (1987) gives an algorithm that finds the smallest convex polygon (probability polygon) containing a proportion (α) of the sample locations. With this approach the sample size bias of the minimum polygon estimator could be eliminated. Hartigan's method has been developed in the context of nonparametric probability density estimation and is a statistically rigorous approach.

7 Home Range Estimation

Even though the minimum convex polygon allows a wide range of shapes, the method often leads to spurious results, when a concave shape is obviously the case. Consider, for example, a terrestrial animal with a home range around the edge of a lake (Fig. 7.3). A minimum convex polygon estimate includes part of the lake, even though a terrestrial animal would not inhabit this area.

The obvious method of correcting the defect of the minimum area polygon method is to make the polygon concave. However, some objective rule has to be made regarding which points are to be on the boundary. Every point in a set of data can be on the boundary of a concave polygon (Fig. 7.4). To understand this, consider the mean of the x's and y's of a set of locations as the center. Then start with any point and begin working in a clockwise direction, connecting the current point to the next point encountered as the hand of the clock pivots around the mean location. After the hand has swept through 360° (i.e., entirely around the clock face), the original data point closest to the mean is encountered again, and the polygon is closed.

In the example shown in Fig. 7.3, the researcher could quite objectively decide whether the lake area in the minimum convex polygon should or should not be excluded. Thus, a concave polygon could be derived on an objective (and thus repeatable) basis. However, a general, objective procedure that is based on biological merits for determining a concave polygon does not exist. One approach is to allow a concave polygon that has acute concave angles greater than 30°. Thus, many of the interior points in Fig. 7.4 would no longer be on the polygon boundary. However, the choice of 30° is no better than some other angle and is not based on biological grounds. Of course, a convex polygon is based on a choice of 180° angle (i.e., a convex polygon), and thus may

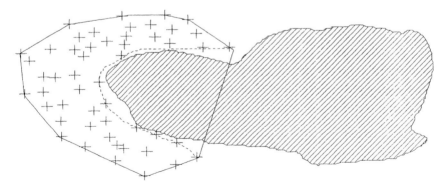

Figure 7.3 Minimum area polygon that includes part of a lake (shaded area) in the home range estimate.

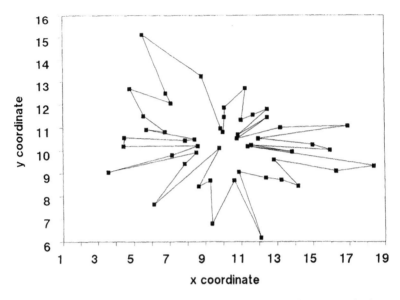

Figure 7.4 Concave polygon that connects all of the locations taken on the animal.

not be any more appropriate than some other angle. A very interesting approach concerned with human perception of shapes is given by O'Callaghan (1974). The input parameters are specified to the computer algorithm, which determines the degree of concavity allowed in the home range polygon. If biologically motivated decision rules could be prescribed, these procedures would offer potential for home range estimation.

Although the minimum area polygon estimate of home range size does not require the data to be uniformly distributed within the polygon, the characteristics of the estimate will depend somewhat on how the data are distributed within the polygon. The minimum area polygon can estimate home range size for data that are uniformly distributed, whereas some of the other estimators to be discussed cannot. Data distributed under a uniform distribution mean that each section of the home range is used equally frequently; that is, there is no center of activity. Samuel and Garton (1985) describe a test of the data against a uniform distribution, using randomly generated locations. This test has been programmed in the SAS code in Listing 1 of Appendix 7. A modification to the procedure proposed by Samuel and Garton (1985) has been made in that the random locations are only generated within the polygon. If the radio-tracking locations are uniformly distributed, then the distances from random points to the nearest radio-tracking location should follow an exponential distribution. The observed distribution of these distances is compared to the expected distribution with the Cramér–von Mises statistic (W^2).

For the data in Table 7.1, the W^2 statistic is 0.157 ($P = 0.132$). Thus, the null hypothesis of a bivariate uniform distribution would not be rejected, even though the data are known to be simulated from a bivariate normal distribution. This result emphasizes the low power of this goodness-of-fit test for small sample sizes. To perform goodness-of-fit tests with reasonable power, much larger sample sizes are required.

The minimum convex polygon area also does not allow for the time correlation of the data; that is, the locations are assumed to be statistically independent. As discussed above, a biased estimate of home range may result if an inadequate sample is taken.

Bivariate Normal Models

Jennrich–Turner Estimator

Hayne (1949) first suggested the use of a circle to estimate home range of animals captured on trapping grids. The approach was greatly improved by Jennrich and Turner (1969), when the circle was generalized to an ellipse. The underlying spatial model assumed in the Jennrich–Turner approach is a bivariate normal probability distribution. That is, location vectors (x_i, y_i) are assumed to be identically and independently distributed according to a bivariate normal model. Thus, animals are assumed to move randomly about their home range, with the most probable location (mode) being the very center—the mean x and mean y locations. Then, a 95% (or any other percentage) ellipse is calculated around the mean location, and the area of this ellipse is an estimate of the animal's home range (Fig. 7.5).

To calculate the home range ellipse from the n pairs of (x_i, y_i) coordinates, the means, variances, and covariances must first be calculated. These quantities are:

$$\bar{x} = \frac{\sum_{i=1}^{n} x_i}{n}$$

$$\bar{y} = \frac{\sum_{i=1}^{n} y_i}{n}$$

Analysis of Wildlife Radio-Tracking Data

$$s_x^2 = \frac{\sum_{i=1}^{n}(x_i - \bar{x})^2}{(n-1)}$$

$$s_y^2 = \frac{\sum_{i=1}^{n}(y_i - \bar{y})^2}{(n-1)}$$

$$s_{xy} = \frac{\sum_{i=1}^{n}(x_i - \bar{x})(y_i - \bar{y})}{(n-1)}$$

Jennrich and Turner (1969) suggest the use of $n - 2$ in the denominator of s_x^2, s_y^2, and s_{xy} to provide unbiased estimates, as opposed to $n - 1$. The covariance matrix, $\hat{\Sigma}$, is defined as

$$\hat{\Sigma} = \begin{bmatrix} s_x^2 & s_{xy} \\ s_{xy} & s_y^2 \end{bmatrix}$$

with the correlation of x and y defined as

$$r = \frac{s_{xy}}{(s_x^2 s_y^2)^{1/2}} = \frac{s_{xy}}{s_x s_y}$$

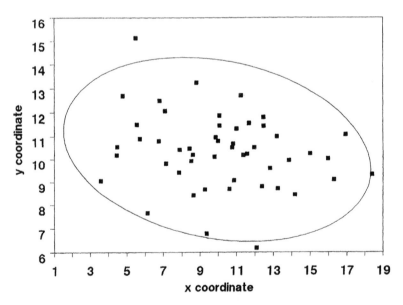

Figure 7.5 Jennrich–Turner bivariate normal ellipse estimator applied to the simulated data of Table 7.1.

7 Home Range Estimation

The $1 - \alpha$ confidence home range area is then calculated as the square root of the determinant of $\hat{\Sigma}$, ($|\hat{\Sigma}|^{1/2}$) times π times the appropriate $\chi^2_{(1-\alpha)(2)}$ statistic for the $1 - \alpha$ level with 2 degrees of freedom. Thus,

$$\hat{A} = \pi \left|\hat{\Sigma}\right|^{1/2} \chi^2_{(1-\alpha)(2)}$$

For $\alpha = 0.05$, a 95% confidence ellipse is obtained, and $\chi^2_{(1-\alpha)(2)} = 5.99$. The determinant $|\hat{\Sigma}|$ equals $s_x^2 s_y^2 - (s_{xy})^2$.

For the data presented in Table 7.1, $\bar{x} = 10.138$, $\bar{y} = 10.347$, $s_x^2 = 11.778$, $s_y^2 = 2.5737$, and $s_{xy} = -1.2167$, with $n - 1$ used in the denominator of s_x^2, s_y^2, and s_{xy}. Thus, $\hat{\Sigma}$ is

$$\hat{\Sigma} = \begin{bmatrix} 11.778 & -1.2167 \\ -1.2167 & 2.5737 \end{bmatrix}$$

with $r = -0.2210$ and $\hat{A} = 0.01011$ ha. This estimate is larger than the estimate of 0.007668 ha from the minimum convex polygon.

The value of $\hat{A}$ only represents the area of the home range estimate. To actually plot the ellipse on a map showing the (x_i, y_i) locations requires the use of the quadratic equation of the ellipse (Batschelet 1981). For the ellipse centered at $(\bar{x}, \bar{y})$, the quadratic equation is

$$s_y^2(x - \bar{x})^2 - 2s_{xy}(x - \bar{x})(y - \bar{y}) + s_x^2(y - \bar{y})^2 = \chi^2_{(1-\alpha)(2)}$$

Before the general case can be considered, two special cases must be checked. If $s_x^2 = s_y^2$, the ellipse is a circle with radius $(\chi^2_{(1-\alpha)(2)}/s_y^2)^{1/2}$. If $s_x^2 < s_y^2$ and $s_{xy} = 0$, the major axis of the ellipse is parallel to the y axis.

Excluding these special cases, the following quantities are calculated:

$$R = [(s_y^2 - s_x^2)^2 + 4(s_{xy})^2]^{1/2}$$

$$a = \left(\frac{(s_y^2 + s_x^2 + R)\chi^2_{(1-\alpha)(2)}}{2}\right)^{1/2}$$

$$b = \left(\frac{(s_y^2 + s_x^2 - R)\chi^2_{(1-\alpha)(2)}}{2}\right)^{1/2}$$

$$\theta = \arctan \frac{-2s_{xy}}{s_y^2 - s_x^2 - R}$$

Here, a and b are the semi-axes ($a > b$), and θ is the angle by which the major axis is inclined versus the X axis ($-90° < \theta < 90°$).

For computer plotting Batschelet (1981:264) uses the parametric equations of the ellipse

$$x = \bar{x} + a \cos \psi \cos \theta - b \sin \psi \sin \theta$$
$$y = \bar{y} + a \cos \psi \cos \theta - b \sin \psi \sin \theta$$

Here, ψ is a variable angle increasing from 0° to 360° in small steps. Batschelet (1981:265) suggests that a convenient step length for computer plotting is 4°. Using this choice, the ellipse is created from 90 points. For the reader desiring a better understanding of the geometry of an ellipse, see the chapter on this subject by Batschelet (1981), and also his chapter on bivariate methods.

The bivariate normal model of Jennrich and Turner (1969) provides a great improvement over the minimum area polygon, because the estimate of home range size is not a function of sample size. That is, the 95% ellipse estimated from 100 data points is expected to be the same size as the 95% ellipse estimated from 500 data points. Thus, home range size is much more comparable from study to study. Koeppl et al. (1975) suggest that the $\chi^2_{(1-\alpha)(2)}$ value should be replaced with the value $\{[2(n-1)]/(n-2)\}F_{(1-\alpha)(2, n-2)}$ where $F_{(1-\alpha)(2, n-2)}$ is an F statistic at level $(1-\alpha)$ and degrees of freedom $(2, n-2)$. The F statistic takes into account the fact that the ellipse center $(\bar{x}, \bar{y})$ is not known exactly, but is estimated from the data. As Table 7.2 shows, very little difference occurs for reasonable sample sizes. However, for sample sizes of less than 40 commonly used in the literature, the difference exceeds 10%. The F statistic estimator is almost two times larger than the χ^2 statistic when $n = 10$ for $\alpha = 0.05$. The disadvantage of using the F statistic of Koeppl et al. is that the expected home range size is now a function of the sample size (Madden and Marcus 1978), just as in the case of the minimum area polygon (Jennrich and Turner 1969). The F statistic would be appropriate if the purpose of estimating A was to have a 95% probability of including the next location

TABLE 7.2

Values of the F Statistic Multiplier Used by Koeppl et al. (1975) Compared to the χ^2 Used by Jennrich and Turner (1969)

n	$F_{(1-0.05)(2, n-2)}$	$\dfrac{2(n-1)}{(n-2)}F$	$\chi^2_{(1-0.05)(2)}$
5	9.55	25.47	5.99
10	4.46	10.04	5.99
20	3.55	7.49	5.99
30	3.34	6.92	5.99
42	3.23	6.62	5.99
62	3.15	6.40	5.99
122	3.07	6.19	5.99
∞	3.00	6.00	5.99

7 Home Range Estimation

taken on the animal inside the home range ellipse, but it is not appropriate for estimating the expected home range size. The sample size is obviously important in determining the quality of the home range estimate, but it should not influence the expected size of the home range. As shown below, sample size should be used to calculate the 95% confidence interval of the home range estimate, which provides an appraisal of the quality of the estimate.

For the data in Table 7.1, the Koeppl et al. estimate is 0.01079 ha. The programs DC80 and McPAAL both use this estimate instead of the Jennrich-Turner estimate provided by HOMER (described in Appendix 5).

Dunn and Brisbin (1982:46) have developed an equation to estimate the confidence interval for the home range estimate $\hat{A}$:

$$P[(2n - 4)\hat{A}/\chi^2_{(1-u)(2n-4)} \leq A \leq (2n - 4)\hat{A}/\chi^2_{(l)(2n-4)}] = 1 - \alpha$$

where l is the area of the tail of the lower χ^2 statistic and u is the area of the tail of the upper χ^2 statistic, so that $l + u = \alpha$ is the probability that the confidence interval does not include the true value. Nonsymmetric confidence intervals ($u \neq l$) provide a shorter interval, but generally a symmetric interval ($u = l = \alpha/2$) is desirable.

Thus, a 95% confidence interval for $\hat{A}$ is ($\alpha = 0.05$)

$$\frac{(2n - 4)\hat{A}}{\chi^2_{(1-\alpha/2)(2n-4)}} < \hat{A} < \frac{(2n - 4)\hat{A}}{\chi^2_{(\alpha/2)(2n-4)}}$$

For the data in Table 7.1, the confidence interval is 0.007760–0.01371 ha.

From these expressions Dunn and Brisbin (1982:46–47) have derived an estimate of the sample size (n) needed to estimate A such that $\hat{A}$ has probability at least $1 - \alpha$ of incurring a relative error less than r:

$$n \geq \frac{z^2_{(1-\alpha/2)}}{r^2} + 2$$

where $z^2_{(1-\alpha/2)}$ is the $(1 - \alpha/2)$ percentage point of the standard normal distribution.

Take, for example (from Dunn and Brisbin 1982:47), $\alpha = 0.05$ and $r = 0.10$; that is, $\hat{A}$ is desired to be within $\pm 10\%$ of the true A with probability 95%. Then

$$n \geq \frac{(1.96)^2}{(0.10)^2} + 2 \approx 386$$

For $r = 0.20$, $n \geq 98$. Thus, these results stress the large sample size needed to adequately determine the home range of an animal. As a general rule of thumb, at least 100 locations are needed to achieve an estimate with a confidence interval of ±20%.

Weighted Bivariate Normal Estimator

Samuel and Garton (1985) have proposed a modification of the Jennrich–Turner ellipse estimate, in which each data point is weighted by its distance from the mean of all of the locations. We follow the matrix notation used by Samuel and Garton in describing this estimator. The weighted distance is calculated in terms of the probability of this particular value occurring this far or farther from the mean location:

$$d_i^* = [(\mathbf{x}_i - \overline{\mathbf{x}^*})'(\mathbf{S}_x^*)^{-1}(\mathbf{x}_i - \overline{\mathbf{x}^*})]^{1/2}$$

where d_i^* is the weighted Mahalanobis distance for location $\mathbf{x}_i$, $\overline{\mathbf{x}^*}$ is the vector of weighted means, and $\mathbf{S}_x^*$ is the variance–covariance matrix of the locations. Several of the asterisks denoting weighted estimates were missing in the equation presented by Samuel and Garton (1985:519) and have been added in the above formula. The weighted mean and variance–covariance matrix are calculated as

$$\overline{\mathbf{x}^*} = \frac{\sum_{i=1}^{n} w_i \mathbf{x}_i}{\sum_{i=1}^{n} w_i}$$

and

$$\mathbf{S}_x^* = \frac{\sum_{i=1}^{n} w_i^2 (\mathbf{x}_i - \overline{\mathbf{x}^*})(\mathbf{x}_i - \overline{\mathbf{x}^*})'}{\sum_{i=1}^{n} w_i}$$

where

$$w_i = \begin{cases} 2/d_i & \text{for } d_i > 2 \\ 1 & \text{for } d_i \leq 2 \end{cases}$$

Thus, points close to the mean are weighted greater than those far away and, of particular importance, outliers are weighted very low, so that they no longer affect the estimate. To make the estimator more stable, points within 2 standard deviations are all given the same weight, with any point beyond 2 standard deviations given less weight, depending on its distance from the mean.

The weighting of the individual data points makes the estimator robust to outliers, hence the estimator is referred to as a robust ellipse estimator. The SAS code to generate this estimate is provided in Listing 3 of Appendix 7.

For the data in Table 7.1, $\overline{x^*} = 10.171$, $\overline{y^*} = 10.368$, $s_x^{*2} = 10.604$, $s_y^{*2} = 1.8918$, and $s_{xy}^* = -0.9476$, giving an area of 0.008240 ha (95% confidence interval 0.006328–0.01118). Thus, the weighted ellipse estimate is smaller than the unweighted estimate.

Multiple Ellipses

Don and Rennolls (1983) have proposed a home range model based on a circular normal distribution that incorporates biological attraction points (termed "nuclei" in their paper). Their model assumes that the coordinates of the attraction points are known, for example, the location of a fox squirrel's (*Sciurus niger*) den tree. Given one or more attraction points, their estimator provides the relative use ("strength") and range (measured by the variance) of the attraction points. Because a circular normal distribution is fit at each of the attraction points, a set of one-dimensional variances is estimated. Although their estimator allows for attraction points, and hence general asymmetry of the home range, the assumption of a circular normal distribution around each of the attraction points seems inconsistent with the goal of a nonsymmetric home range estimator. Don and Rennolls (1983) applied their estimator to gray squirrel (*Sciurus carolinensis*) data, which seems reasonable unless a squirrel's den tree was next to an unusable area, as would be the case if it were on the shore of a lake!

Dunn Estimator

The problem of lack of independence of observations was solved for evenly spaced (temporal) locations by Dunn and Gipson (1977), who developed an estimate of home range from the multivariate Ornstein–Uhlenbeck stochastic process (MOU). The MOU diffusion process describes the motion of a particle on a surface exhibiting a central tendency. The probability distribution of a particle at time $t + \Delta t$ is recognized to be a function of (1) the location of the previous point and (2) the amount of elapsed time. The process is a Markovian process, and thus this home range estimator allows for the time series nature of the data. Locations closely spaced in time are recognized not to provide as much information about home range size as points further apart in time. Relaxing the assumption of independence by modeling the time series nature of the data is a valuable contribution to home range estimation.

Like the Jennrich–Turner method, Dunn's estimator requires the assumption that the animal is using its space as bivariate normal, but the time sequence of the observations is considered; that is, the correlation between the successive fixes along the animal's path is recognized. Two observations close in time do not contribute as much information about the home range as do two observations taken further apart in time. The covariance matrix between two observations taken a fixed amount of time apart is constant, regardless of the absolute time or location. Hence, the home range is considered stable over time and space; that is, the temporal and spatial dimensions are homogeneous.

Input data for the Dunn estimator consist of groups of observations, or "bursts." Initial locations of bursts are assumed to be independent; that is, enough time has elapsed from the end of the previous burst to the start of the current burst that the two observations are statistically independent. As the Dunn estimator is currently developed, the time interval between observations within a burst must be equally spaced in time. That is, the time interval between observations should be the same within (but not necessarily between) bursts. This limitation of the estimator can be removed, but will require significantly more computer time.

An example is shown in Fig. 7.6 comparing the Dunn ellipse with the Jennrich–Turner ellipse. In this case the two estimates are nearly identical (Jennrich–Turner ellipse 55.2 ha; Dunn ellipse, 51.1 ha). The randomization test conducted by HOMER (Appendix 5) was not significant ($P = 0.853$), suggesting that a significant time series correlation was not present in the data. Thus, we would not expect large differences in the estimates from these two home range estimators.

Testing Bivariate Normality

Conceptually, bivariate normal models do not properly describe the movements of most free-living animals. Animals do not move randomly about a central point on a homogeneous surface, but move with purpose to obtain resources available within their home range. The movement assumed with the bivariate normal model is random, except for the tendency to stay to the middle—there is no purpose or predictable direction to the movement. Also, there is no biological reason that an animal is expected to spend the most time at the exact center of its home range. The most likely location may be a den site, which is unlikely to be located at the mean x and y locations, as assumed by bivariate normal models. Further, the elliptical shape assumption will often be violated. The example data given in Fig. 7.3 would produce erroneous estimates if used with any of the ellipse estimators. Smith (1983) provides a χ^2 goodness-of-fit test to evaluate whether the home range data are consistent with the assumption of bivariate normality, although the calculations shown in the paper are incor-

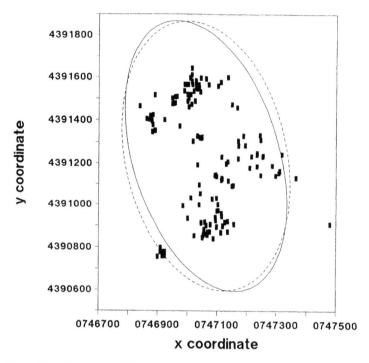

Figure 7.6 Comparison of Dunn (----) and Jennrich–Turner (———) ellipses for a female mule deer summering on the Roan Plateau, Piceance Basin, Colorado.

rect. The test looks at the frequency of locations within a series of concentric ellipses (contours) representing equal probability of occurrence (Fig. 7.7). That is, the test is analogous to a χ^2 goodness-of-fit test for continuous univariate distributions, but with cells replaced by contours. The SAS code for this test is provided in Listing 2 of Appendix 7.

For the simulated data in Table 7.1, the observed frequencies are 12, 8, 11, 9, and 10. These locations can be counted from Fig. 7.7; that is, there are 12 locations in the center ellipse, 8 in the next ring, 11 in the next ring, etc. The four ellipses in Fig. 7.7 are selected to divide the home range up into five equal areas. Therefore, the expected frequencies for each cell under the null hypothesis of a bivariate normal distribution would be 50/5 = 10; that is, the 50 locations should be divided equally into the 0–20, 20–40, 40–60, 60–80, and 80–100 percentage classes. For these data, the null hypothesis is not rejected ($P = 0.911$), suggesting that the data fit a bivariate normal distribution, as is the case.

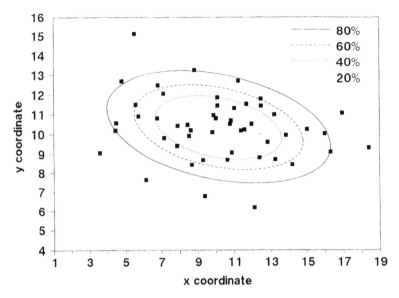

Figure 7.7 Concentric ellipses for probability levels of 20, 40, 60, and 80% are shown plotted over the simulated data from Table 7.1. The goodness-of-fit test suggested by Smith (1983) depends on the number of observed locations within each of the areas defined by the concentric ellipses.

Smith (1983:615) described the χ^2 statistic as having degrees of freedom of one less than the number of cells used to construct the test. Because the mean and the standard deviation of both x and y and the covariance of x and y are used to construct the test, 5 additional degrees of freedom should be subtracted to provide a proper test statistic. As the test is now programmed in Listing 2 of Appendix 7, this correction is not added, and the test will reject less often than it should. A minimum of seven cells would have to be used to construct the test if this correction is made to the program.

Samuel and Garton (1985) provide another test of bivariate normality that is more powerful than the test suggested by Smith. If the data follow a bivariate normal distribution, then the squared probability distances from the mean of the data (i.e., the squared values of d_i described above) should follow a χ^2 distribution with 2 degrees of freedom. Further, a robust version of the test using the d_i^* values can be constructed to accompany the robust estimate of home range size proposed by Samuel and Garton (1985). The SAS code for this test is provided in Listing 3 of Appendix 7.

For the data in Table 7.1, the unweighted goodness-of-fit test gives a W^2

value of 0.116 ($P > 0.15$). The weighted W^2 value is 0.214 ($P = 0.062$). Thus, the weighted ellipse does not appear to fit the observed data as well as the unweighted ellipse.

To illustrate that animal home ranges often do not follow a bivariate normal or bivariate uniform distribution, the following results from 38 mule deer home ranges are reported. The entries are the number of home ranges that met the ($P < 0.10$) criterion:

	Distribution		
Goodness-of-fit result	Bivariate uniform	Bivariate normal	Weighted bivariate normal
$P > 0.10$ (fit)	2	3	1
$P < 0.10$ (does not fit)	36	35	37

Thus, we conclude that for these mule deer, neither the bivariate uniform nor the bivariate normal distribution is a good fit of the data.

One example from this set of data is shown in Fig. 7.8. The impact of a few outlying locations causes the polygon estimate to differ substantially from the ellipse. Further, the effect of these outliers causes the ellipse to extend in the opposite direction from the outliers to compensate for their impact on the shape of the normal distribution. The bivariate normal distribution requires that the utilization distribution be symmetrical, and thus forces the ellipse out into vacant area opposite of the outlying locations. The bivariate normal goodness-of-fit test suggested by Samuel and Garton (1985) rejected the null hypothesis ($P < 0.01$), as did the concentric ellipse test suggested by Smith (1983) ($P < 0.01$). The goodness-of-fit test for the bivariate uniform distribution suggested by Samuel and Garton (1985) also rejected this hypothesis ($P < 0.01$).

One of the pitfalls of testing a set of observations for fit to a distribution is that a small sample size precludes the test's having any power. When the null hypothesis is accepted under these conditions, the conclusion is not that the data fit the distribution, but rather that too few observations may have been taken to reject the null hypothesis.

Although animal locations seldom fit a bivariate normal distribution, the use of the bivariate normal model for home range estimation is still worthwhile. The concept of a probability model to describe animal movements is practical. The ease of calculation, particularly the Jennrich–Turner estimate, and the

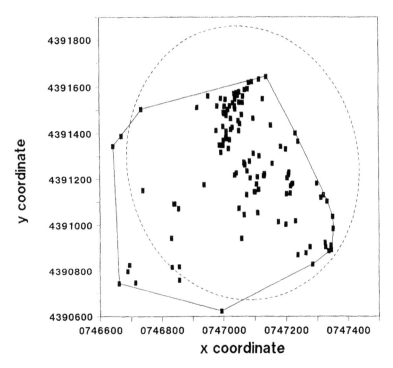

Figure 7.8 Impact of an asymmetrical utilization distribution on the Jennrich–Turner ellipse estimator. The dashed line is the 95% Jennrich–Turner ellipse; the solid line is the minimum convex polygon.

availability of programs (Dunn 1978b) make their application easy compared to other approaches. Also, the many enhancements made by Dunn (1978b) and by Dunn and Brisbin (1982) guarantee the continued use of this model.

Nonparametric Approaches

Grid Cell Counts

Methods utilizing the counts of location within grid cells or a map of an animal's home range (Fig. 7.9) are truly nonparametric approaches to home range estimation. No assumptions are made relative to the shape of the area utilized. The area over which an animal has moved is dissected by a grid of cells, or blocks. The number of animal locations is tabulated for each of these cells, and the sum of the areas of cells containing locations is taken as the estimate of home range area. Several serious problems arise in the practical application of this method. Disjoint areas (areas not connected) tend to be added together to form the home range estimate. Obviously, the animal did not suddenly disappear from one location and reincarnate at the next, but it may have moved across the areas where no locations happened to be taken. Thus, the disjointedness is due to reduced sampling intensity.

One approach to correcting the problem is to assume that cells crossed by the straight line joining two consecutive locations should also be counted as part of the home range area. However, use of a straight line is artificial, and selection of the maximum time interval for which locations will be connected is subjective. Grid cell methods have been utilized for the analysis of movement data from the Cedar Creek natural history area (Siniff and Tester 1965). These estimates avoid the problems mentioned above, because the locations are taken on animals every 45 seconds (Cochran et al. 1965). Thus, the straight line assumption does not appear unreasonable, and the filling in of cells for time intervals of 45 seconds does not seem too subjective. However, in many cases

	2	3	4	5	6	7	8	9	10	11	12	13	14	15	16	17
15				1												
14																
13				1				1								
12					1	2			1		2					
11			1		1	1			3	3	2	1				1
10			1			1	2	2	1	1	1	1	1	1	1	
9			1				1	1		2	1	1			1	
8					1			1					1			
7								1								
6											1					
5																

Figure 7.9 Illustration of how a grid cell home range estimate can be calculated by summing the areas of the cells that contained one or more animal locations. Values are only entered in cells that have one or more locations; remaining cells have zero locations. The top row and the left column are the grid coordinates. For these data the cell size is 1 m^2, giving an estimate of 0.0038 ha. Data are listed in Table 7.1.

the data are sampled (Rongstad and Tester 1969, 1971), and thus time intervals of 1 hour (deer) or 15 minutes (snowshoe hares) are used. Why use these values, as opposed to 30 minutes or 5 minutes? Again, the method raises a question of subjectivity. Also, home range estimates that are direct functions of the sampling intensity of the experiment do not seem to be a reasonable approach.

Rongstad and Tester (1969) also describe procedures for which cells within 2 units of distance of filled cells are also counted in the home range estimate. Again, a subjective estimate of the connecting distance of two cells must be made.

Complete counts of grid cell areas do not allow for removal of the outliers; that is, a 95% confidence area is not defined as described above. However, this criticism is easily overcome, because a plot of cumulative area versus the cumulative number of fixes can be constructed. The area at which 95% (or any other percentage) of the fixes are included is thus a 95% area estimate.

The greatest problem with the grid cell approach is selecting the grid square size, that is, determining the number of grid squares to draw on the map. Choosing too coarse a grid produces coarse estimates and tends to overestimate home range size. Choosing too fine a grid mesh produces underestimates of home range area and aggravates the problems mentioned above. That is, too fine a grid causes many more disjoint areas to be connected than a less fine grid mesh. Choice of grid cell size is an arbitrary decision for which no biologically based, objective procedures are known. Possibilities include setting grid cell size based on the average (or median) distance between consecutive locations. However, the level of arbitrariness in the estimation procedure is now hidden in the rule, rather than being an obvious decision made as part of the data analysis.

Fourier Series Smoothing

Anderson (1982) has presented a nonparametric approach to estimating home range utilizing the bivariate Fourier series. Fourier series have often been used in statistics to smooth data. The function to be smoothed is decomposed into a series of sine and cosine components of different frequencies. For home range estimation the bivariate functions consist of the x–y plot with a "spike" plotted at each (x, y) coordinate where the animal was located. The two-dimensional Fourier transform method smooths these spikes to form a surface representing the animal's use of area. Home range is calculated as the smallest area encompassed by 95% (or any other percentage) of the volume of the Fourier transform surface, just as with the bivariate normal model.

7 Home Range Estimation

Anderson found that the main problem with the use of the Fourier transform method was that areas at the edge of the home range are not estimated very well because of a lack of data. That is, the edges of the home range are where few locations are taken. Therefore, the estimate of home range size is highly variable for large percentages (such as 95%), because the poorly estimated surface at the edge of the home range has to be incorporated into the calculation of volume.

Anderson (1982) suggests that a more reasonable approach is to utilize a 50% volume as an estimate of home range. The impact of the edges is much less, because 50% of the volume lies near the center. A percentage as small as 50% seems to be a rather narrow description of "normal" in Burt's (1943:351) definition of home range. However, there is no biological justification for the use of 95%; this figure has probably been adopted by biologists because of the use of $\alpha = 0.05$ in statistical tests. Thus, the selection of any percentage can probably be justified if the value fulfills the needs of the experiment being conducted.

In comparisons of the Fourier estimator and the Jennrich–Turner bivariate normal ellipse for data simulated from a bivariate normal distribution, the Fourier estimator performed almost as well as the bivariate normal method for a 50% home range estimate (Anderson 1982). Because these data were simulated from a bivariate normal distribution, the bivariate normal estimator has a distinct advantage in this situation. Hence, that the Fourier estimator does so well is encouraging. The means of both estimators were very close to the true home range size determined analytically from the normal distribution. Further, the variance, or range of estimates, was roughly the same for the Jennrich–Turner and Fourier methods. However, we would expect that the Fourier estimate would not perform as well as the Jennrich–Turner if the home range estimator was increased from a 50% home range to a 95% home range. This is because the fringe of the distribution is much more difficult to estimate, as discussed above. The distribution assumptions made for the Jennrich–Turner estimator would greatly increase its effectiveness (because the assumption is true for these data), whereas the Fourier estimator would have only the available data to estimate the distribution (i.e., the additional information about the shape of the distribution is not known to this estimator).

For the data in Table 7.1, Table 7.3 lists the home range estimate (calculated with the McPAAL program) as a function of the percentage of future fixes to be included. For this particular data set, the 95% estimate is greater than the estimates produced by the minimum area polygon (0.007668 ha) and the

TABLE 7.3
Estimates of the Fourier Series Home Range Size (Anderson 1982) Computed with the McPAAL Program for the Data Presented in Table 7.1

Future locations (%)	Home range (ha)
25	0.002636
50	0.005423
75	0.008717
90	0.01125
95	0.01232

Jennrich–Turner (0.01011 ha) estimators. However, many replications of this experiment must be simulated to understand the relationships among these three estimators.

Harmonic Mean

Dixon and Chapman (1980) have proposed the use of a home range estimator based on the harmonic mean of the areal distribution. The center(s) of activity is located in the area(s) of greatest activity. The activity area isopleth is then related directly to the frequency of occurrence of an individual within its home range. Thus, the estimator has some worthwhile features, in that home ranges with two centers of activity can be properly defined, and no shape criteria are imposed on the estimator.

Contours of area are developed from a grid of points, where the observed area is the harmonic mean of the distance from the grid node to the observed radio-tracking locations. The farther a location is from a grid node, the less influence the location has on the mean for the node (Fig. 7.10). The specific equation to calculate the harmonic mean $\bar{d}_{g1.g2}$ for a grid node located at (x_{g1}, y_{g2}) is

$$\bar{d}_{g1.g2} = \frac{1}{\frac{1}{n}\sum_{i=1}^{n}\frac{1}{[(x_i - x_{g1})^2 + (y_i - y_{g2})^2]^{1/2}}}$$

Grid nodes located at centers of activity have the shortest distances and hence the lowest values of $\bar{d}_{g1.g2}$, while those located far from centers of activity have the longest distances and hence highest values of $\bar{d}_{g1,g2}$. The grid of $\bar{d}_{g1,g2}$ values is then contoured to estimate the isopleths of activity.

Study of the above equation immediately exposes one of the problems of the estimator. What if the telemetry location (x_i, y_i) is identical to the grid node location (x_{g1}, y_{g2})? The value of $\bar{d}_{g1,g2}$ is undefined, and a contour map cannot be constructed. A logical response to this problem is to shift the grid some small increment $(\Delta x, \Delta y)$, so that the distance from each grid node to each radio-tracking location is positive. Unfortunately, this action identifies a second problem. The contours constructed from a particular grid are unique to that grid; that is, if the grid origin is shifted, different estimates result. In practice, distances less than 1 unit length are given a value of 1, which also aggravates the problem of the units of dimension used to compute the estimate. Further, the estimator is sensitive to the distance between grid nodes and number of grid nodes. In particular, if the units of distance are converted and a new grid is used, different estimates of home range size (and isopleths of activity) are achieved. Thus, practical application of this estimator is sensitive to both the grid origin and the cell size. However, the procedure can still be useful in defining a map of activity, if the investigator recognizes that the map may vary somewhat in shape and is only an approximation.

For the data in Table 7.1, the DC80 program calculates the the estimates

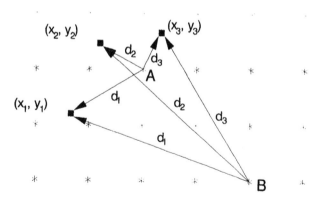

Figure 7.10 Mechanics of the Dixon–Chapman home range estimator. The grid nodes used to compute the estimator are shown as asterisks; animal locations, as squares. For grid node A, the reciprocal of the mean of $1/d_1$, $1/d_2$, and $1/d_3$ is much smaller than for grid node B. Thus, when the grid nodes are used to construct a contour map, the area around grid node A will be included inside a contour with a higher probability of occurrence of the animal than the contour which includes the area around grid node B.

TABLE 7.4
Harmonic Mean Home Range Estimates Computed with the DC80 Program for the Data Presented in Table 7.1[a]

Locations (%)	Home range (ha)
25	0.0006234
50	0.002340
75	0.005590
90	0.008704
95	0.01168

[a] Ten grid divisions were used to compute these estimates.

TABLE 7.5
Harmonic Mean Home Range Estimates Computed with the DC80 Program for the Data Presented in Table 7.1[a]

Locations (%)	Home range (ha)
25	0.001528
50	0.002736
75	0.006626
90	0.009983
95	0.01236

[a] Five grid divisions were used to compute these estimates.

presented in Table 7.4. Ten grid divisions were used. Had only five divisions been used, the estimates presented in Table 7.5 would result.

Similar answers are produced by the McPAAL program, although we have been unable to duplicate estimates with these two programs. Evidently, the estimator is implemented slightly differently in each of these codes. Possibly, the selection of the grid nodes is chosen differently.

Problems Common to All of the Nonparametric Methods

All of the nonparametric methods so far considered have ignored the time series nature of the data. As pointed out earlier, animals do not move about their environment in a random fashion, and thus the lack of statistical independence of the consecutive locations may lead to nonrepresentative home range estimates. The greatest need in estimation of home range is a nonparametric approach (in terms of assumptions about shape) that handles the time series nature of tracking data.

Second, all of the nonparametric methods described here share a problem with the minimum area polygon: None has an associated confidence interval. Thus, a point estimate is produced, but no estimate is provided of its standard error. Not only is the researcher left to wonder about the quality and usefulness of the estimate, but no method is available to determine sample size. Thus, sample size criteria have not been provided in the literature. In contrast, the ellipse-based estimators provide estimates of precision and, hence, necessary sample size can be determined. However, as discussed above, ellipse estimators require strong assumptions about the shape and structure of the use of the home range.

Computer Programs for Home Range Calculation

In the previous sections, we have mentioned various computer codes that will perform the home range calculations discussed. In Table 7.6, we summarize the codes known at this time.

TABLE 7.6
Computer Codes for the IBM Personal Computer and IBM-Compatible Models That Provide Home Range Estimates[a]

Program	Home range estimators
HOMER (Appendix 5)	Minimum convex polygon
	Jennrich–Turner ellipse
	Dunn ellipse
	Siniff-Tester grid cell counts
SAS Listing 1, Appendix 7	Minimum convex polygon
	Test of bivariate uniform distribution
SAS Listing 2, Appendix 7	Jennrich–Turner ellipse
	Test of bivariate normal distribution (Smith 1983)
SAS Listing 3, Appendix 7	Samuel–Garton weighted ellipse
	Test of bivariate normal distribution (Samuel and Garton 1985)
McPAAL	Minimum convex polygon
	Jennrich–Turner ellipse with Koeppl et al. (1975) F statistic
	Fourier series
	Dixon–Chapman
TELEM/PC	Minimum convex polygon
	Jennrich–Turner ellipse
	Dixon–Chapman
DC80	Minimum convex polygon
	Jennrich–Turner ellipse with Koeppl et al. (1975) F statistic
	Dixon–Chapman
HOMERANGE	Minimum convex polygon
	Jennrich–Turner ellipse
	Weighted bivariate normal ellipse
	Dixon–Chapman

[a] Sources for the codes not provided here are given in Table 1.1.

For straightforward home range estimates the McPAAL program is probably the easiest for the novice to use. Graphics are provided on the screen, and printer output can be obtained for the estimates. However, if statistical summaries of the estimates across animals, or other classification variables are desired, the SAS procedures offer considerably more power. The full statistical analysis capacity of SAS is available to perform data summaries and/or visual presentations of the data in x–y plots or histograms, etc.

Extension of Home Range Estimators

Dunn's home range estimator (Dunn and Gipson 1977) allows extensions so that home range can be calculated on more than just the x and y coordinates. For example, suppose that an animal is carrying a thermal sensor so that external body temperature is measured at the same time that its location is determined. Home range can be calculated on the vector (x, y, a_1). Hence, the third dimension of home range is now the range of temperatures utilized by the animal. Instead of generating an ellipse in two dimensions, an ellipsoid is generated in three dimensions. To visualize what an ellipsoid looks like, imagine the two-dimensional ellipse in Fig. 7.5 rotated around the major axis line. An egg-shaped object would result. The third dimension into and out of the paper can be the temperature axis. In this example the three variables all change smoothly in relation to one another. For real data the shape might be quite jagged.

The Jennrich–Turner (1969) estimator can be extended just as the Dunn estimator has been. Of course, the time-dependent nature of the data must be considered. Likewise, Mohr's minimum area polygon, grid cell methods, Dixon and Chapman's (1980) estimator, and Anderson's (1982) Fourier series analysis can all be extended to three dimensions. Just as home range calculations put boundaries on the area that an animal uses, extending the methods to include additional dimensions of temperature, heart rate, etc., can put limits on the range of these auxiliary variables.

Evaluating Home Range Estimators

Part of the confusion about home range estimation methodology has been produced from the lack of rigor in evaluating new estimators. A new estimator is proposed, but the only evidence that it works is an example set of data from an animal for which the true home range is not known. To demonstrate that a home range estimator "works," Monte Carlo simulations are required in which the true home range is known. The bias and the precision of the estimator can then be assessed. Anderson (1982) does a relatively good job of developing the statistical properties of the Fourier estimator. Simulations are presented comparing the Fourier estimator to both the minimum area polygon and the Jennrich–Turner bivariate normal ellipse. The estimator most notably lacking simulations to verify its statistical properties is the Dixon–Chapman harmonic mean estimator. No simulations have been published in the literature that demonstrate whether this estimator will provide unbiased estimates for a known home range.

Part of the problem in demonstrating that an estimator performs well is selecting the underlying distribution of animal movements. We suggest that the solution to this problem is to simulate the two extremes of the reasonable distributions and study the performance of the estimator at the extremes. If an estimator performs well at both ends of the spectrum, then it should perform reasonably through the middle range of distributions. Thus, we recommend that estimators be simulated for a heavily center-weighted distribution such as the bivariate normal, for which the edges of the distribution are not defined, going out to ∞. At the other extreme is the uniform distribution, for which the boundaries of the home range are explicitly delineated and the distribution of locations within the home range are uniformly distributed, that is, no attraction points. To achieve various shapes of home ranges, both convex and concave polygons can be simulated with the uniform distribution, and mixtures of bivariate normal distributions can be combined to produce irregular shapes. Anderson (1982) simulated a mixture of three bivariate normal distributions in his Fig. 6a.

Thus, a range of shapes and utilization distributions can be developed. In each case the true home range size can be calculated analytically, so that the bias and precision of each estimator can be assessed. We feel that such an extensive set of computer simulations is needed to evaluate the commonly used home range estimators. Continued application of the estimators without a better understanding of their statistical properties will only further cloud the issues relative to home range sizes.

Similarity of Home Ranges

An approach common to measuring fidelity and animal association (Chapter 6) is to compare two or more home ranges. In the case of fidelity, home range estimates are compared for a single animal between two or more time periods.

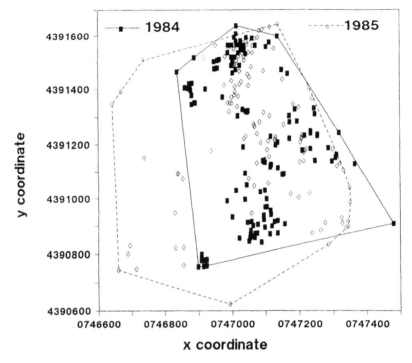

Figure 7.11 Summer home ranges of a female mule deer from the Piceance Basin of western Colorado that was radio-tracked for two consecutive summers. The home range was estimated with the minimum area polygon.

For measuring animal association the home ranges of two or more animals are compared. Continued data collection over several time periods or animals provides a series of home range maps that can be compared for similarity. The approach depends heavily on the assumptions of the home range estimation procedure. Figure 7.11 illustrates this approach for measuring fidelity with data from mule deer summering in the Piceance Basin, Colorado. A similar approach was used by Tierson et al. (1985).

Although not quantitative, Fig. 7.11 adequately illustrates the fidelity of this deer to a specific summering site. However, the influence of a few locations in 1985 greatly affects the estimate of overlap, hence fidelity. As shown in Fig. 7.12, use of a Jennrich–Turner ellipse estimator would change the reader's perception of the overlap of the home ranges. Another problem with estimating home range overlap using the bivariate normal estimators is that the utilization

7 Home Range Estimation

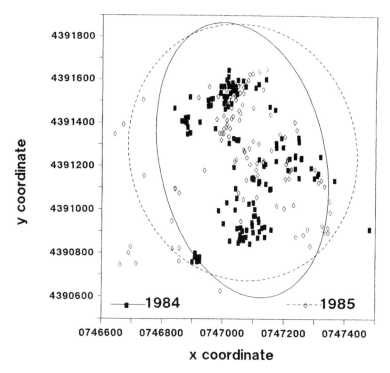

Figure 7.12 Summer home ranges of a female mule deer from the Piceance Basin of western Colorado that was radio-tracked for two consecutive summers. The home range is estimated with the Jennrich–Turner bivariate normal ellipse.

distributions should be taken into account, making the estimate of overlap difficult to calculate for the bivariate normal ellipse. That is, the concentric probability ellipses should be used to estimate overlap: the integral of the utilization function, not just the amount of overlapping area of the two ellipses.

Many applications which have used home range estimates in the literature can be performed without actually calculating a home range estimate. For example, the MRPP test of Mielke and Berry (1982) (see also Zimmerman et al. 1985) can be used to test for changes in an animal's area of utilization, such as is required for testing for fidelity or association. Instead of comparing two home range estimates (where both lack measures of precision), the method tests whether the two or more sets of locations came from a common distribution. The use of the MRPP test was discussed in Chapter 6 as a method of testing for fidelity to an area.

Preferred Home Range Estimator

There is no single preferred estimator of home range. All estimators have their faults. However, if a method must be used, there are some criteria by which the most appropriate method may be selected:

1. *Time series randomization test*: If each point is independent, the average distance between consecutive points should be the same, regardless of the order in which they are entered (Schoener 1981). Swihart and Slade (1985) have extended this test by making additional assumptions about the underlying statistical distribution of the locations.
2. *Shape criterion*: Apply a χ^2 goodness-of-fit test to determine whether the home range shape is close to the expected shape (Samuel and Garton 1985, Smith 1983).
3. *Sample size*: The large sample size required for nonparametric estimators may preclude their use in a particular study. Estimators requiring additional assumptions may be preferred.

Usefulness of the Home Range Concept

Home range is a concept that is frequently abused. Some of the problems go back to its definition as the area used by the animal in its normal activities (Burt 1943:351). What is use? What is normal? Normal is frequently defined as 95% of the time, but with no biological justification. New, more complex estimators, such as the kernel methods used by Worton (1989), make these questions even more difficult as the number of home range estimators increases.

An approach which eliminates the need for a home range estimate is to use the actual location estimates to test the hypothesis of interest. For example, home range size may be desired because it may be related to an animal's energetic costs. That is, the larger the home range, the more energy expended in searching for various life requisites. However, the distance between consecutive locations (corrected for time differences) would also measure energetic costs and does not suffer from the many assumptions required to obtain a home range estimate. Comparison of distance moved for two groups of animals seems preferable to comparing home range estimates. The use of the MRPP test described above and in Chapter 5 is another example of how the problems of home range estimates can be avoided by applying statistical techniques to the raw data rather than comparing home range estimates.

Finally, before computing home range, the investigator should consider what will be learned. Often, home range estimates are the only product from a

7   Home Range Estimation

tracking study, because no hypothesis was originally designed to be tested. Home range estimates are a poor substitute for good experimental protocol. No information about why an animal moves is provided in a home range estimate. Wildlife biologists need more than just descriptive data from "binoculars that see in the dark." Wildlife science has moved beyond this point in its quest for knowledge.

Summary

1. The concept of a home range may be useful to biologists. However, implementation of the concept in a rigorous statistical format is not easy. Most of the benefits of the concept are lost when actual estimates of home range size are constructed, because of the many subjective decisions that must be made to construct the estimate. Estimates lack objective criteria in their construction, and thus provide little biological insight. Further, the statistical properties of most of the commonly used estimators are unknown, and Monte Carlo simulations are needed to compare the estimates when the true home range size is known.
2. Three basic models have been used to develop home range estimators: bivariate normal ellipses, convex polygons, and nonparametric models based on grids.
3. Estimators based on the bivariate normal ellipse require stronger assumptions about the internal use of the home range than the other models. The animal is expected to spend the most time near the center of the home range, but very little time at the periphery. Three bivariate normal estimators are those by Jennrich and Turner (1969), Koeppl et al. (1975), and Dunn and Gipson (1977).
4. Neither the elliptical nor convex polygon models allow home ranges with concave shapes.
5. The convex polygon suffer from its relationship between sample size and home range size: As sample size increases, the expected home range size also increases.
6. Nonparametric models tend to have lower precision than the parametric models, because fewer assumptions are made to develop an estimate. Prominent nonparametric models are the Fourier estimator by Anderson (1982) and the harmonic mean estimator by Dixon and Chapman (1980).
7. Because home range is a function of time as well as space, sampling procedures must sample time in a representative fashion. Home range

estimates should be reported in the context of the time frame of the sample. Further, the autocorrelation of the animal's consecutive locations must be considered if a random sample of the animal's time is not taken. Only the estimator developed by Dunn and Gipson (1977) assumes any autocorrelation of locations.

8. Extension of telemetry data to include physiological parameters means that additional dimensions are included in the analysis. However, most of the methods discussed for a two-dimensional space can be extended to three dimensions, and so many times the analyses are straightforward.

9. A serious limitation of most home range estimators is the lack of a precision estimate and hence a confidence interval estimator. Statistical estimates that lack precision estimates provide a false sense of security to the user. Further, statistical tests of the equivalence of two home range estimates cannot be constructed without measures of precision.

10. To test whether two or more animals are using the same area in the same fashion, or whether one animal is using an area in the same way through time, the biologist should be testing whether the distribution of locations is the same. Statistical tests that test for changes in the bivariate probability density function of the locations should be used. A promising candidate for this approach is the MRPP test by Mielke and Berry (1982).

11. Often, home range estimates are developed because the researcher has not carefully designed the research being conducted. The hypotheses to be tested were not formulated prior to data collection, so home range estimates provide a quantitative summary of the data. Careful design of research can usually avoid the pitfalls of home range estimation, because a statistical analysis based on the raw data can be developed to test important hypotheses. Home range estimates are not used in the analysis, and thus the assumptions they require are not needed, and the biases inherent in the technique are avoided.

References

Anderson, D. J. 1982. The home range: a new nonparametric estimation technique. Ecology 63:103–112.

Batschelet, E. 1981. Circular Statistics in Biology. Academic Press, New York. 371 pp.

Burt, W. H. 1943. Territoriality and home range concepts as applied to mammals. J. Mammal. 24:346–352.

Cochran, W. W., D. W. Warner, J. R. Tester, and V. B. Kuechle. 1965. Automatic radio-tracking system for monitoring animal movements. BioScience 15:98–100.

Conover, W. J. 1971. Practical Nonparametric Statistics. Wiley, New York. 462 pp.

Dixon, K. R. and J. A. Chapman. 1980. Harmonic mean measure of animal activity areas. Ecology 61:1040–1044.

Don, B. A. C. and K. Rennolls. 1983. A home range model incorporating biological attraction points. J. Anim. Ecol. 52:69–81.

Dunn, J. E. 1978a. Computer programs for the analysis of radio telemetry data in the study of home range, Stat. Lab. Tech. Rep. No. 7. Univ. of Arkansas, Fayetteville. 73 pp.

Dunn, J. E. 1978b. Optimal sampling in radio telemetry studies of home range. Pages 53–70 *in* H. H. Shugart, Jr. ed. Time Series and Ecological Processes. SIAM, Philadelphia, PA.

Dunn, J. E. and I. L. Brisbin, Jr. 1982. Characterizations of the multivariate Ornstein-Uhlenbeck diffusion process in the context of home range analysis, Stat. Lab. Tech. Rep. No. 16. Univ. of Arkansas, Fayetteville. 72 pp.

Dunn, J. E. and P. S. Gipson. 1977. Analysis of radio telemetry data in studies of home range. Biometrics 33:85–101.

Eddy, W. F. 1977. A new convex hull algorithm for planar sets. ACM Trans. Math. Software 3:398–403.

Ford, R. G. and D. W. Krumme. 1979. The analysis of space use patterns. J. Theor. Biol. 76:125–155.

Hartigan, J. A. 1987. Estimation of a convex density contour in two dimensions. J. Am. Stat. Assoc. 82:267–270.

Hayne, D. W. 1949. Calculation of size of home range. J. Mammal. 30:1–18.

Jennrich, R. I. and F. B. Turner. 1969. Measurement of non-circular home range. J. Theor. Biol. 22:227–237.

Kenward, R. E. 1987. Wildlife Radio Tagging. Academic Press, San Diego, CA. 222 pp.

Koeppl, J. W., N. A. Slade, and R. S. Hoffmann. 1975. A bivariate home range model with possible application to ethological data analysis. J. Mammal. 56:81–90.

Macdonald, D. W., F. G. Ball, and N. G. Hough. 1980. The evaluation of home range size and configuration using radio tracking data. Pages 405–424 *in* C. J. Amlaner, Jr. and D. W. Macdonald eds. A Handbook on Biotelemetry and Radio Tracking. Pergamon Press, Oxford, England.

Madden, R. and L. F. Marcus. 1978. Use of the F distribution in calculating bivariate normal home ranges. J. Mammal. 59:870–871.

Mielke, P. W., Jr. and K. J. Berry. 1982. An extended class of permutation techniques for matched pairs. Commun. Stat. 11:1197–1207.

Mohr, C. O. 1947. Table of equivalent populations of North American small mammals. Am. Midl. Nat. 37:223–249.

O'Callaghan, J. F. 1974. Computing the perceptual boundaries of dot patterns. Comput. Graphics Image Process. 3:141–162.

Rongstad, O. J. and J. R. Tester. 1969. Movements and habitat use of white-tailed deer in Minnesota. J. Wildl. Manage. 33:366–379.

Rongstad, O. J. and J. R. Tester. 1971. Behavior and maternal relations of young snowshoe hares. J. Wildl. Manage. 35:338–346.

Samuel, M. D. and E. O. Garton. 1985. Home range: a weighted normal estimate and tests of underlying assumptions. J. Wildl. Manage. 49:513–519.

Schoener, T. W. 1981. An empirically based estimate of home range. J. Theor. Biol. 20:281–325.

Siniff, D. B. and J. R. Tester. 1965. Computer analysis of animal movement data obtained by telemetry. BioScience 15:104–108.

Smith, W. P. 1983. A bivariate normal test for elliptical home-range models: biological implications and recommendations. J. Wildl. Manage. 47:613–619.

Swihart, R. K. and N. A. Slade. 1985. Testing for independence of observations in animal movements. Ecology 66:1176–1184.

Tierson, W. C., G. F. Mattfield, R. W. Sage, and D. F. Behrend. 1985. Seasonal movements and home ranges of white-tailed deer in the Adirondacks. J. Wildl. Manage. 49:760–769.

VanWinkle, W. 1975. Comparison of several probabilistic home-range models. J. Wildl. Manage. 39:118–123.

Worton, B. J. 1989. Kernel methods for estimating the utilization distribution in home-range studies. Ecology 70:164–168.

Zimmerman, G. M., H. Goetz, and P. W. Mielke, Jr. 1985. Use of an improved statistical method for group comparisons to study effects of prairie fire. Ecology 66:606–611.

CHAPTER

8

Habitat Analysis

Habitat use is a critical facet in the management of a wildlife species. Habitat provides food and cover essential for the population to survive. Thus, interest in habitats is shown in the analysis of wildlife tracking data.

Four questions are often asked about wildlife and habitat relationships:

1. What is the availability of the habitat to the wildlife population, that is, how much of each habitat type is available?
2. What is the degree of utilization of each habitat by the wildlife population?
3. What is the preference for each habitat type by the wildlife population?
4. Which of the habitat types is critical for the wildlife population to survive?

In this chapter each of these aspects (availability, utilization, preference, criticalness) of habitat is discussed relative to wildlife populations.

Availability

Availability of each habitat type is usually determined from specific areas on a habitat map. Availability consists of the amount of area of each habitat type that is available to the wildlife population. Unfortunately, what the biologist determines is available habitat and what the animal perceives as available habitat may be two different things. For instance, because of territoriality, the animal may select only a portion of a specific area, even though to the biologist the area appears to be suitable habitat. Possibly noise pollution from a nearby development or disturbance by occasional vehicles may stop an animal's use of an area. Even the presence of another species may preclude a specific area from being used. In the following presentation we assume that availability can be measured using a map, but the reader must remember that gross errors in conclusions may result if this assumption is incorrect. One of the methods discussed (Johnson 1980) handles the problem of undefinable habitat boundaries.

Measurement of the size of each habitat on a map can be performed in a number of ways. One standard method is to use a planimeter to measure the map to obtain areas of each habitat type. A second approach is to accurately weigh the map, cut it into pieces representing each habitat type, and weigh each of the pieces to obtain the percentage of the map contained in a particular habitat type. These percentages are converted to area of each habitat type by converting to the proportion of the total area.

If the map has been digitized for computer processing, numerous computer programs are available to estimate the size of each habitat type, including commercially available GIS systems. Some public domain GIS systems include MOSS (mapping and overlay statistical system) (Data Systems Support Group 1979), WRIS (Russell et al. 1975a,b), and AREAS (Wehde et al. 1980). These programs operate on maps using two different methods: the grid cell approach and the polygon approach. In the grid cell approach a map is divided into grid cells. The dominant habitat type for that cell is coded as the habitat type of that cell. The sum of the cell areas for each habitat type then provides the measurement of the area of that habitat type. Obviously, the accuracy of this method depends to a large degree on the fineness of the scale at which the grid cells are defined (the smaller the better, but also the more difficult).

In the polygon approach the habitat boundaries are digitized; that is, a sequence of (x, y) coordinates is used to represent the habitat boundary line. These line segments are then connected to form polygons for each portion of habitat. The sum of the areas of the polygons of each habitat type then provides the area of each habitat type. Some systems, such as MOSS, can convert maps back and forth between the two representations.

If a habitat map is not available, more advanced approaches are required. LANDSAT data are available on computer tapes from which habitat maps can be developed. This methodology is beyond the scope of this presentation (see LaPerriere et al. 1980 for an example). The National Cartographic Information Center (NCIC 1979) has surface elevations available for computer processing

8 Habitat Analysis

that may prove useful in developing a habitat map. Photos produced on special films are also available from which a habitat map may be constructed.

Whatever the approach used, the above methods provide the areas of each habitat type. In theory, these methods provide exact measurements, within limitations of the map, of the area of each habitat type. An approach suggested by Marcum and Loftsgaarden (1980) estimates the area of each habitat type. Random points are placed on the map and habitat types are determined for each of these points. Because a finite number of points is used, this method has a sampling error associated with the habitat area estimate and therefore the statistical analysis must be treated differently, as we discuss later.

Utilization

Radio-tracking data provide examples of utilization of habitat by the members of a population. Locations taken for each animal are classified as to the habitat type in which they occurred. Thus, estimates are developed for the percentage of time each animal spends in a particular habitat type.

Note that the usual problems of sampling occur here again. First, a representative (random) sample of the population of animals must be selected for radio-marking. Each of the radio-marked animals is then sampled over time to obtain estimates of the percentages of time each spends in various habitat types. In order to assume that the habitat analysis will be representative of the wildlife population, representative samples must be taken at both levels. In addition, the statistical tests described require that all observations must be independent.

The methodology to determine the habitat type of a radio-tracking location varies from simple to complex. Possibly, the habitat type can be recorded in the field, either at the time the location is taken or at a later time when the observer travels to the location. A second approach is to physically plot the location on a habitat map and record the habitat type from the map.

All of the above approaches are probably labor intensive. A computer, however, can be used to obtain the habitat type directly from a digitized habitat map. Usually, if a digitized map has been used to obtain habitat availability, the same map can be used to determine the habitat type of radio-tracking locations.

Just as availability was measured from either a cell map or a polygon map, likewise, the habitat type of a radio-tracking location can be determined from either type of map. In general, the polygon approach is preferred, in that the distance to the nearest habitat boundary can be determined and, hence, the probability of the location's being misclassified can be estimated. That is, as is pointed out in Chapter 4, each location is an estimate with an associated confidence ellipse. When this confidence ellipse includes a habitat boundary, the probability that the location may be misclassified in the wrong habitat type increases. A disadvantage in using cell maps is that many cells do not consist of only one habitat. Also, locations may often be placed in the wrong cell. Thus, polygon maps provide the best approach for assessing the reliability of the habitat type associated with a fix.

Preference

The above sections describe how habitat availability is determined and its utilization estimated. Next, preference by the wildlife population for a particular habitat needs to be considered. The question of preference is really whether the animal population selects some habitat types more than others and thus spends more time in these habitats (and hence less time in the remaining habitats) than would be expected based on the availability of each habitat type. We will use the term "avoidance" to refer to habitat types where the animal spends less time than expected, even though the animal may not actually be avoiding these areas. If one habitat type is preferred, then more time must be spent in this habitat type than expected by chance alone. Thus, if one habitat type is preferred, then one or more of the remaining habitat types must be avoided because of constraints on time. Alldredge and Ratti (1986) have compared four methods of testing for the selection of habitat types from availability and use data. We discuss each of these approaches and then summarize their conclusions concerning which of these methods is appropriate for radio-tracking data.

χ^2 Analysis

A standard approach to testing the hypothesis of preference has been presented by Neu et al. (1974). A χ^2 test is performed to test for the goodness of fit of utilized habitat to available habitat types. The two null hypotheses tested by the χ^2 test, as described by Alldredge and Ratti (1986), are H_{01}, in which usage occurs in proportion to availability, considering all habitats simultaneously, with the option of testing the second hypothesis, H_{02}, that usage occurs in proportion to availability, considering each habitat separately.

A simple example will illustrate the technique. Suppose five habitat types (A–E) have been defined on a 10-km^2 study area. The areas of each habitat type and the proportion of the study area contained in each habitat (availability) are given in Table 8.1.

8 Habitat Analysis

TABLE 8.1
Habitat Area and Proportion of the Total Area
(Availability) for Each of Five Habitats on a
Hypothetical 10-km² Study Area

Habitat type	Area (km²)	Proportion	Percent
A	5	0.5	50.0
B	3	0.3	30.0
C	1	0.1	10.0
D	0.5	0.05	5.0
E	0.5	0.05	5.0
Total	10	1.0	100.0

TABLE 8.2
Numbers and Percentages of Hypothetical Radio-Tracking Locations
in Each Habitat Type (Utilization) for Three Animals.

Habitat type	Animal 1		Animal 2		Animal 3	
	Number	%	Number	%	Number	%
A	302	46.5	162	49.1	381	45.3
B	180	27.7	91	27.6	263	31.2
C	69	10.6	33	10.0	78	9.3
D	49	7.5	16	4.8	65	7.7
E	50	7.7	28	8.5	55	6.5
Total	650	100.0	330	100.0	842	100.0

Suppose that three radio-marked animals are tracked during a 1-month period in this study area. The number of locations in each habitat type for each animal (utilization) has been determined and is shown in Table 8.2. The sample size of the χ^2 test is determined by the number of locations for an animal. Therefore, the χ^2 table for animal 1 is calculated as shown in Table 8.3. Note that the expected values are calculated as the proportion of area made up by the habitat type in the study area times the total number of locations for the animal. That is, the expected number of locations in each habitat type, if no preference or avoidance is shown by the animal, would be the proportion of locations that fell in the habitat type if the locations were picked at random. Hence, this χ^2 test is just a goodness-of-fit test of the locations to the expected distribution of locations if no preference or avoidance is shown by the animal. In this example,

TABLE 8.3
Calculation of the χ^2 Statistic (Test of Preference) for Animal 1[a]

Habitat type	Observed	Expected	χ^2 Contribution	G Contribution
A	302	0.5 × 650 = 325	1.6277	−22.166
B	180	0.3 × 650 = 195	1.1538	−14.408
C	69	0.1 × 650 = 65	0.2462	4.121
D	49	0.05 × 650 = 32.5	8.3769	20.118
E	50	0.05 × 650 = 32.5	9.4231	21.539
Total	650	1.0 × 650 = 650	20.8277	18.408

[a] Calculation is from Table 8.2, with expected values determined from Table 8.1.

the conclusion is that the animal is showing preference (and hence also avoidance) of some habitat, since the probability of observing such a large χ^2 statistic (20.828) is $P = 0.0003$, with 4 degrees of freedom. The degrees of freedom for the table is the number of habitat types (k) minus 1. Thus, we reject H_{01} described above. We also include the maximum likelihood χ^2 statistic (G) for comparison to the usual Pearson χ^2 statistic (see Sokal and Rohlf 1981 for a discussion of these two methods of calculating χ^2 values). The formula for calculating the G χ^2 statistic is similar to that for a χ^2 statistic:

$$G = 2 \sum_{i=1}^{k} O_i \ln\left(\frac{O_i}{E_i}\right)$$

The G statistic is also distributed as χ^2, and has 4 degrees of freedom.

The main contribution of the work by Neu et al. (1974) is the methodology to determine which specific habitat types are avoided or preferred, given that the χ^2 test rejects the null hypothesis for all habitat types considered together. They construct confidence intervals for the proportion of times an animal uses each habitat type. The interval used is

$$\hat{p}_i - z_{\alpha/2k}\left[\frac{\hat{p}_i(1-\hat{p}_i)}{n}\right]^{1/2} \le p_i \le \hat{p}_i + z_{\alpha/2k}\left[\frac{\hat{p}_i(1-\hat{p}_i)}{n}\right]^{1/2}$$

where $\hat{p}_i$ is the proportion of locations in habitat type i, and $z_{\alpha/2k}$ is the upper standard normal variate corresponding to a probability tail area of $\alpha/2k$. The $2k$ denominator under α is used because multiple confidence intervals are being computed simultaneously. To keep the Type I error level low, Bonferroni normal statistics are used with $k = 5$, the number of habitat types. For the example above, the confidence intervals are shown in Table 8.4 for $\alpha = 0.10$, with

8 Habitat Analysis

TABLE 8.4
Bonferroni Confidence Intervals for the Proportion of Time Spent in Each Habitat Type for Animal 1

Habitat type	Number of locations	$\hat{p}_i$	Confidence interval
A	302	0.46462	$0.41909 \leq p_A \leq 0.51014$
B	180	0.27692	$0.23608 \leq p_B \leq 0.31777$
C	69	0.10615	$0.078039 \leq p_C \leq 0.13427$
D	49	0.07538	$0.051288 \leq p_D \leq 0.099482$
E	50	0.07692	$0.052602 \leq p_E \leq 0.10124$

$z_{(1-\alpha/2k)} = 2.327$. Exact values of the z statistic can be computed with the SAS function *PROBIT*. To determine whether a habitat is avoided or preferred, the confidence interval is checked for overlap with the availability proportion of the corresponding habitat (Table 8.1). If the confidence interval includes the availability proportion, then the hypothesis of no preference or avoidance of this habitat type cannot be rejected. However, if the lower bound of the interval exceeds the availability proportion, then the animal has shown preference for this habitat type. For example, in Table 8.4, the lower bound for habitat type D is 0.051288, which exceeds the expected proportion of 0.05 in Table 8.1. Thus, the animal shows preference for habitat type D and thus we reject H_{02} for habitat type D. In contrast, the upper bound for habitat type A is 0.51014 (Table 8.4) and, hence, is greater than the expected value of 0.5. Thus, the null hypothesis H_{02} for habitat type A cannot be rejected. In summary, we conclude that habitat types D and E are preferred, with data lacking to reject the null hypothesis for habitat types A, B, and C. Additional details of this procedure are clarified by Byers et al. (1984).

Similar analyses would be conducted for animals 2 and 3. Thus, a set of preference and avoidance statements for each animal is developed. From these statements, conclusions about the population must be drawn. Because of animal variability, different animals will prefer and/or avoid different habitats. Results for the three animals in Table 8.2 are shown in Table 8.5.

The χ^2 test of preference requires that the observations of an animal's location be independent. Hence, it is not reasonable to inflate the sample size of the test by including two observations near one another in time. Lack of independence of the observations in the test will result in too many Type I errors; that is, the null hypothesis is rejected when in reality it should be accepted.

TABLE 8.5
Conclusions on Preference and/or Avoidance of Three Animals for the Five Habitat Types[a]

Animal	χ^2 Statistic	Degrees of freedom	Probability	Habitat				
				A	B	C	D	E
1	20.8277	4	<0.001				Prefer	Prefer
2	8.7313	4	0.068					
3	21.0942	4	<0.001	Avoid		Prefer		
Combined	50.6532	12	<0.001					

[a] The overall conclusion for the three animals from Table 8.2 combined is formed by summing the individual statistics and the degrees of freedom.

To get an overall test of the presence of habitat selection by the population, the three χ^2 statistics and their degrees of freedom in Table 8.5 can be summed. Thus, the total χ^2 statistic is 50.6532 with 12 degrees of freedom, and we conclude that there is significant habitat selection for these three animals ($P < 0.001$). The three χ^2 values can be summed together into one overall value because they can be assumed to be statistically independent; that is, the animals are moving independently of each other. The sum of two independent χ^2 statistics is also χ^2 distributed with degrees of freedom equal to the sum of the degrees of freedom. If two of the animals had been observed to move together and generally in close proximity to one another, then the assumption of independence of animals would be false, and the technique of summing the χ^2 values would be inappropriate.

Another approach that may occur to the reader is that the data could be pooled over animals and then analyzed with the χ^2 test. We believe that this approach is usually invalid with radio-tracking data because of the question of independence of observations within animals as opposed to among animals. Presumably, the biologist has gathered a large number of observations for each of the radio-marked animals. χ^2 tests can be conducted for each animal, with the among-animal variability maintained, as is shown in Table 8.5. In contrast, if the data for each animal are pooled to give the total number of locations in each habitat type, the among-animal variability is lost. As a worst-case situation, the effects of opposite choices of preference and avoidance may cancel one another. For instance, suppose animal 1 prefers habitat type A and avoids habitat type B. Animal 2 does the opposite. If approximately equal numbers of observations are taken on each animal, then pooling the data may result in the conclusion that the population uses each habitat type equally. In reality, different animals use habitats differently. But it is possible that the correct population inference has been drawn.

In some cases pooling of data may be justified, as when few observations are taken on many animals. Such was the case presented by Neu et al. (1974) for aerial counts of moose in different habitat types. For radio-tracking studies in which few observations are taken on many animals, pooling of the data is appropriate, and the techniques presented above for each individual can be applied to the pooled data.

Marcum–Loftsgaarden Analysis

In the Neu et al. (1974) analysis the expected habitat utilization is determined from measured map areas; that is, there is no sampling error. Although the procedure by Marcum and Loftsgaarden (1980) was not discussed by Alldredge and Ratti (1986), the approach is important when habitat availability has been estimated rather than determined exactly, as in the χ^2 goodness-of-fit test procedure. To estimate habitat availability, the habitat type of random locations is determined. A test of independence of the random locations and the animal locations is desired. The χ^2 table for this test is constructed as a $2 \times k$ table for each animal. The expected value for each cell is then determined as

$$\text{expected} = \frac{\text{row } \sum \times \text{column } \sum}{\text{table } \sum}$$

The degree of freedom is $(2 - 1)(k - 1) = k - 1$. As with the goodness-of-fit test of Neu et al. (1974), the sum of the χ^2 statistics and degrees of freedom can be used to test for population habitat selection.

An example with the data from Table 8.2 will illustrate the procedure. Instead of the known values for availability presented in Table 8.1, consider that 300 random locations are selected in the study area. The following results:

Habitat type	Number of locations
A	151
B	94
C	31
D	14
E	10

Now a χ^2 test of independence is constructed for animal 1 as

Random locations	151	94	31	14	10
Telemetry locations	302	180	69	49	50

The χ^2 value for the table is 10.332, with 4 degrees of freedom ($P = 0.035$). The G statistic is 11.242 ($P = 0.024$), and is calculated from the formula

$$G = 2\left[\left(\sum_{i=1}^{2}\sum_{j=1}^{5} O_{ij} \ln O_{ij}\right) - \left(\sum_{i=1}^{2} R_i \ln R_i + \sum_{j=1}^{5} C_j \ln C_j\right) + N \ln N\right]$$

where R_i is the ith row sum, C_j is the jth column sum, and N is the sum of the observations in the table. Thus, the null hypothesis of availability and use occurring in the same proportions is rejected. This is the same conclusion that was reached for the goodness-of-fit procedure, but is not as strong because estimates of availability take some of the power out of the test.

Heisey's Analysis

Heisey (1985) proposes the use of log-linear models for the analysis of habitat selection experiments, an extension of the χ^2 analysis. Manly's measure of selection (Manly et al. 1972, Manly 1974) can be used in log-linear models to test resource selectivity. Animals can be cross-classified by factors such as age and sex, providing a convenient method for examining how selectivities depend on these factors and a rigorous method for combining data from multiple animals to draw conclusions concerning the selectivity of the population of animals. Also, Heisey's method provides estimates of the selection coefficients for each of the factors included in the model.

Because Heisey's paper does not provide the specifics of how to conduct the test with a common statistical package, we present the input to BMDP (Dixon 1983) for the example data from his paper.

```
/ problem    title is 'Availability -- Use data from Heisey paper'.
/ input      variables are 4.
             format is '(3f2.0,f4.0)'.
             cases are 16.
/ variables  names are habitat, deer, time, freq.
/ table      indices are habitat, deer, time.
             count is freq.
             initial are 66,20,9,5,65,13,13,9,66,20,9,5,65,13,13,9.
```

8 Habitat Analysis

```
/ category  names(1) are aspen, clearcut, plantation, spruce.
            codes(1) are 1 to 4.
            names(2) are '68','342'.
            codes(2) are 1, 2.
            names(3) are 'mid-day','morn-even'.
            codes(3) are 1, 2.
/ fit       all.
/ end
 1 1 1 18
 2 1 1  2
 3 1 1  0
 4 1 1  0
 1 2 1 29
 2 2 1  1
 3 2 1  4
 4 2 1  0
 1 1 2 43
 2 1 2 33
 3 1 2  5
 4 1 2  0
 1 2 2 46
 2 2 2 29
 3 2 2  4
 4 2 2  2
```

Johnson's Analysis

Johnson (1980) developed a method to test for habitat preference or avoidance that does not assume that habitat areas are precisely known. His method is appropriate for the case in which errors in habitat classification are possible. For example, suppose radio-marked dabbling ducks are tracked on the shore of a large lake. Typically, these ducks stay near the marsh edge, in good cover, but occasionally they do venture some distance out into the lake. The question is then how much of the open water of the lake should be called available dabbling duck habitat? Depending on the answer, the results of the χ^2 test will indicate preference and avoidance or else no habitat selection.

Johnson's (1980) test uses the ranks of the habitat availability figures to test for preference. That is, the observations used in the test are the ranks of habitat use for each animal. Thus, the sample size of the test is the number of animals. The null hypothesis tested is that the population uses each habitat type proportionally to its availability. Because only ranks are used, the precise area of each habitat type is not needed, as the expected use is not calculated based on this proportion. Thus, the method is less sensitive to availability. Johnson's method does not test for habitat selection for each animal, but rather uses each animal as an observation to test for a preference by the population. A limitation of the test is that there must be more animals than the number of habitat types, although this limitation seems reasonable, given that the sample size is the number of animals. Because of this limitation, however, the data in Tables 8.1 and 8.2 cannot be used to demonstrate this test. In general, a large number of animals (>30) is required to meet the assumptions of normality required by this test.

The specific hypotheses tested with Johnson's procedure (Alldredge and Ratti 1986) are H_{01}, the relative selections for all habitats are equal, with the follow-up, H_{02}, the relative selection for habitat i equals that for habitat j. Thus, this procedure also provides two levels of tests. First, an overall test of preference is performed. However, the second set of hypotheses does not determine which habitat type is being selected relative to its availability, but only the relative ranking of use of each habitat type.

One conceptual difficulty with Johnson's test is that the results for an animal with ten locations and an animal with 100 observations are given equal weight. That is, each is summarized as the vector of ranks of different habitat types.

Also, the test appears to have low power relative to the χ^2 methods presented above; that is, the test is less likely to reject a false null hypothesis for the same amount of data. This is part of the price that must be paid in order to reduce the sensitivity of the analysis to the habitat availability estimates. Stronger assumptions are made about habitat availability in the χ^2 analysis, with a corresponding increase in power. If these assumptions are incorrect, the likelihood of incorrect conclusions increases.

Program PREFER has been provided by Johnson (1980) to perform this analysis. The program operates on an IBM personal computer or IBM-compatible model and is available from Tony Frank, Great Lakes Fisheries Laboratory, Duluth, Minnesota.

Friedman Test

The Friedman (1937) test can be used to test for differences in the percentage of availability and the percentage of utilization of each habitat to determine selection. The differences are ranked for each animal to compute the test statis-

8 Habitat Analysis

tic. For this test the animals are "blocks," and the treatments are "habitats." The specific hypotheses tested (Alldredge and Ratti 1986) are H_{01}, the ranks of the differences in selection and availability are the same for all habitats, with the follow-up, H_{02}, the rank of the difference in selection and availability is the same for habitat i and habitat j. The second hypothesis is equivalent to a multiple comparison test for analysis of variance. Computing of the Friedman test is described by Conover (1980).

For the data in Tables 8.1 and 8.2, the following table of differences of the percentage of selection minus the percentage of availability results:

			Habitat		
Animal	A	B	C	D	E
1	-3.5	-2.3	+0.6	+2.5	+2.7
2	-0.9	-2.4	0.0	-0.2	+3.5
3	-4.8	+1.2	-0.7	+2.7	+1.5

From this table the table of ranks within animals (from smallest to largest) results. We will use the notation R_{ah} to denote the rank for the ath animal ($a = 1, \ldots, A$ animals; in this case 3) and the hth habitat ($h = 1, \ldots, H$ habitats; in this case 5).

		Habitat			
Animal	A	B	C	D	E
1	1	2	3	4	5
2	2	1	4	3	5
3	1	3	2	5	4
Sum = R_h	4	6	9	12	14

Thus, habitat type E has the highest sum of ranks over animals, suggesting that it is selected more often than the other habitats in relation to its availability. These sums of ranks are denoted as

$$R_h = \sum_{a=1}^{A} R_{ah}$$

The Friedman statistic is calculated as

$$\chi^2 = \left[\frac{12}{HA(H+1)} \sum_{h=1}^{H} R_h^2 \right] - 3A(H+1)$$

$$= \frac{12}{5(3)(5+1)} (4^2 + 6^2 + 9^2 + 12^2 + 14^2) - 3(3)(5+1)$$

$$= \frac{12}{90}(473) - 54$$

$$= 9.0667$$

This value is χ^2 with $H - 1$ degrees of freedom. Thus, a value of 9.0667 with 4 degrees of freedom is marginally significant ($P = 0.0595$). One problem that the Friedman test has in common with Johnson's procedure is that each animal is compared as if the sample sizes were identical. Such is usually not the case.

Another method to test for differences in use and availability data is the Quade (1979) test. The hypotheses tested are identical to those by the Friedman test described above, so we will not pursue this test further. However, it is one of the tests also simulated by Alldredge and Ratti (1986).

Differences between Animals

The Johnson and Friedman procedures assume that animals are the sample, and thus assume that animals select the same habitat. This assumption can be tested with a simple χ^2 test of independence. We will illustrate the procedure with the data in Table 8.2. A χ^2 test is constructed as

	Habitat					
Animal	A	B	C	D	E	Total
1	302	180	69	49	50	650
2	162	91	33	16	55	330
3	381	263	78	65	55	842
Total	845	534	180	130	133	1822

The expected value for animal 1, habitat A, is just the row total multiplied by the column total divided by the total number of locations, that is, $(650 \times 845)/1822 = 301.45$. The contribution to the χ^2 statistic is (observed − expected)^2/expected. Thus, for the data in Table 8.3, $\chi^2 = 7.846$ with (number of habitats − 1) × (number of animals − 1) = 8 degrees of freedom ($P = 0.449$). For comparison, the maximum likelihood statistic, G, is 8.119 ($P = 0.422$).

8 Habitat Analysis

The SAS (SAS Institute Inc. 1985) code to perform this test is shown here.

```
/********************************************************************/
/*    SAS procedure to test for differences between habitat         */
/*    use for 2 or more animals. For each animal, the number        */
/*    of locations in each habitat type is entered. The data        */
/*    are then used in a chi-square test of independence to         */
/*    test for differences between animals. The output is           */
/*    the chi-square statistic and its probability level.           */
/********************************************************************/
data;
    keep animal habitat count;
    array hab{5} hab1-hab5;
    input animal hab1 -- hab5;
    do habitat=1 to 5;
        count=hab{habitat}; output;
    end;
cards;
1 302 180 69 49 50
2 162  91 33 16 28
3 381 263 78 65 55
;
proc freq; weight count; tables animal*habitat / chisq; run;
```

If the available habitat for the three animals is identical, then the above test detects differences in animal preferences. That is, it will detect whether different animals are using the habitat differently. If the availabilities are not the same for the animals, that is, they are in different study areas, then the test still looks for differences in habitat use between animals, but the reason can be differences in either animal selection or animals because of what was available to them.

Selecting a Test

In the previous sections four tests have been demonstrated in some detail. Which is appropriate for a particular set of data? The first criterion is whether the availability values are known exactly. If not, and if only approximate rankings can be determined, then the test by Johnson (1980) is appropriate. If the availability values can be estimated from random locations, the Marcum and Loftsgaarden (1980) procedure should be considered.

A second criterion for considering which test to use is whether the individual animals are considered to have the same selection criteria. That is, can

197

the animals that are used to estimate use be assumed to have the same preferences? From computer simulations, D. Hewitt (personal communication) has demonstrated that differences in individual preference can cause a much higher Type I error rate of the Friedman test than would be expected. Thus, use of this test would reject the null hypothesis much more often than expected if the animals differed in their habitat selection. Such differences between animals would also influence the Johnson test, because animals are assumed to be "blocks." Such a problem does not occur for the χ^2 goodness-of-fit procedure, because the test is performed for each individual. As shown in Table 8.5, conclusions might be different for each animal.

If differences between animals can be explained by covariates, then the methods suggested by Heisey (1985) are appropriate. However, his method still assumes that the animals are identical in their selection of habitat at the most basic level.

From their Monte Carlo simulation study, Alldredge and Ratti (1986) made no strong recommendations, but the Neu et al. (1974) procedure generally did well compared to the others, as expected, because this procedure requires the most data. Because of the extension to log-linear models suggested by Heisey (1985), we prefer the χ^2 procedure.

Critical Habitat

The above tests of habitat preference tell us nothing about whether habitats utilized by the animals are critical to their continued survival and reproduction. If we modify habitat composition to make it identical to use data, will the animal's fitness be increased? We believe that much of the selection shown by animals is neutral relative to their fitness, because individual animals often show differences in their preferences. Probably, only strong preferences that are consistent across animals actually benefit fitness. Undoubtedly, few if any habitat preferences are negative to fitness in the long term, that is, actually harm the animal's fitness, because these behaviors would have been selected against and eliminated from the population. Short-term preferences for habitats with negative value may be caused by predation pressure or intraspecific behavioral interactions. Thus, we do not believe that preference for a habitat type provides much evidence that the habitat type is necessary to the animal's survival and reproduction. The only approach to determining the criticalness of a habitat type is to keep the animal out of it and then monitor its survival and reproduction. Thus, tests of criticalness require perturbation studies, with large sample sizes, because population parameters are being monitored.

8 Habitat Analysis

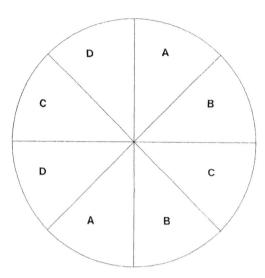

Figure 8.1 Conceptual design for a pen study to demonstrate whether a particular habitat is critical to survival and reproduction of a group of animals. Animals can be kept from using one or more of the eight available segments by closing off access from the central point, the necessary perturbation needed to show that a habitat type is critical to the animals' continued survival and reproduction. Survival and reproduction are monitored as a function of the available habitat.

Consider the pen study illustrated in Fig. 8.1. The eight segments of the enclosure represent habitat types A–D. A test of whether habitat type A is critical to an animal's survival would be to place a representative sample of animals in the enclosure, with habitat A initially being available. Once it is determined that the population is surviving within the enclosure, the gates are shut on habitat A, making it unavailable. If the population continues to survive and reproduce, we conclude that habitat A is probably not critical, at least for the short time span sampled. However, if the population begins to decline or animals decline in body condition, habitat A must be a critical requirement. Of course, a temporal control would be needed to demonstrate that the loss of habitat A was the cause of the population decline, and not a temporal effect (see Chapter 2).

Obviously, such a simple-minded experiment is not reasonable. The differences between two of the habitats might not be sufficiently limiting in the short term to be detected, so the experiment would lack power to detect meaningful biological differences. For example, the difference between A and B might be the proportion of some nutrient, which is important in long-term survival, but causes little effect over the duration of the study. Replications and better controls would be necessary, as well as demonstration that the enclosures themselves were not influencing the results. The point to emphasize is that determination of the criticalness of a habitat type cannot be determined by simply monitoring habitat utilization in the field. Demonstration that a habitat type is preferred does not mean it is critical. A particular habitat type is critical only if its loss affects an animal's fitness, and thus the population level. Keep in mind that the habitat preference analyses discussed in this chapter are just another correlational analysis, another application of telemetry as "binocs that see in the dark."

Accuracy of Radio-Tracking Locations

Habitat studies require a precise estimate of an animal's location so that it can be correctly placed in a habitat type. However, a number of locations will still be on the boundary between two habitats; that is, the confidence ellipse includes the habitat boundary. Several possibilities exist for handling this problem.

If the number of locations is small, they could be eliminated. No information would be available from these locations because an unambiguous habitat type would not be available.

A second possibility would be to assume that the location has equal probability of being in either habitat, and so randomly assign the animal to one of the two habitats. This approach is definitely not appropriate, as the 50–50 assignment rule would lower the power of the tests for preference. If the animal is actually selecting one of the two habitats along the boundary, then this approach would tend to misclassify locations.

Finally, we could just use the habitat type of the estimated location and ignore the fact that the confidence ellipse included the habitat boundary. This approach would probably be acceptable as long as the habitat types do not influence the radio-tracking system. However, such effects have been demonstrated (Hupp and Ratti 1983), because a heavily forested habitat causes greater bounce problems than an open habitat. Thus, a potential bias may result.

The effect of error in the estimates of location is to lower the power of the tests of habitat selection, as shown by White and Garrott (1986). They used Monte Carlo simulations to determine the impact of radio-tracking triangulation errors on the power of the χ^2 goodness-of-fit test to detect habitat selection. Their results indicated that the power of χ^2 tests were greatly decreased by decreasing precision of triangulation bearings. For example, power of the χ^2 test decreased by a factor of 2.5 as the bearing standard deviation doubled, or decreasing the standard deviation of bearings by 50% meant that only 71% as many locations were needed to obtain the same power to detect habitat selection. These researchers suggested that habitat selection experiments should be simulated on a computer prior to fieldwork, in order to verify that adequate accuracy of location estimates is available for the expected sample size to meet study objectives.

Home Range Approach

Another approach to estimating habitat utilization involves the proportion of each habitat type that is in the animal's home range. However, we believe this approach does not merit use in place of the methods described in this chapter.

As discussed in Chapter 7, home range estimates are seldom satisfactory. Given the lack of a perfect home range estimate, errors due to home range estimates are then injected into the habitat analysis. That is, any errors in the home range model are included in the habitat analysis. The methods described above do not have this problem.

Second, the home range approach generally assumes that the area within the home range is uniformly utilized by the animal. However, most home range models are based on assumptions other than uniform utilization (cf. with ellipse models). Thus, habitat at the edge of the home range should not be treated with the same weight as habitat at the center of the home range. Although it is theoretically possible to integrate the weighted area of the home range to get habitat utilization, the result would be equivalent to the methods described in this chapter (assuming a perfect home range estimator). Therefore, the effort is not necessary.

Another application of home range estimates to habitat selection experiments is to use the home range map to determine what is available to the animal. However, an animal's home range represents some prior selection; that is, the animal has already selected a particular area. Depending on study objectives, this prior selection may be tolerable. If the objective is to determine habitat use during a particular portion of the day, when selection within the home range is expected, then use of the home range may be acceptable (if the problems of estimating home range can be overcome). However, if the objective of the study is to evaluate habitat selection over the life span of the animal, the prior selection of a particular area by the animal will bias the results.

Sample Size

Sample sizes for the χ^2 goodness-of-fit test should be based on the individual Bonferroni statistics, because the biologist wants to be able to detect preference/avoidance for specific habitats, not a general result across all habitats. For this calculation M. Samuel (personal communication) has supplied the following SAS (SAS Institute Inc. 1985) code to calculate either the power or the sample size of a particular comparison.

```
* ------------------------------------------------------------- *
|      This procedure calculates the sample size and/or power for  |
|   the Bonferroni multiple comparison tests.                      |
|      The procedure requires:                                     |
|          1) either ES or DELTA to be specified.                  |
|          2) either (NU or NA) or R to be specified.              |
|          3) ALPHA, K1, and K2 to be specified.                   |
|      The remaining missing values will automatically be solved   |
|   by the procedure.                                              |
|      Thanks to Michael D. Samuel for supplying this SAS code.    |
* ------------------------------------------------------------- *;
data done;
  retain zero 0;
  input id $ nu pu na pa alpha beta r k1 k2 es delta;
  if delta ne '.' and es = '.' then es = delta * sqrt(pa*(1-pa));
  if alpha ne '.' then zalpha = probit(1-alpha/(2*k1));
  if beta ne '.' then zbeta = probit(1-beta/k2);
  if nu = '.' and pu = '.' then pu = pa + es;
  if nu = '.' or na = '.' then do;
   rad = sqrt(pu*(1-pu) + pa*(1-pa));
   cap_a = (zalpha*rad + zbeta*rad)**2;
   n_star = cap_a * (1.0 + sqrt(1 + 4*abs(pu-pa)/cap_a))**2
            / (4*(pu-pa)**2);
   if nu = '.' then
      if r = '.' then
         nu = na*n_star/(2*na-n_star);
      else do;
         radr = sqrt(pu*(1-pu) + pa*(1-pa)/r);
         cap_a = (zalpha*radr + zbeta*radr) ** 2;
         nu = cap_a * (1.0 + sqrt(1 + 2*(r+1)*abs(pu-pa)/(r*cap_a)))**2
              / (4*(pu-pa)**2);
      end;
```

```
          if na = '.' then
             if r = '.' then
                na = nu*n_star/(2*nu-n_star);
             else na = r*nu;
          output done;
          end;
          if beta = '.' then do;
          r = na / nu;
          if pu = '.' then pu = pa + es;
          if pu > pa then pu = pa + es
          else pu = pa - es;
          zbeta = (sqrt(nu*(pu-pa)**2 - ((r+1)/r)*abs(pu-pa) +
                       ((r+1)/r)**2/(4*nu))
                   - probit(1-alpha/(2*k1)) * sqrt(pu*(1-pu) + pa*(1-pa)/r))
              / sqrt(pu*(1-pu) + pa*(1-pa)/r);
          power = probnorm(zbeta);
          beta = 1 - power;
          output done;
          end;
        * input variables: id nu pu na pa alpha beta r k1 k2 es delta; cards;
        cc0    . . . .075 .1 .1 2 4 2 . .15
        cc1    . . . .305 .1 .1 2 4 2 . .15
        cc2    . . . .420 .1 .1 2 4 2 . .15
        cc3    . . . .200 .1 .1 2 4 2 . .15
        proc print data=done;
          var id nu pu na pa alpha beta r k1 k2 es delta zalpha zbeta cap_a
                n_star power;
        run;
```

As noted in the section on the accuracy of the telemetry locations, the power will be influenced by the amount of error in classifying the animal's habitat type.

A quick calculation can be performed using the methods from Sokal and Rohlf (1981) presented in Chapter 3. Another possibility is to use the variance on the ratio of use to availability suggested by Hobbs and Bowden (1982) to compute the necessary sample size required for a useful estimate of the preference ratio.

Summary

1. Habitat analysis consists of comparisons of availability and use to determine preference/avoidance of particular habitats. Because the true

availability can never be known, these analyses are only approximations, and might be quite misleading.

2. The χ^2 test of habitat selection suggested by Neu et al. (1974) is the most powerful test when habitat availability is known and still allows the test to be conducted for individual animals.
3. Johnson's (1980) test assumes that habitat availability is not precisely known, and so is more appropriate when habitat availability is undefinable.
4. The tests by Johnson (1980) and by Friedman and Quade (Alldredge and Ratti 1986) assume that animals within the population all prefer the same habitats. This assumption is often shown to be incorrect when the χ^2 test procedure is used for each animal.
5. The effect of errors in animal locations on habitat selection tests is to lower the power of the test to detect preferences.
6. Even if the null hypothesis of no selection is rejected, no conclusions can be made on how critical the selected habitats are to the animal's fitness (i.e., survival and reproduction). The need for the habitat can only be determined through manipulations, and thus is not determined by observing habitat use and comparing it to availability.

References

Alldredge, J. R. and J. T. Ratti. 1986. Comparison of some statistical techniques for analysis of resource selection. J. Wildl. Manage. 50:157–165.

Byers, C. R., R. K. Steinhorst, and P. R. Krausman. 1984. Clarification of a technique for analysis of utilization-availability data. J. Wildl. Manage. 48:1050–1053.

Conover, D. 1980. Practical Nonparametric Statistics, 2nd ed. Wiley, New York. 493 pp.

Data Systems Support Group. 1979. MOSS user's manual version II, WELUT-79/07. U.S. Fish and Wildl. Serv., Fort Collins, CO. Not consecutively paginated pp.

Dixon, W. J. (ed.). 1983. BMDP Statistical Software 1983. Univ. of Calif. Press, Berkeley. 734 pp.

Friedman, M. 1937. The use of ranks to avoid the assumption of normality implicit in the analysis of variance. J. Am. Stat. Assoc. 32:675–701.

Heisey, D. M. 1985. Analyzing selection experiments with log-linear models. Ecology 66:1744–1748.

Hobbs, N. T. and D. C. Bowden. 1982. Confidence intervals on food preference indices. J. Wildl. Manage. 46:505–507.

Hupp, J. W. and J. T. Ratti. 1983. A test of radio telemetry triangulation accuracy in heterogeneous environments. Pages 31–46 in D. G. Pincock ed. Proc. 4th Int.

Wildl. Biotelemetry Conf. Applied Microelectronics Institute and Technical Univ. of Nova Scotia, Halifax.

Johnson, D. H. 1980. The comparison of usage and availability measurements for evaluations of resource preference. Ecology 61:65–71.

LaPerriere, A. J., P. C. Lent, W. C. Gassaway, and F. A. Nodler. 1980. Use of LANDSAT data for moose-habitat analyses in Alaska. J. Wildl. Manage. 44:881–887.

Manly, B. F. J. 1974. A model for certain types of selection experiments. Biometrics 30:281–294.

Manly, B. D. J., P. Miller, and L. M. Cook. 1972. Analysis of a selective predation experiment. Am. Nat. 106:719–736.

Marcum, C. L. and D. O. Loftsgaarden. 1980. A nonmapping technique for studying habitat preferences. J. Wildl. Manage. 44:963–968.

National Cartographic Information Center (NCIC). 1979. Digital Terrain Tapes. U.S. Dep. of the Interior, Geological Survey, Washington, D.C. 12 pp.

Neu, C. W., C. R. Byers, J. M. Peek, and V. Boy. 1974. A technique for analysis of utilization-availability data. J. Wildl. Manage. 38:541–545.

Quade, D. 1979. Using weighted rankings in the analysis of complete blocks with additive block effects. J. Am. Stat. Assoc. 74:680–683.

Russell, R. M., D. A. Sharpnack, and E. L. Amidon. 1975a. WRIS: A resource information system for wildland management, Res. Pap. PSW-107/1975. Pacific Southwest Forest and Range Experiment Station, U.S. Forest Service, Berkeley, CA. 12 pp.

Russell, R. M., D. A. Sharpnack, and E. L. Amidon, 1975b. Wildland resource information system: user's guide, Gen. Tech. Rep. PSW-10/1975. Pacific Southwest Forest and Range Experiment Station, U.S. Forest Service, Berkeley, CA. 36 pp.

SAS Institute Inc. 1985. SAS® Language Guide for Personal Computers, Version 6 Edition. SAS Institute Inc., Cary, NC. 429 pp.

Sokal, R. R. and F. J. Rohlf. 1981. Biometry, 2nd ed. Freeman, San Francisco, CA. 859 pp.

Wehde, M. E., K. J. Dalsted, and B. K. Worcester. 1980. Resource applications of computerized data processing: the AREAS example. J. Soil Water Conserv. 35:36–40.

White, G. C. and R. A. Garrott. 1986. Effects of biotelemetry triangulation error on detecting habitat selection. J. Wildl. Manage. 50:509–513.

CHAPTER

9

Survival Rate Estimation

The rates at which animals die are critical parameters of wildlife populations. Estimates of survival (i.e., 1 − mortality) are needed to manage a population in an adequate fashion. In this chapter methods will be developed for the estimation of survival rates from radio-tracking data. Even more important are the construction of tests to compare estimates of survival. As discussed throughout, radio-tracking should be used to test hypotheses. No hypothesis in wildlife management is more basic than comparing the survival of two populations of animals in response to a treatment.

Trent and Rongstad (1974) recognized that transmitters could be used to estimate survival rates of cottontail rabbits (*Slyvilagus floridanus*). The method they used is similar to the Mayfield estimator for nest survival (Mayfield 1961, 1975, Miller and Johnson 1978, Johnson 1979, Hensler and Nichols 1981, Bart and Robson 1982). Heisey and Fuller (1985) extended the method proposed by Trent and Rongstat by developing a maximum likelihood estimation procedure and likelihood ratio tests between groups of animals. Bunck (1987) provided a recent summary of techniques.

In contrast to attaching transmitters to animals, other methods of identifying individual animals have been used to estimate survival rates. Survival rates of the North American waterfowl population are estimated from the recovery of leg bands. The statistical procedures for estimating survival rates from band recovery data are developed thoroughly by Brownie et al. (1985). These models are more general than needed for radio-tracking data in that the fate of radio-marked animals is almost always known, whereas leg-banded birds are often never seen again. That is, with transmitters the researcher knows how many animals are still alive. With bands the researcher doesn't know whether an animal has died and the band was not found and reported or the animal is still alive. Thus, banding data require a more complex framework for the estimation of survival rates than do radio-tracking data.

In this chapter we first discuss some of the basics about estimating finite survival rates and their relationship to the binomial distribution. Next, we describe the use of SURVIV, a powerful program for the analysis of wildlife radio-tracking data. Then we consider two other approaches to the estimation of survival from wildlife radio-tracking data: the MICROMORT program (Heisey and Fuller 1985) and the Kaplan–Meier method. Finally, we will consider other methods developed for use in the medical profession to estimate survival as a function of different medications or other treatments.

Binomial Distribution

The finite rate of survival for a period (say, 1 year) can be estimated with a sample of radio-marked animals. The number of animals still alive at the end of the year (n_1) divided by the number of animals alive at the start of the year (n_0) gives an estimate of survival. Thus,

$$\hat{S} = \frac{n_1}{n_0} \qquad (9.1)$$

with the variance of $\hat{S}$ estimated by

$$\text{Var}(\hat{S}) = \frac{\hat{S}(1 - \hat{S})}{n_0} \qquad (9.2)$$

because n_1 is a binomially distributed variable. That is, n_0 radio-marked animals start the year. At the end of the year, n_1 animals are still alive, while $n_0 - n_1$ animals died. There are two mutually exclusive classes at the end of the year: alive and dead. For the moment, problems of transmitter failure, animals leaving the study area, etc., will be ignored.

The formula for the variance of $\hat{S}$ given above is for large samples, that is, $n_0 > 30$. Small sample confidence intervals can be constructed by cumulating the probabilities in the tails of the binomial distribution. Hollander and Wolfe (1973:22–24) provide procedures for both large and small sample sizes for putting confidence intervals on $\hat{S}$.

9 Survival Rate Estimation

To illustrate these calculations, we will compute the survival and a 95% confidence interval for the 61 treatment mule deer fawns shown in Fig. 9.1. Of the 61 fawns, 19 survived through the winter, and four had transmitter failures and will be discarded. Therefore, we take $n_0 = 57$ and $n_1 = 19$. Thus, $\hat{S} = 19/57 = 0.3333$. The variance of $\hat{S}$ is

$$\text{Var}(\hat{S}) = \frac{0.3333 \times (1 - 0.3333)}{57}$$

$$= 0.003899$$

The 95% confidence interval for $\hat{S}$ is computed as

$$\hat{S} \pm z_\alpha \sqrt{\text{Var}(\hat{S})}$$

where z_α is the z statistic for an appropriate probability level. For a 95% confidence interval, $\alpha = 0.05$ and $z_{0.05} = 1.96$. Thus, the 95% interval is 0.2110–0.4557.

For confidence intervals calculated by the procedures developed from the binomial distribution to be correct, the animals must be independent observations. That is, the survival of one animal cannot depend on the survival of another. If survival rates should be dependent, then the assumptions behind the binomial distribution are no longer met and the procedures are not valid for the data.

The binomial distribution allows estimation of both point and interval estimates, as shown above, plus testing for differences between two or more rates. That is, to test the null hypothesis that $S_1 = S_2$, a simple χ^2 test can be constructed from the 2 × 2 table consisting of the numbers of animals alive and dead at the end of the specified time interval in each population. Consider the example from Fig. 9.1, in which 61 animals are radio-marked on the treatment area, with 19 still alive at the end of the winter and four transmitter failures, and 59 animals radio-marked on the control area, with 19 still alive at the end of the year and two transmitter failures. So that the two survival rates will not be equal, we will assume that the animals carrying the failed transmitter lived through the interval. The two survival rates are then $23/61 = 0.3770$ and $21/59 = 0.3559$. Then the question is whether these two rates are significantly different. The χ^2 table is

	Alive	Dead
Treatment	23	38
Control	21	38

```
Treatment 191.075 Male   34.2 125.5 03DEC82 Lived    .
Treatment 191.085 Female  .   129.0 04DEC82 Lived    .
Treatment 191.600 Female 32.6 123.0 03DEC82 Died    12JAN83
Treatment 191.610 Female 31.8 122.0 03DEC82 Lived    .
Treatment 191.620 Female 34.0 126.0 03DEC82 Died    13DEC82
Treatment 191.630 Female 37.0 125.5 03DEC82 Lived    .
Treatment 191.640 Female 29.9 119.0 03DEC82 Died    15DEC82
Treatment 192.010 Male    .   119.0 04DEC82 Died    18MAY83
Treatment 192.020 Male    .   134.0 04DEC82 Lived    .
Treatment 192.140 Male   36.0  .    03DEC82 Died    17DEC82
Treatment 192.150 Female 31.0 118.0 02DEC82 Lived    .
Treatment 192.160 Female 34.5 121.0 01DEC82 Died    11MAY83
Treatment 192.170 Female 33.0 127.5 01DEC82 Lived    .
Treatment 192.180 Female 35.5 128.0 01DEC82 Lived    .
Treatment 192.190 Female 32.5 124.0 01DEC82 Lived    .
Treatment 192.200 Female 32.5 120.0 02DEC82 Died    31DEC82
Treatment 192.220 Female 26.7 116.0 02DEC82 Died    18DEC82
Treatment 192.230 Female 32.9 126.5 02DEC82 Died    04APR83
Treatment 192.270 Female 30.2 121.0 01DEC82 Died    05JAN83
Treatment 192.280 Male   31.0 118.5 02DEC82 Died    30MAR83
Treatment 192.300 Male   41.0 117.5 30NOV82 Died    31DEC82
Treatment 192.320 Male   33.9 125.0 02DEC82 Censored 01MAY83
Treatment 192.340 Male   37.5 128.0 03DEC82 Censored 01MAY83
Treatment 192.360 Male   33.0 120.5 03DEC82 Died    12JAN83
Treatment 192.370 Male   45.6 133.0 02DEC82 Died    06JAN83
Treatment 192.380 Male   37.0 129.5 01DEC82 Died    28DEC82
Treatment 192.390 Male   41.5 131.0 01DEC82 Censored 01MAY83
Treatment 192.400 Male   45.5 134.0 30NOV82 Died    15MAR83
Treatment 192.410 Male   38.5 128.0 29NOV82 Censored 01MAY83
Treatment 192.420 Male   32.5 119.0 03DEC82 Died    06JAN83
Treatment 192.450 Male   31.5 126.0 01DEC82 Died    10JAN83
Treatment 192.470 Male   40.0 121.5 30NOV82 Died    01MAR83
Treatment 192.480 Female 34.2 129.0 01DEC82 Lived    .
Treatment 192.490 Female 26.9 120.0 02DEC82 Died    28FEB83
Treatment 192.500 Female 32.4 123.0 02DEC82 Died    28APR83
Treatment 192.510 Female 31.4 124.0 02DEC82 Died    10JAN83
Treatment 192.530 Female 23.9 124.0 02DEC82 Died    20DEC82
Treatment 192.540 Female 36.4 130.0 03DEC82 Lived    .
Treatment 192.550 Female 30.5 119.0 02DEC82 Lived    .
```

Figure 9.1 Survival data from mule deer fawns radio-collared at the start of the winter in the Piceance Basin, Colorado. Treatment animals were on a study area near an oil shale development, while control animals were seldom near human activity. Radio frequencies were in the 191–193 MHz band. Body weight (kg) and body length (cm) are in columns 4 and 5. The next column is the day the animal was captured, with the next column the fate of the animal. The last column is the day the animal died or the radio failed, with a single dot indicating that the animal survived the winter. (*Figure continues.*)

9 Survival Rate Estimation

Treatment	192.560	Female	32.5	118.5	03DEC82	Died	07JAN83
Treatment	192.570	Female	32.5	112.5	03DEC82	Lived	.
Treatment	192.590	Female	27.5	.	03DEC82	Died	13DEC82
Treatment	192.600	Female	33.4	128.5	02DEC82	Died	13DEC82
Treatment	192.610	Female	32.1	121.0	02DEC82	Lived	.
Treatment	192.620	Male	38.5	127.5	30NOV82	Died	05JAN83
Treatment	192.630	Male	45.1	132.0	01DEC82	Died	22JAN83
Treatment	192.640	Male	30.8	124.0	02DEC82	Died	11MAY83
Treatment	192.650	Male	41.4	131.0	02DEC82	Died	04FEB83
Treatment	192.660	Male	33.5	126.0	29NOV82	Died	27DEC82
Treatment	192.670	Male	34.5	120.5	02DEC82	Died	03MAR83
Treatment	192.680	Female	26.5	113.0	02DEC82	Died	28DEC82
Treatment	192.690	Male	42.0	126.0	01DEC82	Lived	.
Treatment	192.700	Male	40.5	125.5	01DEC82	Died	14FEB83
Treatment	192.710	Male	41.5	129.5	30NOV82	Lived	.
Treatment	192.720	Female	35.0	122.0	30NOV82	Died	21MAR83
Treatment	192.740	Female	32.0	125.0	30NOV82	Lived	.
Treatment	192.750	Female	34.0	125.0	30NOV82	Lived	.
Treatment	192.760	Female	37.6	125.5	01DEC82	Lived	.
Treatment	192.770	Female	38.5	119.0	30NOV82	Died	04FEB83
Treatment	192.780	Female	24.5	108.0	30NOV82	Died	28MAR83
Treatment	192.790	Female	31.0	113.0	30NOV82	Died	16DEC82
Control	191.350	Female	35.5	128.0	05DEC82	Lived	.
Control	191.360	Female	33.0	122.0	05DEC82	Died	30MAR83
Control	191.370	Female	34.3	123.0	05DEC82	Lived	.
Control	191.400	Female	29.6	117.5	05DEC82	Died	11FEB83
Control	191.410	Female	32.5	118.5	05DEC82	Lived	.
Control	191.420	Male	24.9	114.0	09DEC82	Died	16DEC82
Control	191.430	Male	40.8	129.0	09DEC82	Died	29MAR83
Control	191.440	Male	29.1	114.0	09DEC82	Lived	.
Control	191.450	Female	33.5	122.5	08DEC82	Lived	.
Control	191.460	Female	32.0	124.5	08DEC82	Lived	.
Control	191.510	Female	33.5	125.5	08DEC82	Lived	.
Control	191.520	Female	31.4	123.0	08DEC82	Died	25MAR83
Control	191.530	Female	32.2	119.0	08DEC82	Died	11FEB83
Control	191.540	Female	39.0	129.0	06DEC82	Lived	.
Control	191.560	Female	34.9	122.0	06DEC82	Lived	.
Control	191.590	Female	30.0	121.0	06DEC82	Died	15MAR83
Control	191.660	Female	37.8	135.5	07DEC82	Lived	.
Control	191.680	Female	34.0	126.0	08DEC82	Lived	.
Control	191.690	Female	35.5	127.5	07DEC82	Lived	.
Control	191.710	Female	30.4	120.0	08DEC82	Died	14JAN83
Control	191.730	Female	32.8	117.0	07DEC82	Died	31JAN83
Control	191.970	Male	28.0	112.0	08DEC82	Died	22MAR83
Control	191.980	Female	28.0	120.0	07DEC82	Died	30MAR83

(*Figure continues.*)

Analysis of Wildlife Radio-Tracking Data

Control	191.990	Female	27.5	112.0	08DEC82	Died	23DEC82
Control	192.030	Female	34.0	127.5	05DEC82	Died	18DEC82
Control	192.050	Male	34.3	127.5	05DEC82	Died	20APR83
Control	192.060	Male	37.6	129.5	05DEC82	Died	16FEB83
Control	192.070	Male	32.0	117.0	08DEC82	Died	04APR83
Control	192.090	Female	28.0	118.5	10DEC82	Died	08APR83
Control	192.110	Male	31.5	124.5	10DEC82	Died	22MAR83
Control	192.120	Male	36.5	130.0	09DEC82	Died	17MAR83
Control	192.130	Male	29.0	116.0	08DEC82	Died	06JAN83
Control	192.290	Male	35.4	123.5	05DEC82	Died	10MAR83
Control	192.310	Male	37.5	124.0	06DEC82	Died	27DEC82
Control	192.330	Male	29.3	118.5	05DEC82	Died	03MAR83
Control	192.350	Male	32.5	118.0	05DEC82	Died	17DEC82
Control	192.430	Male	33.0	120.5	07DEC82	Died	27APR83
Control	192.440	Male	33.6	121.0	06DEC82	Censored	15MAY83
Control	192.460	Male	39.0	127.0	07DEC82	Died	16DEC82
Control	192.580	Male	36.2	125.5	08DEC82	Died	07MAR83
Control	192.620	Male	33.8	117.0	05DEC82	Died	11MAR83
Control	192.630	Male	35.2	123.0	07DEC82	Died	18DEC82
Control	192.640	Male	36.6	123.0	06DEC82	Lived	.
Control	192.650	Male	40.5	128.0	06DEC82	Lived	.
Control	192.660	Male	34.0	126.0	06DEC82	Died	16MAR83
Control	192.670	Male	38.2	127.0	09DEC82	Died	11MAR83
Control	192.680	Female	32.0	122.0	08DEC82	Lived	.
Control	192.690	Male	22.8	115.5	09DEC82	Died	16DEC82
Control	192.700	Male	36.0	126.0	09DEC82	Lived	.
Control	192.710	Male	38.6	131.0	06DEC82	Died	15APR83
Control	192.720	Male	33.5	122.0	08DEC82	Lived	.
Control	192.730	Male	36.5	125.5	06DEC82	Died	13JAN83
Control	192.730	Male	33.5	125.5	08DEC82	Lived	.
Control	192.740	Male	36.3	127.0	08DEC82	Lived	.
Control	192.750	Male	33.2	120.0	07DEC82	Died	15DEC82
Control	192.760	Male	36.6	122.0	08DEC82	Censored	12MAY83
Control	192.770	Male	35.0	125.0	08DEC82	Died	01MAR83
Control	192.780	Male	.	116.0	08DEC82	Died	01MAR83
Control	192.790	Male	40.0	133.0	08DEC82	Died	06APR83

Figure 9.1 (*Continued*)

with the calculated $\chi^2_{(1)} = 0.058$. Based on this χ^2 value, the probability of observing a value this large, or larger if the two survival rates are truly equal, is $P = 0.81$. Thus, because this value is much greater than the 0.05 level, we conclude that survival rates did not differ between the treatment and control areas and do not reject the null hypothesis. We did not use a 2 × 2 table correction to simplify the example, because the test is more appropriate without the correction (D'Agostino et al. 1988). A more extensive treatment of testing for differences in probabilities is given by Conover (1971:150–154) and illustrates the procedure for testing more than two survival rates.

Parameter Estimation by Numerical Methods

The general model for estimating survival rates using radio-tracking data is structured similarly to that of Brownie et al. (1985), consisting of a set of multinomial distributions tied together by common parameters to describe the expected cell probabilities. A multinomial is a binomial extended to more than two classes, that is, dead in year 1, dead in year 2, dead in year 3, or still alive after year 3. The model is best illustrated by considering a simple example. Assume that 40 and 50 animals are radio-marked on years 1 and 2, respectively. The fate of these 90 transmitters is shown in Table 9.1, based on a 3-year battery life. Thus, the disappearance of the animal after 3 years is assumed to be due to battery failure. The expected cell probabilities for the simulated data are shown in Table 9.1, including the cell labeled "battery failure."

Brownie et al. (1985) used the cell probabilities of models such as that in Table 9.1 to construct a likelihood function and derive analytical estimates of the unknown parameters (in the example S_1, S_2, S_3, and S_4). Several difficulties can occur with this approach. First, the parameter estimates are not constrained in a bounded region. Thus, survival rates greater than unity occasionally occur. Second, if a sparse data matrix happens to occur (i.e., certain critical cells have zeros), the analytical estimators may be undefined, even though the logical estimate of survival for that cell might be unity (cf. North and Cormack 1981).

These drawbacks can be corrected by numerically maximizing the likelihood function, with the parameter space restricted to the range of admissible parameter values. Several other advantages are incurred by using numerical procedures. Sets of parameters can be constrained to a common value. Thus, in the simple example, S_1 and S_2 can be constrained to a common value, and likewise S_3 and S_4. Thus, the parameter space can be reduced from 4 to 2, and if this model is still appropriate for the data, better estimates of the parameters are achieved.

The appropriateness of estimating common parameter values can easily be determined using numerical methods, because likelihood ratio tests can be constructed from the values of the maximum likelihood functions. Thus, if the null hypothesis of $H_0: S_1 = S_2$ and $S_3 = S_4$ is to be tested, first the likelihood

TABLE 9.1
Simulated and Expected Recovery History of 90 Transmitter Collars Placed on Animals in Years 1 and 2[a]

Recovery	Year collared	Number collared	Year collars returned				Battery failure
			1	2	3	4	
Simulated	1	40	12	8	13	—	7
	2	50	—	13	17	7	13
Expected	1	N_1	$N_1(1-S_1)$	$N_1S_1(1-S_2)$	$N_1S_1S_2(1-S_3)$	—	$N_1S_1S_2S_3$
	2	N_2	—	$N_2(1-S_2)$	$N_2S_2(1-S_3)$	$N_2S_2S_3(1-S_4)$	$N_2S_2S_3S_4$

[a] The true survival rates for years 1–4 were 0.6, 0.7, 0.5, and 0.6, respectively.

function is maximized with all four parameters individually estimated and then maximized again with $S_1 = S_2$ and $S_3 = S_4$. Then a likelihood ratio test is constructed with the two likelihood values. The numerical optimization approach allows a general approach.

Generally, no models have been developed that allow the same survival parameters for two sexes at one age class, but different survival rates at a later age class. Consider, for example, the estimation of survival of a cervid species; the survival of male and female young may be considered identical to develop a parsimonious model. However, when the age class is reached in which sex differences become obvious (such as antlers), the survival rates should be different. Thus, models to estimate the survival rates of cervid species should allow common survival rates for the young animals of both sexes, but different survival rates for the sexes at later age classes. Tests of the hypothesis of identical survival of the young should be made before this hypothesis is accepted, however. The advantage of the numerical techniques described here is that such a complicated model can be constructed fairly easily.

The big drawback to the numerical optimization approach is the actual construction of the likelihood functions. SURVIV, the computer program described in the next section, is available to construct the likelihood function from the algebraic expressions for the expected cell probabilities, perform optimization (with constraints), and construct likelihood ratio tests between models. However, the user must understand thoroughly the construction of the expected cell probabilities to utilize this program.

A second drawback of the numerically derived estimates is that the small sample bias of the estimators cannot be corrected as is done with analytical estimates (cf. Brownie et al. 1985:16). Monte Carlo simulation results presented by White (1983) suggest that constraining the range of the estimate to its admissible range may partially offset the bias of the estimators.

Program SURVIV

The practical application of the numerical optimization approach described above requires that a "powerful" computer program be available. The chore of constructing the matrix of expected cell probabilities is difficult in itself, and additional complications due to complex input for a computer program make the task nearly insurmountable. SURVIV (White 1983) was written to handle the numerical optimization task, with input to SURVIV being straightforward and simple. The advantages of numerical estimation should not be negated by a computer program that requires complex coding to enter model specifications and observed data. The simplicity of the input to SURVIV is demonstrated by the following example.

SURVIV uses the procedures MODEL, ESTIMATE, and TEST to perform the numerical estimation calculations. PROC MODEL constructs the likelihood function from algebraic expressions describing the cell probabilities. The generality and ease of model specification is shown by the input in Fig. 9.2 for the data in Table 9.1. The PROC MODEL statement sets various options and alerts the program that the observed and expected cell probabilities are to follow. The COHORT card sets the number of animals for the first multinomial and is followed by the three multinomial cells, with the observed value separated from the expected value by a colon. The entry of the expected cell probability is the feature of SURVIV that makes the numerical approach described in this chapter feasible. The parameters to be estimated are denoted by the S(I) notation in the algebraic expression. This algebraic expression must be FORTRAN 77 compatible, because the program manipulates this code to construct a FORTRAN 77 subroutine to evaluate the likelihood function.

If the sum of the observed returns (i.e., the numbers before the colons on the cell probability cards) does not equal the number of animals in the cohort (specified on the COHORT card), the program generates an additional cell, with the probability being 1 minus the sum of the previous cells in the cohort. The number of observed animals for this cell is the total in the cohort minus the sum of the observed animals in the previous cells of the cohort. The technique of generating the last cell of the multinomial as the complement of the previous cells is particularly useful for models, such as banding models, in which the algebraic expression for the animals never observed again is difficult to write out, except as the complement of the sum of the previous cells.

PROC MODEL continues to read COHORT, cell probability, and LABEL cards until the next PROC card is encountered. At this time the FORTRAN 77 compiler is called to compile the likelihood function subroutine, the system linker is called to link the binary code into memory, and control is passed to the next procedure. Because some machines lack a dynamic linking capability, a two-pass approach is used. The first pass constructs the FORTRAN 77 subroutine. The program is terminated, the FORTRAN 77 compiler is called to compile the model subroutine, and the newly created binary is placed in the binary file of the program, which is then executed again with an input flag, indicating that the correct model is already in the core image. Thus, processing of the MODEL procedure can be bypassed. The main disadvantage of the two-pass approach is that only one general model can be processed in a particular

```
PROC TITLE RADIO-TRACKING EXAMPLE FOR JWM, 1983, VOL. 47, PAGES 716-728;
PROC MODEL NPAR=4 /* SIMPLE RADIO-TRACKING EXAMPLE */;
   COHORT = 40 /* NUMBER OF ANIMALS COLLARED IN YEAR 1*/;
   12:(1.-S(1));
    8:S(1)*(1.-S(2));
   13:S(1)*S(2)*(1.-S(3));
    7:S(1)*S(2)*S(3);
   COHORT = 50 /* NUMBER OF ANIMALS COLLARED IN YEAR 2*/;
   13:(1.-S(2));
   17:S(2)*(1.-S(3));
    7:S(2)*S(3)*(1.-S(4));
   13:S(2)*S(3)*S(4);
   LABELS;
    S(1)=SURVIVAL RATE IN YEAR 1;
    S(2)=SURVIVAL RATE IN YEAR 2;
    S(3)=SURVIVAL RATE IN YEAR 3;
    S(4)=SURVIVAL RATE IN YEAR 4;
PROC ESTIMATE NAME=GENERAL
   /* ALL PARAMETERS INDIVIDUALLY ESTIMATED */ ;
   INITIAL; S(1)=0.6; S(2)=0.7; S(3)=0.5; S(4)=0.6;
PROC ESTIMATE NAME=CONSTRAIN
   /* SETS OF 2 PARAMETERS CONSTRAINED */ ;
   INITIAL; S(1)=0.6; S(2)=0.7; S(3)=0.5; S(4)=0.6;
   CONSTRAINTS; S(1)=S(2); S(3)=S(4);
PROC TEST
   /* GENERATE LIKELIHOOD RATIO TEST OF ABOVE 2 MODELS */ ;
PROC STOP /* SIGNAL END OF ANALYSIS. */ ;
```

Figure 9.2 Input to program SURVIV to analyze the data and the model in Table 9.1. Statements are separated by semicolons, while observed recoveries are separated by colons from the algebraic expression for the expected cell probabilities.

run. Replacing the binary image in core when the likelihood subroutine is created allows multiple data sets to be analyzed in one run.

PROC ESTIMATE is called to make the actual parameter estimates. The INITIAL statement (Fig. 9.2) signifies that initial estimates of some parameters are provided [the default is $S(I) = 0.5$]. The first time ESTIMATE is called in Fig. 9.2, only boundary constraints are provided; that is, the admissible range of the parameter values are set. The default interval for a parameter is (0, 1). This call to PROC ESTIMATE evaluates the model with four individual values of $S(1)$, $S(2)$, $S(3)$, and $S(4)$.

The second call to PROC ESTIMATE (line 21 in Fig. 9.2) evaluates the model with S(1) = S(2) and S(3) = S(4), that is, the reduced model in which two sets of parameters are assumed equal. The CONSTRAINTS statement is used to specify these equalities, and the degrees of freedom of the model are automatically reduced by 2.

PROC TEST performs a likelihood ratio test between all pairs of models called in PROC ESTIMATE, given that the degrees of freedom available for each model are not equal. In this simple example only one test is performed, because only one pair of models is available: the general model with four parameters and the reduced model with two parameters. The null hypothesis tested is $H_0: S_1 = S_2$ and $S_3 = S_4$ versus the alternative hypothesis H_a of at least one of the equalities not true.

Other procedures in SURVIV useful for performing survival rate estimation experiments are SIMULATE and SAMPLE SIZE. PROC SIMULATE performs Monte Carlo simulation of a model entered via PROC MODEL. Use of PROC SIMULATE allows the researcher to determine the power of hypothesis tests and the expected confidence interval length of interval estimates based on numbers of marked animals and estimated survival rates. PROC SAMPLE SIZE performs sample size estimation for banding experiments as described by Brownie et al. (1985:186–193). Both of these procedures are designed to aid the researcher in performing an optimal experiment to estimate survival rates. PROC SIMULATE is also useful in studying the operating characteristics of a model and the associated estimates, such as determining the effect of boundary constraints on parameter estimates. A complete user's manual for SURVIV is given in Appendix 6. The source code is available from the authors or by dialing the SESAME bulletin board (see Chapter 1).

In the remainder of this section, we further develop the methods used by SURVIV and relate them to the statistical theory of categorical data analysis. Readers not interested in this material should skip to the next section. The procedures used by SURVIV are identical to log-linear analysis theory, and the χ^2 tests are identical to G tests or maximum likelihood χ^2 tests (Sokal and Rohlf 1981). To demonstrate this, we construct the G test for the example data in Table 9.1 of constant survival for all years and cohorts, that is, $H_0: S_1 = S_2 = S_3 = S_4$, versus the alternative that at least one of the equalities is not true.

First, we must calculate the goodness-of-fit statistic for the complete model. Because potential differences between cohorts are ignored in the complete model, the six cells collapse into four. For example, the 78 animals alive at the start of the second year are the (40 − 12) that lived through the first year, plus the 50 that were captured at the start of the second: (40 − 12) + 50 =

9 Survival Rate Estimation

78. Of these 78, eight of the 40 from the first year died, and 13 of the 50 captured the second year died. This gives a total of 21 animals that died the second year. Similar calculations are used to obtain the following table:

Year	Alive at start	Died during year	Lived through year	Survival estimate
1	40	12	28	28/40 = 0.700
2	78	21	57	57/78 = 0.731
3	57	30	27	27/57 = 0.474
4	20	7	13	13/20 = 0.650

The G statistic for this complete model is

$$G = 2\left[\sum_{i=1}^{4} O_i \ln\left(\frac{O_i}{E_i}\right)\right]$$

This G statistic equals zero, because the model is completely specified. That is, for cells containing the number of animals that lived, the expected value is identical to the observed value. The term $\ln(O_i/E_i)$ is then equal to zero. Likewise, for cells containing the number of animals that died, the expected value is also identical to the observed value. Thus, $G = 0$ for the complete model.

Now we must calculate the goodness-of-fit statistic for the reduced model, that is, under the null hypothesis. The estimate of S is

$$\hat{S} = \frac{28 + 57 + 27 + 13}{40 + 78 + 57 + 20}$$

where 28 of 40 animals lived the first year, 57 of 78 the second, etc., from the above table. The G statistic is calculated

Year	Alive at start	Died during year			Lived through year		
		Observed	Expected	Contribution to G	Observed	Expected	Contribution to G
1	40	12	14.359	−2.154	28	25.640	2.464
2	78	21	28.000	−6.041	57	50.000	7.469
3	57	30	20.462	11.480	27	36.54	−8.168
4	20	7	7.179	−0.177	13	12.821	0.181
Total	195	70	103.000	3.108	125	92.000	1.946

```
Submodel      Name       Log-likelihood    NDF    G-O-F
--------      ----       --------------    ---    -----
   1         GENERAL       -12.253604       2    0.3807
   2         ALL_SAME      -17.306328       5    0.0139

Likelihood Ratio Tests Between Models
General     Reduced                      Degrees    Pr(Larger
Submodel    Submodel     Chi-square      Freedom    Chi-square)
--------    --------     ----------      -------    -----------
GENERAL     ALL_SAME       10.105           3         0.0177
```

Figure 9.3 Output from Program SURVIV for the example data in Table 9.1.

and after multiplying the total G by 2, we find $G = 10.108$. This is the same value as calculated by SURVIV in the test of GENERAL versus ALL_SAME in the output shown in Fig. 9.3. We leave to the reader the problem of testing whether the two cohorts have identical survival.

Example

Elk were radio-marked by Los Alamos National Laboratory personnel from 1978 to 1980, with the transmitter collars having a 3-year battery life. Initially, animals were collared only in the eastern Jemez Mountains, but in 1980 a second herd was studied in the western Jemez Mountains. The age and sex of the animals collared by location and year and the collar returns are shown in Table 9.2 and the corresponding input to SURVIV is shown in Appendix 8.

SURVIV can be used to test the hypothesis of interest to the researchers involved in this research. The first question was: Is the survival of western Jemez elk different than that of eastern Jemez elk? This hypothesis involves testing the equality of the 1980 survival rates for the western and eastern animals; that is, the survival parameters for western Jemez elk are set equal to corresponding parameters for the eastern Jemez elk in the CONSTRAINTS section of PROC ESTIMATE, and the results of this model are tested against the general model, in which all parameters are estimated individually.

Note that some of the survival rates are not estimable for this data set (e.g., adult male survival for 1978) because no animals were captured. SURVIV can handle such data because either the parameter can be left out of the cell probability statements or the parameter can be fixed to a particular value and not estimated. The latter choice is preferred, because when the parameters are constrained to be equal in the reduced models, estimates can be generated. This procedure is used in the input to SURVIV to analyze the elk data in Table 9.2.

9 Survival Rate Estimation

TABLE 9.2
Recoveries of Transmitter Collars by Age and Sex Classes and Year Attached to Elk in the Jemez Mountains, New Mexico, by Los Alamos National Laboratory Personnel[a]

Year	Location	Sex	Age	N	Year returned[b] 1978	1979	1980	1981	Lived
1978	East Jemez	Male	Adult	0	0	0			0
1979	East Jemez	Male	Adult	1		0	1		0
1980	East Jemez	Male	Adult	1			0	0	1
1980	West Jemez	Male	Adult	1			1	0	0
1978	East Jemez	Male	Yearling	1	1	0			0
1979	East Jemez	Male	Yearling	1		1	0		0
1980	East Jemez	Male	Yearling	2			1	1	0
1980	West Jemez	Male	Yearling	5			0	4	1
1978	East Jemez	Male	Calf	5	1	1			3
1979	East Jemez	Male	Calf	2		0	1		1
1980	East Jemez	Male	Calf	1			0	0	1
1980	West Jemez	Male	Calf	0			0	0	0
1978	East Jemez	Female	Adult	8	0	0			8
1979	East Jemez	Female	Adult	4		2	0		2
1980	East Jemez	Female	Adult	7			1	1	5
1980	West Jemez	Female	Adult	5			0	0	5
1978	East Jemez	Female	Yearling	0	0	0			0
1979	East Jemez	Female	Yearling	0		0	0		0
1980	East Jemez	Female	Yearling	4			0	1	3
1980	West Jemez	Female	Yearling	0			0	0	0
1978	East Jemez	Female	Calf	2	0	0			2
1979	East Jemez	Female	Calf	4		0	0		4
1980	East Jemez	Female	Calf	2			0	0	2
1980	West Jemez	Female	Calf	2			0	1	1

[a] Blanks in a cell indicate that these cells are before or after the 2-year life assumed for the transmitters.

[b] Years are from March 1 to February 28.

The sequence of five models was tested from these data (six models if YEAR_SAME is included). The most general model allows for individual survival rates for each age and sex class, year, and geographic area, making a total of 36 parameters. Only 25 of these parameters are estimable from the data in Table 9.2. The next most complex model assumed that survival was the same in both geographic areas. This model was tested against the most general model by a likelihood ratio test, and the null hypothesis of survival equal across geographic areas was not rejected ($P = 0.19$). Then the hypothesis of survival equal between years was tested; again, it was not rejected ($P = 0.60$). The next hypothesis was that survival was the same for age and sex classes within the categories of hunted and nonhunted animals. That is, male calves and all female age classes are pooled as one survival rate and adult and yearling males as one survival rate, leaving only two survival parameters to estimate. This highly reduced model was not significantly different from the previous model ($P = 0.29$). The last model of all survival rates constant was rejected by the likelihood ratio test against the previous model ($P < 0.01$). Therefore, we conclude that survival rates are different between adult and yearling males and all other age classes, certainly a reasonable conclusion, given that only adult and yearling males were extensively hunted.

The above results must be interpreted within the framework of statistical hypothesis testing; that is, we "accept" the null hypotheses, but cannot prove them true. Thus, the simple model with only two survival rates is probably not the true model for the elk population, but sufficient data are lacking to justify a more complex model. Therefore, this model is accepted as the model consistent with the observed data. The estimated survival rates and their associated standard errors are 0.56 ± 0.10 and 0.91 ± 0.03, respectively, for the hunted and nonhunted segments of the collared elk population.

However, a more extensive data set may indicate that survival rates do differ between age and sex classes other than hunted and nonhunted. Certainly, the small sample sizes in this example are not adequate for good estimates of survival, as indicated by the large standard errors of the above two estimates.

More Complex Applications of Program SURVIV

SURVIV can also be used to estimate instantaneous survival rates. The finite survival rate (probability of survival for a time period of observed recoveries) can be replaced with the appropriate algebraic expression containing the instantaneous survival rate. Rather than an estimate and a standard error of a finite rate, the estimate and standard error of the instantaneous rate would be obtained. Boundary constraints on the instantaneous rate would differ from the default interval of (0, 1) appropriate for a finite rate.

Another extension that can be developed with SURVIV is to model survival rates as a function of one or more covariates. Suppose that survival is postulated to be a function of winter severity, denoted as W. A simple linear model might be postulated:

$$S_i = \beta_0 + \beta_1 W_i$$

where S_i is survival in year i, W_i is the winter severity for year i, and β_0 and β_1 are the parameters of the linear model relating survival and winter severity. Then, in the input to program SURVIV, every survival parameter will be replaced with the winter severity model.

However, the simple linear model is too rudimentary to actually be used, because survival could easily exceed 1 or be less than 0, depending on the values of β_0 and β_1. Thus, we might use a logistic model, such as that described by Burnham et al. (1987), to model desert tortoise survival (*Gopherus agassizii*). This model would be constrained between the values of β_0 and β_2:

$$S_i = \frac{\beta_2}{1 + \exp(-\beta_0 - \beta_1 W_i)}$$

where β_0 is the intercept of the curve for $W_i = 0$, β_1 is the slope, and β_2 is the asymptote for $W_i \to \infty$. Although a more complex model, survival can be constrained to the interval (0, 1). This extension to SURVIV is identical to a logistic regression, to be discussed later in this chapter. The advantage to using SURVIV is that more complex model testing can be formulated, for example, tests of the equality of the regression model parameters between age classes of animals or geographic areas.

Assumptions Required for Use of Program SURVIV

In the methods presented for SURVIV, the failure of radio-tracking transmitters attached to experimental animals is assumed to occur only at the predicted time of battery depletion. Although failure with today's commercially available equipment is less common than earlier, failure still does occur. How transmitter failures are handled in the above analysis depends on the cause of failure. If the failure is truly a random event (i.e., failure was not caused by some factor attributable to the animal's survival characteristics), then the animal can be discarded from the study and the number of radio-marked animals can be reduced by 1. However, often the failure of a transmitter is due to the animal's mortality, such as when the animal is poached and the transmitter is destroyed by the poacher. In this case the time of transmitter failure should be taken as the time of the animal's death. Obviously, the researcher is never sure which of the above cases is the true one and thus influences the data analysis and results when he chooses between the alternatives. A small percentage of transmitter failures will not change the results greatly. However, significant numbers of transmitter failures lead to subjective decisions and could lead to questions about the results of the research. The only solution to this problem seems to be to use high-quality equipment, so that random failures do not occur and time of death can therefore be taken as the time of transmitter failure. Reliability studies of transmitters might also be performed with doubly radio-tagged animals. Note that transmitter failure is not the same problem as that of band loss (Nelson et al. 1980), because transmitter failure is known to occur, while the loss of a band is never identified. The failure of a transmitter can be placed in a specified time interval, and the problem is rather one of excluding the animal from the experiment, or assuming that the transmitter failure time is also the animal's time of death.

A subtle way in which significant numbers of transmitter failures can bias estimates of survival is shown in Fig. 9.4. Animal 1 died before the transmitter failed. Animal 2 died after the transmitter failed, so its death would never be detected. Animal 3 never died, but the transmitter failed. Animals 4 and 5 died, but their transmitters never failed for the interval considered reliable for the transmitters. Animal 6 and its transmitter both survived the interval. Thus, the true survival rate for these six animals is 2/6 = 0.33. If an animal is discarded from the analysis if the transmitter fails during the interval, then the estimated survival rate would be 1/4 = 0.25, since animals 2 and 3 would be discarded because of transmitter failure. Animal 1 would have been included in the analysis, because the researcher would not have detected the transmitter failure and would not have known to reject the data for this animal. Thus, unbiased esti-

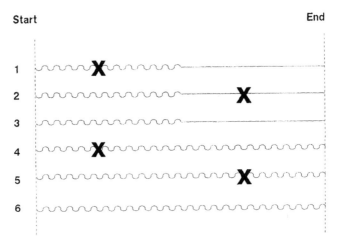

Figure 9.4 Example of how transmitter failures can bias the estimates of survival obtained from radio-tracking data. Wavy lines indicate the periods when transmitters are working, straight lines are when they have failed, and Xs indicate the death of the animal wearing the transmitter.

mates can only be achieved if the transmitters are known to work throughout the interval being considered. Had the transmitters been left in the field, the data for animal 1 would have been discarded also, so that an unbiased estimate of survival would still be obtained, that is, 1/3 = 0.33. The bias in survival for this example is because the probability of a transmitter failure is greatest for the animals that live throughout the interval, with a smaller probability of transmitter failure being detected for animals that die early in the interval.

A second common problem occurs when the transmitter fails, but it happens to be retrieved later, when the animal is killed by a hunter or trapped as part of further research. This animal was originally discarded from the analysis because of the failure. The temptation is to now include the animal in the analysis as a survivor for the interval. To do so biases the analysis, because had the animal been killed by a predator, the transmitter would never have been recovered. Thus, only survivors will be added, and hence survival will be biased high.

Radio-marked animals that move great distances from the population being studied present an additional complication. If they are known to have moved off the area, their survival rates may not be representative of the population and, hence, should be discarded from the study. If the animal is not located again (i.e., the biologists do not know the fate of the animal), the animal can then be classified as a transmitter failure, and the above discussion applies.

Pollock and Raveling (1982) discuss the problem of heterogeneous survival rates of the members of the population. They found that the band recovery models of Brownie et al. (1985) provide unbiased estimates of the average survival rate and the average recovery rate if year-to-year recovery and survival rates are all independent. They argue that such a fact is very unlikely, because an individual with a high survival probability in 1 year will probably continue to have a high survival probability the next year.

Estimates of survival rates obtained from radio-tracking data with the techniques described in this chapter would represent the average rate for heterogeneous populations, with the degree of bias depending on the correlation between survival rates among years. The ability of the researcher to separate the population based on age, sex, or any other identifiable characteristics of the animals helps in examining the causes of heterogeneity. The example of survival rate estimation for New Mexico elk presented above illustrates the procedure to test for differences in survival rates based on the geographic area where the animal was transmitter-collared. Tests of differences in survival rates of radio-marked animals could also be constructed based on the geographic area of recovery or previous areas utilized. For example, Pollock and Raveling

(1982) mention the case of waterfowl banded at the same location utilizing different migration corridors. During migration, radio-marked birds could be followed to determine a category for separating the marked animals, and a test of the equality of survival rates for migration corridors could be conducted. Note that this approach will not work for waterfowl that are only banded, because the model requires both survival and recovery rates to be estimated. Recovery rates for radio-marked birds are known, except if the proportion of transmitter failures becomes significant as discussed above.

Fundamentally, ecological/evolutionary theory predicts that individuals in a population will have heterogeneous survival rates, and therefore variation in fitness. The variability of survival rates could not be too large, or the proportion of the population with low survival rates would already be extinct (unless that segment of the population is just going extinct!), because it would be unable to compete. However, a corresponding increase in reproduction and recruitment could allow a segment of a population to have a much lower survival rate than the remainder of the population that maintains itself with low reproduction and high survival. Thus, two segments of a population could have equal fitness, even if survival and reproduction were quite different. The results of Pollock and Raveling (1982) show that the variance of the distribution of population survival rates must be fairly large to obtain substantial bias in survival estimates. We suspect that the natural variability of survival rates in most populations is not large enough to bias survival estimates.

Methods Incorporating Time Until Death

The amount of time until an individual animal dies (T) also contributes information on survival. For example, suppose that all the animals in both treatments of an experiment died. However, for treatment A the average time survived was 50 days, whereas for treatment B the average time survived was only 10 days. Naturally, the researcher would suspect that treatment B was a much more severe treatment.

The times that animals survive are commonly presented in three types of functions: the survivorship function [$S(t)$], the probability density function [$f(t)$], and the hazard function [$h(t)$]. All provide similar information, and a unique specification of one of these functions means that the other two are also uniquely specified. Transformations among the three functions are mathematically easy to perform. The following descriptions of these three functions are adapted from Lee (1980).

The length of times that animals survive can be described with the survivorship function, $S(t)$. The survivorship function is a function describing the probability that an individual animal survives until time t. Let T be the length of time that an individual survives. Mathematically, the function is

$$S(t) = Pr(T \geq t)$$

The survivorship function is estimated as the proportion of animals surviving longer than t:

$$\hat{S}(t) = \frac{\text{Number of animals surviving longer than } t}{\text{Total number of animals}}$$

$S(t)$ is always a level or decreasing function of time, because animals either continue to live or else die and decrease the function. At $t = 0$, $S(t) = 1$, because no animals have died yet. At $t = \infty$, $S(t) = 0$, because all animals die eventually. An example of a survivorship function is given in Fig. 9.5 for the case where the probability of dying is constant with time.

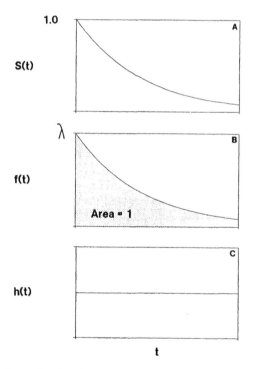

Figure 9.5 The survivorship function (A), probability density function (B), and hazard function (C) for the circumstance when the risk of mortality is constant through time. For a constant hazard function the survival function declines exponentially through time.

The survivorship function can be used to construct the probability density function. As with any continuous random variable, the survival time T has a probability density function defined as the limit of the probability that an animal dies in the short interval t to $t + \Delta t$ per unit width Δt (Lee 1980). The probability density function is expressed as

$$f(t) = \lim_{\Delta t \to \infty} \frac{Pr[\text{an animal dying in the interval } (t, t + \Delta t)]}{\Delta t}$$

or mathematically,

$$f(t) = \lim_{\Delta t \to \infty} \frac{Pr(t \leq T < t + \Delta t)}{\Delta t}$$

The probability density function $f(t)$ is estimated as the proportion of animals dying in an interval per unit width:

$$\hat{f}(t) = \frac{\text{Number of animals dying in the interval beginning at time } t}{[(\text{Total number of animals})(\text{Interval width})]}$$

The graph of $f(t)$ is called the density curve (Lee 1980), with an example given in Fig. 9.5 for a constant hazard function. Note that the area between the density curve and the t axis is equal to 1.

The hazard function $h(t)$ gives the probability that an animal dies during a very short interval, i.e., that T is in the range $(t, t + \Delta t)$. The hazard function can be described as:

$$h(t) = \lim_{\Delta t \to 0} \frac{Pr[\text{an animal alive at } t \text{ dies in the time interval } (t, t + \Delta t)]}{\Delta t}$$

or mathematically as

$$h(t) = \lim_{\Delta t \to 0} \frac{Pr(t \leq T < t + \Delta t \mid T \geq t)}{\Delta t}$$

The hazard function is a function of the animal's environment. For example, deep snow and cold weather would increase a mule deer fawn's chances of dying, and hence the hazard function would elevate during periods of inclement weather. Likewise, warm weather and spring green-up would lower the hazard function. The hazard function plotted in Fig. 9.5 shows a constant hazard; that is, the probability of dying is the same in all time intervals $(t, t + \Delta t)$. A constant hazard function leads to a probability density function known as the exponential distribution. Other examples of hazard functions for a human population and a wildlife population are given in Fig. 9.6.

The relationships among the survivorship function, the probability density

9 Survival Rate Estimation

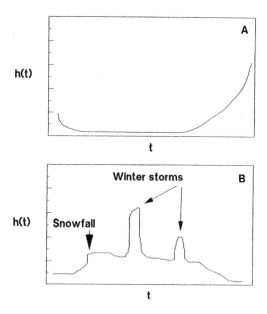

Figure 9.6 Hypothetical hazard functions that might typify the survival of humans (A) through the time period of their life (from Lee 1980) and the survival of mule deer fawns through winter.

function, and the hazard function are mathematically straightforward. Readers not interested in these relationships may skip this section without loss of continuity. The following discussion is taken from Lee (1980). An expression that relates all three functions is

$$h(t) = \frac{f(t)}{S(t)} \qquad (9.3)$$

The cumulative distribution function can be derived from the survivorship function as

$$F(t) = 1 - S(t) \qquad (9.4)$$

so that the probability density function can be defined from the survivorship function as

$$f(t) = \frac{d}{dt}[1 - S(t)] = -S'(t) \qquad (9.5)$$

Substituting these relationships,

$$h(t) = -\frac{S'(t)}{S(t)} = -\frac{d}{dt} \log_e S(t) \qquad (9.6)$$

229

In words, the "derivative of the logarithm of the survivorship function times -1 is equal to the hazard function" (Lee 1980:17). The above equation can be integrated from 0 to t (knowing that $S(0) = 1$) to obtain

$$-\int_0^t h(x)\, dx = \log_e S(t) \tag{9.7}$$

or

$$S(t) = \exp\left[-\int_0^t h(x)\, dx\right] \tag{9.8}$$

Thus, Eq. 9.8 can be substituted into Eq. 9.3 to obtain the relationship between $h(t)$ and $f(t)$:

$$f(t) = h(t) \exp\left[-\int_0^t h(x)\, dx\right] \tag{9.9}$$

The message from the above mathematics is that specification of any one of the three functions $h(t)$, $f(t)$, or $S(t)$, uniquely specifies the remaining two functions. Usually, a graph of one of the three is more informative than the other two in evaluating the "reasonableness" of the survival model. This phenomenon is most clear to us when we inspect the graphs in Fig. 9.5. The survivorship function $S(t)$ seems reasonable for overwintering mule deer fawns, but the constant hazard function $h(t)$ does not seem reasonable when considered relative to the stochastic environment in which mule deer fawns and most other wildlife species exist.

Program MICROMORT

Heisey and Fuller (1985) extended the Mayfield (1961, 1975) survival estimator for nesting success to radio-tracking data. The following material is taken from their work. The basic estimator is the daily survival rate during an interval of 1 or more days:

$$\hat{s}_i = \frac{x_i - y_i}{x_i} \tag{9.10}$$

where x_i is the total number of transmitter days in the interval and y_i is the total number of deaths, respectively, in interval i (Trent and Rongstad 1974). Then $\hat{s}_i$ is the daily survival rate for the interval. This estimator is the maximum

9 Survival Rate Estimation

likelihood estimator of s_i (Heisey and Fuller 1985). The probability of survival for interval i is then the product of the daily survival rates:

$$\hat{S}_i = \hat{s}_i^{L_i} \tag{9.11}$$

where L_i is the length in days of interval i. Note that this estimator assumes that survival is identical for each day during the interval; that is, the hazard function is constant during the interval. The length of the intervals for $i = 1, \ldots, I$ need not be the same. Then the survival rate, S^*, over all I intervals is estimated as the product of the $\hat{S}_i$'s:

$$\hat{S}^* = \prod_{i=1}^{I} \hat{S}_i \tag{9.12}$$

$\hat{S}^*$ is also a maximum likelihood estimator.

This estimator is then extended to multiple causes of death. The daily mortality rate due to a particular cause of death is the probability that an animal alive at the beginning of a day in interval i dies during the day due to this cause (identified as j). The maximum likelihood estimator of this probability, m_{ij}, is $\hat{m}_{ij} = y_{ij}/x_i$, where y_{ij} is the number of deaths in interval i due to cause j. The probability that an animal dies from cause j during the interval i (M_{ij}) is the sum of the probabilities that it survives to a particular day, and then dies on that day from cause j. This result is expressed mathematically as

$$M_{ij} = m_{ij} + s_i m_{ij} + s_i^2 m_{ij} + \ldots + s_i^{(L_i-1)} m_{ij}$$

$$= \left(\frac{m_{ij}}{1 - s_i}\right)(1 - s_i^{L_i})$$

From this basic development the method results in procedures to estimate survival and test for the equality of the competing causes of death.

The strong assumption required to employ the MICROMORT program is that intervals can be defined in which the daily survival rate is constant, that is, that the hazard function is constant over the interval. Certainly, this can be an arbitrary process and could lead to different estimates and test results if care is not taken in selecting the intervals used. Most temperate climate wildlife populations are greatly affected by weather, particularly winters. For example, consider the survival of mule deer in northwest Colorado. Winter storms cause increases in mortality (White et al. 1987). White et al. demonstrated large fluctuations in the survival rate through the winter. Prescribing intervals over the winter that would have constant daily survival would be nearly impossible. Rather, a more natural estimator of the overwinter survival rate is the binomial estimator described above, that is, the number of deer that die during the winter divided by the number that started the winter. Similar conclusions were presented by Kurzejeski et al. (1987) for the survival of wild turkey hens (*Meleagris gallopavo silvestris*) in northern Missouri.

The importance of the assumption of constant survival over the interval on the survival estimate is shown with a small example. Suppose that we want to estimate survival for a 100-day period starting on January 1, and 100 animals are radio-marked. In scenario 1, suppose a bad storm hits on January 2, and 50% of the animals die on January 5, 5 days into the interval. The remainder survive the interval. The number of deaths (y) for this scenario is 50, and the number of transmitter days (x) is 50 animals × 100 days + 50 animals × 5 days = 5250 days. Then the survival estimate is

$$\hat{s} = \frac{5250 - 50}{5250}$$
$$= 0.99048$$

The survival estimate for the entire interval is $0.99048^{100} = 0.3841$.

Now, consider a second year (scenario 2), during which the bad storm hits on day 92 of the winter interval. Again, 50% of the animals die 3 days later (day 95), so that $y = 50$, and the number of transmitter days (x) is 50 animals × 100 days + 50 animals × 95 days = 9750. The daily survival estimate is

$$\hat{s} = \frac{9750 - 50}{9750}$$
$$= 0.99487$$

with the interval estimate $0.99487^{100} = 0.5980$.

Intuitively, we would say that the survival estimate is 0.5 for the interval for both scenarios, because in both cases half of the radio-marked population did not survive the winter. MICROMORT will seldom give the same estimate as would a simple binomial approach developed from Eqs. 9.1 and 9.2.

Kaplan–Meier Method

The Kaplan–Meier product limit estimator (Kaplan and Meier 1958) is a simple extension of the binomial estimator presented above. Further, this estimator is derived from the survival function in a direct fashion. Suppose that n_0 animals are alive at the start of the interval for which survival is to be estimated. At time T_1 the first animal dies. Up to time T_1 the survival estimate has been 100%, or $\hat{S}(t) = 1$. At time T_1 $\hat{S}(t)$ drops to $(n_0 - 1)/n_0$. Survival con-

9 Survival Rate Estimation

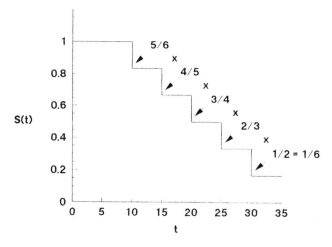

Figure 9.7 Kaplan–Meier estimates for the simple example described in the text.

tinues at this rate until a second animal dies at time T_2. Then $\hat{S}(t)$ becomes $(n_0 - 2)/n_0$. This process continues until the last mortality in the interval, leaving n_1 animals alive. Then the survival estimate would be

$$\hat{S}(t) = \frac{n_1}{n_0}$$

Thus, the survival estimate for the interval is identical to the estimator derived for the binomial distribution described in Eq. 9.1.

A simple example of the estimator will be used to explain how it works, with the estimated survivorship function plotted in Fig. 9.7. Consider the following small radio-tracking study. Six animals ($n_0 = 6$) are monitored. All but one dies. Let $r(T_j)$ represent the number of animals at risk of dying (i.e., alive at the start of the interval) in the interval (T_{j-1}, T_j) and $d(T_j)$ be the number of animals that died during the interval (T_{j-1}, T_j), for $j = 1, \ldots, n_0$.

j	T_j	Status	$r(T_j)$	$d(T_j)$
1	10	Died	6	1
2	15	Died	5	1
3	20	Died	4	1
4	25	Died	3	1
5	30	Died	2	1
6	35	Study ended	1	0

The formal definition of the Kaplan–Meier product limit estimator is:

$$\hat{S}(t) = \prod_{j=1}^{n_0} \frac{r(T_j) - d(T_j)}{r(T_j)} \qquad (9.13)$$

The survival estimate at time T_1 is

$$\hat{S}(T_1) = \frac{r(T_1) - d_1}{r(T_1)}$$

$$= 5/6$$

At time T_2 the estimate is

$$\hat{S}(T_2) = \frac{r(T_1) - d(T_1)}{r(T_1)} \times \frac{r(T_2) - d(T_2)}{r(T_2)}$$

$$= \frac{6-1}{6} \times \frac{5-1}{5}$$

$$= 2/3$$

This product multiplication continues until we obtain the final estimate of 1/6, exactly the same estimate that would be obtained from the binomial estimator. A graphic description of this estimator is given in Fig. 9.7.

The variance of $\hat{S}(T_j)$ is approximated by

$$\text{Var}[\hat{S}(T_j)] = \hat{S}^2(T_j) \sum_j \frac{d(T_j)}{r(T_j)[r(T_j) - d(T_j)]} \qquad (9.14)$$

known as Greenwood's formula (Cox and Oakes 1984). Thus, the variance of $\hat{S}(T_1)$ is

$$\text{Var}[\hat{S}(T_j)] = (5/6)^2 \times \frac{1}{6(6-1)}$$

$$= 0.02315$$

and the variance of $\hat{S}(T_2)$ is

$$\text{Var}[\hat{S}(T_j)] = (2/3)^2 \times \left[\frac{1}{6(6-1)} + \frac{1}{5(5-1)}\right]$$

$$= 0.03704$$

The variance of $\hat{S}(T_6)$, that is, the variance of the survival estimate for the entire interval, is $\hat{S}(T_6) = 0.02315$. Note that this is the same estimate of the variance that is obtained from the binomial distribution for $\hat{S} = 1/6$ with $n_0 = 6$. Cox and Oakes (1984:51) provide an alternate, simpler estimate:

$$\text{Var}[\hat{S}(T_j)] = \frac{[\hat{S}(T_j)]^2[1 - \hat{S}(T_j)]}{r(T_j)} \qquad (9.15)$$

This estimator is better in the tails of the distribution.

Approximate $(1 - \alpha)$ percentage of confidence intervals are constructed for $\hat{S}(t)$ with the formula

$$\hat{S}(t) \pm z_\alpha \sqrt{\text{Var}[\hat{S}(t)]}$$

where $z_\alpha = 1.96$ for $\alpha = 0.05$.

The main advantages of the Kaplan–Meier product limit estimator are that no assumptions are needed about the hazard function and that animals can be added to the population at risk or removed from the population at risk at any time. Thus, most radio-tracking studies do not mark all of the animals simultaneously, so that animals are being added into the marked population through time. This is described as the staggered entry of the transmitters into the marked population (Pollock et al. 1989a). Likewise, generally a few of the transmitters fail, so that censoring occurs. The product limit estimator allows for these incidents by either adding or removing animals from the population at risk $[r(T_j)]$.

For example, consider control and treatment animals for the fawn mule deer data presented in Fig. 9.1. The first animal died 8 days into the study (control animal 192.750). BMDP program P1L (Benedetti et al. 1983) or PROC LIFETEST of SAS can be used to compute the Kaplan–Meier estimates for these data if all animals enter the experiment at the same time. However, we present the SAS program in Fig. 9.8 to compute the estimates for staggered entry as described by Pollock et al. (1989a). The program computes the Kaplan–Meier product limit estimate for each time occasion and generates a simple plot of the survivorship function. Staggered entry and censoring are allowed in this program, and both of the variance estimates described above are computed.

Output from this SAS program was used to construct the survivorship functions for control and treatment animals plotted in Fig. 9.9. In addition, confidence intervals for the two survivorship functions are included. Note that the confidence intervals are narrower at the beginning of the study, when more animals were at risk of mortality. Confidence intervals expand (become wider) as the number of animals at risk decreases. To maintain narrow confidence intervals at the end of the time interval, new animals would have to be captured and their survival monitored, that is, new animals included in the study.

The Kaplan–Meier product limit estimator can also be extended to test for differences between two populations of animals, such as shown in Fig. 9.9. The most common method is the log rank test, formally described by Cox and Oakes (1984: 104–106). Lee (1980: 129) gives a more understandable description for the mathematically unsophisticated.

```
*------------------------------------------------------------*
|  SAS program to compute Kaplan-Meier product limit         |
|  estimates and compute log rank test between 2 groups      |
|                                                            |
|  Data are read from observations listed in Fig. 9.1.       |
|                                                            |
|  Variable  Contents                                        |
|  --------  ----------------------------------              |
|  group     Control or Treatment group                      |
|  id        Transmitter frequency                           |
|  sex       Sex                                             |
|  weight    Weight at capture                               |
|  length    Total body length at capture                    |
|  captured  Start of sampling period                        |
|  status    Fate of animal (lived, died, radio failed)      |
|  returned  End of sampling period (until death,            |
|            transmitter failure, or 15 June)                |
|                                                            |
*------------------------------------------------------------*;
title 'Survival of mule deer fawns, Piceance Basin, CO';
* The following data step reads in the observed data.;
data fawns;
 infile 'fig9.1';
 informat captured returned date.;
 length group $ 9;
 input group $ id $ sex $ weight length captured status $ returned;
 format captured returned date.;
 label group='Control or Treatment group'
       id='Transmitter frequency'
       sex='Sex'
       weight='Weight at capture (kg)'
       length='Total body length at capture (cm)'
       captured='Start of sampling period'
       status='Fate of animal (lived, died, radio failed)'
       returned='End of sampling period';

* This DATA step converts the file from captured and returned
  on the same record, to separate records for each.
  All animals are processed.;
data temp;
   set fawns;
```

Figure 9.8 SAS program to compute the Kaplan-Meier estimate for both control and treatment animals for the data in Fig. 9.1. In addition, a log rank test was also computed and the results were placed in the SAS LOG file. (*Figure continues.*)

```
   label group='Group membership'
         day='Days since start of study'
         type='Type of record';
   keep group day type;
   type='Captured';
   * Convert date to number of days since start of study,
     which is taken as November 15, 1982;
   day=captured-mdy(11,15,82);
   output;
   if status='Lived' then return;
   else if status='Censored' then type='Censored';
   else type='Died';
   day=returned-mdy(11,15,82);
   output;
proc sort; by day;

* This DATA step generates input for the Kaplan-Meier estimates
  and log rank test calculations.;
data kminput;
   retain norisk1 nodead1 nocensr1 noadded1 0 firstday;
   retain norisk2 nodead2 nocensr2 noadded2 0;
   array norisk{1:2} norisk1 norisk2;
   array nodead{1:2} nodead1 nodead2;
   array nocensr{1:2} nocensr1 nocensr2;
   array noadded{1:2} noadded1 noadded2;
   set temp; by day;
   if group='Control' then groupid=1;
   else groupid=2;
   if first.day then firstday=day;
   if type='Captured' then noadded{groupid}=noadded{groupid}+1;
   else if type='Died' then nodead{groupid}=nodead{groupid}+1;
   else if type='Censored' then nocensr{groupid}=nocensr{groupid}+1;
   if last.day then do;
      time=day;
      keep time norisk1 nodead1 nocensr1 noadded1
                norisk2 nodead2 nocensr2 noadded2;
      if time > 0 then output;
      do i=1 to 2 by 1;
         norisk{i}=norisk{i}+noadded{i}-nodead{i}-nocensr{i};
         nodead{i}=0; nocensr{i}=0; noadded{i}=0;
      end;
   end;
```

(Figure continues.)

```
* This data step computes the Kaplan-Meier Estimator for each group,
  plus calculates the log rank test between groups.;
data logrank;
   title2 'Kaplan-Meier Estimates for Control and Treatment';
   retain sumd2 0 sumEd1 0 sumEd2 0 svar1d2 0 svar2d2 0;
   retain Shat1 Shat2 1 Grensum1 Grensum2 0;
   set kminput end=lastobs;
   if _N_=1 then do;
      sumd2=0; sumEd1=0; sumEd2=0; svar1d2=0; svar2d2=0; end;
   * Compute the Kaplan-Meier Product Limit Estimator for
      each group, so that plots can be generated;
   if norisk1 > 0 then
      Shat1=Shat1*(1 - nodead1/norisk1);
   if norisk2 > 0 then
      Shat2=Shat2*(1 - nodead2/norisk2);
   if norisk1 > 0 & (norisk1 - nodead1) > 0 then
      Grensum1 = Grensum1 + nodead1/(norisk1*(norisk1-nodead1));
   if norisk2 > 0 & (norisk2 - nodead2) > 0 then
      Grensum2 = Grensum2 + nodead2/(norisk2*(norisk2-nodead2));
   seShat1 = sqrt(Shat1*Shat1*Grensum1);
   if seShat1 > 0 then do;
      lci1 = Shat1 - 1.96*seShat1;
      uci1 = Shat1 + 1.96*seShat1;
      end;
   else do;
      lci1=.;
      uci1=.;
      end;
   seShat2 = sqrt(Shat2*Shat2*Grensum2);
   if seShat2 > 0 then do;
      lci2 = Shat2 - 1.96*seShat2;
      uci2 = Shat2 + 1.96*seShat2;
      end;
   else do;
      lci2=.;
      uci2=.;
      end;
   keep time Shat1 Shat2 lci1 lci2 uci1 uci2;
```

Figure 9.8 (*Continued*)

9 Survival Rate Estimation

```
    output;
    sumd2 = sumd2+nodead2;
    if (norisk1 + norisk2) > 0 then do;
        sumEd1 = sumEd1 + (nodead1 + nodead2)*norisk1/(norisk1 + norisk2);
    sumEd2 = sumEd2 + (nodead1 + nodead2)*norisk2/(norisk1 + norisk2);
        if (norisk1 + norisk2) > 1 then
            svar1d2 = svar1d2 + (nodead1 + nodead2)*norisk1*norisk2*
                    (norisk1 + norisk2 - nodead1 - nodead2)
                   /((norisk1 + norisk2)**2*(norisk1 + norisk2 - 1));
        svar2d2 = svar2d2 + (norisk1*norisk2*(nodead1 + nodead2))
                / (norisk1 + norisk2)**2;
    end;
    if lastobs then do;
        * Put the log rank test results out in the LOG file listing.;
chisq1 = (sumd2 - sumEd1)**2/svar1d2;
        chisq2 = (sumd2 - sumEd1)**2/svar2d2;
        chisq3 = (sumd2 - sumEd1)**2 * (1/sumEd1 + 1/sumEd2);
        put sumd2= sumEd1= svar1d2= svar2d2=;
        put chisq1= chisq2= chisq3=;
        probchi1 = 1 - probchi(chisq1,1);
        probchi2 = 1 - probchi(chisq2,1);
        probchi3 = 1 - probchi(chisq3,1);
        put probchi1= probchi2= probchi3=;
        end;
proc print;
proc plot;
   plot Shat1*time='C' Shat2*time='T'
   lci1*time='-' uci1*time='-' lci2*time='+' uci2*time='+'
   / overlay;
* Generate plot with SAS/GRAPH procedures;
goptions device=ega;
symbol1 c=gold i=join v=C;
symbol2 c=gold i=join v=T;
symbol3 c=white i=join v=;
proc gplot;
   plot Shat1*time=1 Shat2*time=2
   lci1*time=3 uci1*time=3 lci2*time=3 uci2*time=3
   / overlay;
run;
```

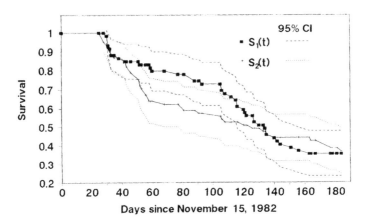

Figure 9.9 Estimates from the Kaplan–Meier model of the survivorship functions from the fawn mule deer data in Fig. 9.1. The dashed lines are 95% confidence intervals.

To compare two survivorship functions, we need to distinguish between the number of animals at risk and the number of animals dying for each group. Thus, we will add a subscript so that $r_1(T_j)$ and $r_2(T_j)$ represent the number of animals at risk of dying for groups 1 and 2 at time T_j, respectively, and $d_1(T_j)$ and $d_2(T_j)$ represent the number of animals dying for groups 1 and 2 at time T_j, respectively. The values $r(T_j)$ and $d(T_j)$ will represent the number of animals at risk of dying and the number of animals that died at time T_j for the combination of both groups. That is, the sum of both groups has no group designator subscript. Cox and Oakes (1984) present three tests to test the hypothesis that the two survivorship functions are identical. Each test is an approximate χ^2 statistic with 1 degree of freedom. All of the tests assume the conditional independence of the number of animals dying, that is, that the fates of the individuals in the population are independent of each other. Second, all of the tests assume that the observed values $d_1(T_j)$ and $d_2(T_j)$ are asymptotically normally distributed.

The first test is constructed as

$$\chi^2 = \frac{\left[\sum_{j=1}^{n_0} d_2(T_j) - \sum_{j=1}^{n_0} \frac{d(T_j)r_2(T_j)}{r(T_j)}\right]^2}{\sum_{j=1}^{n_0} \frac{d(T_j)r_1(T_j)r_2(T_j)[r(T_j) - d(T_j)]}{r(T_j)^2[r(T_j) - 1]}} \qquad (9.16)$$

9 Survival Rate Estimation

The second test is slightly more conservative, that is, less likely to make a Type I error and hence more likely to make a Type II error. The second test is constructed as

$$\chi^2 = \frac{\left[\sum_{j=1}^{n_0} d_2(T_j) - \sum_{j=1}^{n_0} \frac{d(T_j)r_2(T_j)}{r(T_j)}\right]^2}{\sum_{j=1}^{n_0} \frac{d(T_j)r_1(T_j)r_2(T_j)}{r(T_j)^2}}$$

Note that the only difference between the two tests is the term in the denominator. The third test is even more conservative, and is constructed as

$$\chi^2 = \frac{\left[\sum_{j=1}^{n_0} d_2(T_j) - \sum_{j=1}^{n_0} \frac{d(T_j)r_2(T_j)}{r(T_j)}\right]^2}{\left[\frac{1}{\sum_{j=1}^{n_0} \frac{d(T_j)r_1(T_j)}{r(T_j)}} + \frac{1}{\sum_{j=1}^{n_0} \frac{d(T_j)r_2(T_j)}{r(T_j)}}\right]^{-1}}$$

We will illustrate these three tests with the following simple example:

Obser-vation j	Time T_j	Group 1		Group 2		Total	
		$r_1(T_j)$	$d_1(T_j)$	$r_2(T_j)$	$d_2(T_j)$	$r(T_j)$	$d(T_j)$
1	4	10	1	10	0	20	1
2	5	9	0	10	1	19	1
3	8	9	1	9	0	18	1
4	10	8	0	9	1	17	1
5	12	8	1	8	0	16	1
6	15	7	0	8	1	15	1
7	16	7	1	7	0	14	1
8	20	6	1	7	1	13	2
9	24	5	1	6	0	11	1
10	25	4	0	6	1	10	1

In this simple example an animal in group 1 dies every fourth day, and an animal in group 2 dies every fifth day. We monitor survival until day 25, with ten animals monitored for each group. The χ^2 statistics for each of the tests are 0.0735, 0.0727, and 0.0725, respectively. Thus, because these values are all much lower than the critical value of 3.84, we conclude that we cannot show any significant differences in the two survivorship functions.

The log rank procedure allows for staggered entry of animals into the experiment, as well as censoring. PROC LIFETEST of SAS will conduct the log rank test if all the animals enter the experiment at the same time (Pollock et al. 1989b). To aid in understanding the test, the SAS program in Fig. 9.8 is included. This program will compute the log rank test for the control and treatment groups shown in Fig. 9.1 and handles data with staggered entry, as described by Pollock et al. (1989a).

Medical Survival Analysis

Survival data from radio-marked animals can be considered identical to those from clinical studies, in which human subjects report back to the physician periodically, or to laboratory studies, in which the experimental animals can be monitored through time. Numerous statistical techniques used by the medical field may be applicable to the analysis of some radio-tracking survival studies.

As an example of the types of problems that can be handled with medical techniques, consider the problem of testing two different methods of attaching transmitters to animals on the survival of the animals. Transmitter collars are put on two groups of animals (N_1 animals in group 1 and N_2 in group 2) at the start of the project. These radio-marked animals are monitored through time, to obtain $N_1 + N_2$ survival times. The hypothesis to be tested is the following: Did one group of animals exhibit greater survival than the other?

Numerous approaches to this problem have been developed. Three texts on this subject are those by Lee (1980), Cox and Oakes (1984), and Kalbfleisch and Prentice (1980). The book by Lee is somewhat less technical mathematically and, hence, more readable by those not inclined to mathematical presentations. We suggest that readers needing a thorough coverage of this type of survival analysis should consider these sources of information. The methods presented are excerpted from these texts to provide the reader with some understanding of methods commonly used to analyze medical survival data.

One issue to keep in mind while examining the medical survival estimation techniques is that medical research usually presumes that survival is a function of the age of the individual at risk and that the time at which the experiment takes place is not that important. For example, medical studies are often conducted over a period of years, but any year effect is not incorporated into the analysis. Rather, the age of the patient at the time the treatment was administered is included as a covariate. The assumption that time is not important, while age is important, is generally untrue for wildlife studies. That is, most wildlife populations experience major changes in survival related to environmental conditions (particularly winters, but also disease outbreaks, predator population changes, etc.), making time (the year effect) important. In contrast, survival is relatively constant once an individual is mature in most wildlife populations (cf. Brownie et al. 1985), making age unimportant once the animal reaches maturity. For wildlife studies the relative magnitude of the effects of year-to-year fluctuations of the environment affect survival rates more than the differences in survival rates of different-aged, but mature, animals.

A second consideration in using medical analyses to estimate wildlife survival rates is that "smooth" hazard functions can be postulated. For example, the risk of mortality may gradually increase with age for human subjects, and a smooth function can be fit to the data. Typical distribution functions include the exponential, Weibull, lognormal, gamma, logistic, and log logistic distributions (Lee 1980, Kalbfleisch and Prentice 1980). Because of the stochastic nature of time effects on wildlife populations, such assumptions generally do not appear to be reasonable. The constant hazard function results in the exponential distribution, and thus does not appear likely to fit actual wildlife data. The result is that many statistical programs designed for medical use must be applied carefully to wildlife research. An example of such a program is PROC LIFEREG of SAS.

A third subtle difference in the approach of medical analyses to estimating survival rates is the time period being considered. Medical studies usually must somewhat arbitrarily define the endpoint of the study; that is, does the new cancer treatment increase survival over the next 5 years? If the study period is too short, no differences may be found between the treatment and control patients. However, if the interval is too long, other causes of mortality than the one being studied will affect survival and tend to bring the survival rates of treatment and control patients to the same value, that is, zero. Thus, medical analyses often stress differences in the survival functions, $S(t)$, between the treatment and control groups. In contrast, most wildlife population studies are evaluating survival over annual increments. A logical set of intervals for any treatment is generally the annual cycles that wildlife populations experience. Thus, the parameter of interest in many wildlife studies is the annual finite survival rate. Researchers are more interested in testing for differences in the annual survival rates than in actual differences in the survival functions. This difference in analyses explains why many wildlife studies use the binomial distribution and programs such as SURVIV to perform the analysis, whereas medical researchers choose methods such as the Kaplan–Meier procedure that test for differences in the survival curves.

Cox Model

Cox's (1972) regression method uses explanatory or independent variables to predict survival times. The explanatory variables may consist of discrete indicator variables, such as the type of transmitter attachment for the example above, or continuous variables, such as weight and body length of the animal, taken as a measure of its condition.

The problem is to identify a relationship of T_i (survival time of the ith animal) or a function of T_i, say, $g(T_i)$ and the $(x_{1i}, x_{2i}, \ldots, x_{pi})$ independent variables that can be expressed in a regression function:

$$T_i = f_1(x_{1i}, x_{2i}, \ldots, x_{pi})$$

or

$$g(T_i) = f_2(x_{1i}, x_{2i}, \ldots, x_{pi})$$

Cox (1972) proposed a general nonparametric model appropriate for the analysis of survival data with and without censoring. He utilizes the hazard function (the animal's instantaneous probability of death). When survival times are continuously distributed and the possibility of ties (i.e., two deaths at exactly the same time) can be ignored, the hazard function is modeled as

$$h_i(t) = h_0(t) \exp(\beta_1 x_{1i} + \beta_2 x_{2i} + \cdots + \beta_p x_{pi})$$

$$= h_0(t) \exp\left(\sum_{j=1}^{p} \beta_j x_{ji}\right)$$

where $h_0(t)$ is the hazard function of the underlying distribution (arbitrary) when all of the x variables are ignored, that is, all x's equal zero and the β's are regression coefficients. The model can be generalized so that

$$\sum_{i=1}^{n} \beta_j x_{ji}$$

can be replaced by any known function of x_i's and β's. Thus, the Cox model is similar to an analysis of covariance or regression. Tests are developed for the coefficients (β's) to determine whether the jth explanatory variable affects survival.

The critical assumption of the Cox model regression approach is that the functions of the x_i's and β's act only on the underlying hazard function $h_0(t)$ in

9 Survival Rate Estimation

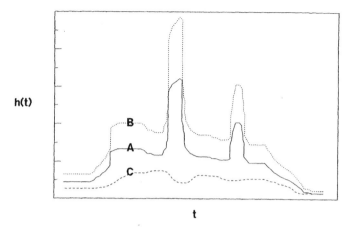

Figure 9.10 Hazard functions A and B are proportional, because $h_A(t)$ multiplied by a constant will give $h_B(t)$. In contrast, $h_A(t)$ cannot be multiplied by a constant to obtain $h_C(t)$.

a multiplicative manner. This assumption is generally stated that the individual hazard functions are proportional to each other, and this model is often termed the "proportional hazards" model. In biological terms this means that the hazard functions for the groups of animals being compared have the same shape. However, the peaks would be higher and the valleys deeper for the hazard function of one group than another. Examples of hazard functions that are proportional or not proportional are presented in Fig. 9.10.

The details of the estimation and testing procedures developed by Cox are not provided here. He utilizes maximum likelihood methods; hence, a statistically rigorous approach exists. The interested reader is referred to work by Lee (1980), Cox and Oakes (1984) or Kalbfleisch and Prentice (1980).

Programs for Cox's model analysis are available for SPSS (Emrich et al. 1981), SAS (Harrell 1980a), and BMDP program P2L (Hopkins 1983).

Logistic Regression

Another approach (also developed by Cox 1970) to analyzing the experiment described above and testing for differences in transmitter attachment methods would be linear logistic regression (Lee 1980:355–359). In this analysis the survival data are treated as dichotomous—an animal is either alive or dead. Hence, the problem of how to utilize the data from animals whose transmitters failed is identical to the analysis in SURVIV. The advantage of logistic regression is that continuous variables are easily incorporated into the analysis.

Define S_i as the probability of survival (success) of the ith individual. The probability of death (failure) is thus $(1 - S_i)$. A straightforward least-squares technique would model survival linearly as a function of independent variables, such as

$$S_i = \sum_{j=1}^{p} \beta_j x_{ji}$$

or as the ratio of survival rate to mortality rate:

$$\frac{S_i}{1 - S_i} = \sum_{j=1}^{p} \beta_j x_{ji}$$

These methods of treating a dichotomous dependent variable as if it were continuous have several limitations, including that the dependent variable is not normally distributed and that the variance around the dependent variable is not homogeneous over its range of values. Therefore, no method of estimation that is linear for the dependent variable will, in general, be fully efficient. Also, the predicted $\hat{S}_i$'s are not always going to be in the interval $0 < \hat{S}_i < 1$. A more appropriate model is the linear logistic model proposed by Cox (1970).

In the logistic regression model the dependence of the probability of success on independent variables is assumed to be

$$S_i = \frac{\exp\left(\sum_{j=1}^{p} \beta_j x_{ji}\right)}{1 + \exp\left(\sum_{j=1}^{p} \beta_j x_{ji}\right)}$$

and

$$1 - S_i = \frac{1}{1 + \exp\left(\sum_{j=1}^{p} \beta_j x_{ji}\right)}$$

where, again, β_j's are unknown coefficients. These equations appear to be complicated, until the logarithm of $S_i/(1 - S_i)$ is used:

$$\lambda_i = \log_e\left\{\frac{S_i}{1 - S_i}\right\} = \sum_{i=1}^{p} \beta_j x_{ji}$$

The quantity

$$\lambda_i = \log_e\left\{\frac{S_i}{1 - S_i}\right\}$$

is called the logistic transform of S_i, and the above equation is called a linear logistic model. Another name for λ_i is "log odds."

To estimate the β_i's, Cox (1970) constructs a maximum likelihood estimator, so that a rigorous statistical analysis is possible. The likelihood is constructed from the binomial distribution, so that logistic regression may be considered an extension of the basic binomial estimator given in Eq. 9.1. To see this, consider the following likelihood:

$$L = \binom{n_0}{n_1} S^{n_1} (1 - S)^{n_0 - n_1}$$

where n_0 animals are radio-tracked, and n_1 of them survive the monitoring interval. We are interested in estimating the parameter S, for which the estimator in Eq. 9.1 is the maximum likelihood estimator. Suppose that we want to determine whether S is a function of the animal's weight W. We could propose the linear model

$$\hat{S}_i = \beta_0 + \beta_1 W_i$$

but the model is somewhat absurd, because the values of $\hat{S}_i$ can be less than zero or greater than 1. To correct this anomaly, we might propose a logistic model:

$$\hat{S}_i = \frac{\exp(\beta_0 + \beta_1 W_i)}{1 + \exp(\beta_0 + \beta_1 W_i)}$$

where the values of $\hat{S}_i$ are constrained to the interval (0, 1). When this new function of β's and W to predict S is replaced in the likelihood equation, the following results:

$$L = \prod_{i=1}^{n_1} \left[\frac{\exp(\beta_0 + \beta_1 W_i)}{1 + \exp(\beta_0 + \beta_1 W_i)} \right] \cdot \prod_{i=1}^{n_0 - n_1} \left[1 - \left(\frac{\exp(\beta_0 + \beta_1 W_i)}{1 + \exp(\beta_0 + \beta_1 W_i)} \right) \right]$$

Survival is now modeled as a function of the covariate (weight), and the analogy to analysis of covariance or regression is evident. A set of tests to determine whether $\beta_j = 0$ has been developed. This test determines the importance of the covariate in predicting the fate of an animal. The approach can be generalized to many covariates and their influence on the survival of the population.

When logistic regression is used with continuous variables predicting the survival probability of individuals, then the S_i values are either 0 or 1. At this extreme of sparseness, the log odds values require either the logarithm of 0 or division by 0 to be performed. Because both are undefined operations, the computer procedures do not actually take the log odds values, but work from the exponential forms of the equations.

Programs for the analysis of the logistic regression are provided in BMDP program PLR (Engelman 1983), SAS PROC LOGIST (Harrell 1980b), and PROC CATMOD (SAS Institute Inc. 1985). In addition, a subprogram for SPSS is available from Reese (1982).

Discriminant Analysis

The reader familiar with discriminant analysis may have noted the similarity of the linear logistic analysis to linear discriminant analysis. Both approaches were suggested by Lee (1980) as possibilities for analyzing the fates of the experimental subjects.

Press and Wilson (1978) compared the logistic regression method to the discriminant analysis and found that if the independent variables are normal with identical covariance matrices, discriminant analysis is preferred. Under nonnormality, the logistic regression method is preferred. In particular, if the independent variables are dichotomous (discrete), discriminant analysis does not perform as well as does logistic regression. Therefore, we recommend that logistic regression be used instead of discriminant analysis for survival problems.

An extension of the logistic regression approach is to partition the likelihood function into intervals (Efron 1988). The term used to describe this approach is "partial logistic regression." If these intervals are made as short as those used in the Kaplan–Meier method, then the two approaches are identical. However, if shorter intervals are used, then the approach has greater power than the Kaplan–Meier method. Both censored data and staggered entry data can be used with partial logistic regression. The method can be computed with either standard logistic regression methods or with SURVIV. Efron (1988) suggests that the subjective decisions on how the data are partitioned have little effect on the final estimates. Further, covariates can be incorporated in the analysis just as in standard logistic regression. Thus, this new procedure appears to offer significant advantages over both standard logistic regression and the Kaplan–Meier approach.

χ^2 Analysis

If most or all of the explanatory variables are dichotomous (such as age or sex), then the appropriate analysis of the fates (alive or dead) of the radio-marked animals is a χ^2 contingency table or a log-linear analysis for a model with three or more factors. As with discriminant analysis, these analyses are presented in

9 Survival Rate Estimation

many statistical texts, so they will not be discussed here. A text we have found helpful is that by Freeman (1987). He compares logistic regression and loglinear models and presents numerous segments of SAS code to generate the analyses discussed.

Censoring

One of the recurrent problems with radio-tracking is the limited battery life of the transmitters. This problem is identical to the problem in human survival studies, in which the patient, although still alive, stops coming to see the physician and thus chooses not to continue to be a part of the study. Hence, the record of this individual is "censored." Likewise, when the transmitter stops transmitting, the survival time of the animal is censored. In the use of the word "censored" here, we are not throwing out data, but losing observations because of problems with equipment or a lack of time to allow all animals in the study to die.

The Cox model analysis allows for censoring of survival times and thus provides a reasonable and objective procedure to handle the problem of censoring. In contrast, when the data are treated as strictly dichotomous (alive or dead), the problem of censoring becomes more acute. Animals that, due to transmitter failure, can no longer be located usually cannot objectively be put into the "alive" or "dead" category. The discussion of transmitter failure relative to SURVIV applies, with the same problems of interpreting the results.

Example Analysis

The mule deer fawn survival data from the Piceance Basin of Colorado (Fig. 9.1) will be used to illustrate the logistic regression approach. Fawns were radio-marked in December and the data were used to estimate their survival through the winter until their migration in late April. The purpose of the study was to compare the survival of two groups of animals: one in the treatment area where future oil shale development posed potential impacts (both positive and negative) and the other a control population.

Figure 9.11 provides the SAS code used to test whether size (measured by length and weight), sex, and treatment area affected survival using logistic regression. The variable STATUS (lived or died) is used as input to PROC CATMOD. Animals with transmitters that failed are not included in the analysis.

Results of this analysis are shown in Fig. 9.12. GROUP is not significant

```
/**************************************************/
/*  SAS code to demonstrate the use of PROC CATMOD to  */
/*  perform a logistic regression with the data in     */
/*  Fig. 9.1.                                          */
/**************************************************/
title 'Logistic Regression of Piceance Mule Deer Fawns';
* The following data step reads in the observed data.;
data fawns;
 infile 'fig9.1';
 informat captured returned date.;
 length group $ 9;
 input group $ id $ sex $ weight length captured status $
returned;
 format captured returned date.;
 label group='Control or Treatment group'
       id='Transmitter frequency'
       sex='Sex'
       weight='Weight at capture (kg)'
       length='Total body length at capture (cm)'
       captured='Start of sampling period'
       status='Fate of animal (lived, died, radio
       failed)'
       returned='End of sampling period';
proc catmod;
   where status^='Censored';
   direct weight length;
   model status=group sex weight length / ml nogls;
run;
```

Figure 9.11 Fawn survival data for radio-marked mule deer fawns collared in December 1982 in the Piceance Basin, Colorado, set up in a SAS job to evaluate the effect of sex, length, and weight on survival probabilities with logistic regression. The variables WEIGHT and LENGTH are continuous, so must be included in the DIRECT statement of PROC CATMOD. GROUP and SEX are discrete, so are not included in the DIRECT statement. No WEIGHT statement is used because each observation represents only one fawn. The WHERE statement eliminates all animals with transmitter failures from the analysis.

9 Survival Rate Estimation

```
         MAXIMUM LIKELIHOOD ANALYSIS OF VARIANCE TABLE
    Source                    DF    Chi-Square     Prob
    ----------------------------------------------------
    INTERCEPT                  1       4.23      0.0396
    GROUP                      1       1.47      0.2249
    SEX                        1      13.05      0.0003
    WEIGHT                     1       2.14      0.1432
    LENGTH                     1       1.04      0.3082
    LIKELIHOOD RATIO         103     116.95      0.1642


              ANALYSIS OF MAXIMUM LIKELIHOOD ESTIMATES
                                       Standard    Chi-
Effect           Parameter  Estimate    Error     Square    Prob
-----------------------------------------------------------------
INTERCEPT            1      13.4858    6.5542     4.23    0.0396
GROUP                2      -0.2874    0.2368     1.47    0.2249
SEX                  3      -1.0794    0.2989    13.05    0.0003
WEIGHT               4      -0.1289    0.0880     2.14    0.1432
LENGTH               5      -0.0665    0.0653     1.04    0.3082
```

Figure 9.12 A portion of the output from the SAS job in Fig. 9.11. Only the tests and parameter estimates are presented.

($P = 0.2249$), but SEX is a significant predictor of survival ($P = 0.0003$). Neither WEIGHT ($P = 0.1432$) nor LENGTH ($P = 0.3082$) is a significant predictor of survival.

Summary

1. Finite survival rates can be modeled statistically as a binomial variable. Extensions to multiple periods extend the model to a multinomial distribution. These distributions require that each animal be independent of the remainder of the population and that each have an identical probability of survival; that is, no heterogeneity of survival rates exists.
2. SURVIV is based on maximum likelihood estimation of survival rates from one or more multinomial distributions. This program is a general purpose program to estimate survival rates and test them among one or more populations.
3. Survival data can be presented in three forms: the survivorship function, the probability density function, and the hazard function. Each

function uniquely specifies the other two, so that the same information is provided in each.
4. MICROMORT (Heisey and Fuller 1985) provides a method of estimating survival rates and examining causes of mortality. The main disadvantage of this method is that survival is assumed to be constant over user-defined intervals. The definition of these intervals is often subjective, and thus selection of intervals may lead to widely different estimates and conclusions. However, as is usually the case with statistical models, a more powerful method results when stronger assumptions are used to build the model.
5. The Kaplan–Meier product limit estimator is a nonparametric approach to modeling the survivorship function. The method is not as powerful as parametric methods that make stronger assumptions about the data, but is much more objective to apply.
6. Analysis of survival data from animals with transmitters is conceptually the same as medical studies in which the patient periodically reports back to the hospital. Thus, the statistical methods used in medical studies also generally apply to radio-tracking studies. Two of the important methods are the Cox model regression and logistic regression.

References

Bart, J. and D. S. Robson. 1982. Estimating survivorship when the subjects are visited periodically. Ecology 63:1078–1090.
Benedetti, J., K. Yuen, and L. Young. 1983. P1L 19.1 Life tables and survival functions. Pages 557–575 in W. J. Dixon ed. BMDP Statistical Software. Univ. of Calif. Press, Berkeley.
Brownie, C., D. R. Anderson, K. P. Burnham, and D. S. Robson. 1985. Statistical inference from band recovery data—A handbook, 2nd ed., Res. Publ. No. 131. U.S. Fish and Wildl. Serv., Washington, D.C. 305 pp.
Bunck, C. M. 1987. Analysis of survival data from telemetry projects. J. Raptor Res. 21:132–134.
Burnham, K. P., D. R. Anderson, G. C. White, C. Brownie, and K. H. Pollock. 1987. Design and analysis methods for fish survival experiments based on release-recapture. Am. Fish. Soc. Monogr. 5:1–437.
Conover, W. J. 1971. Practical Nonparametric Statistics. Wiley, New York. 462 pp.
Cox, D. R. 1970. Analysis of Binary Data. Methuen, London. 142 pp.
Cox, D. R. 1972. Regression models and life tables (with discussion). J. R. Stat. Soc. B 34:187–220.

Cox, D. R. and D. Oakes. 1984. Analysis of Survival Data. Chapman & Hall, New York. 201 pp.

D'Agostino, R. B., W. Chase, and A. Bélanger. 1988. The appropriateness of some common procedures for testing the equality of two independent binomial populations. Am. Stat. 42:198–202.

Efron, B. 1988. Logistic regression, survival analysis, and the Kaplan–Meier curve. J. Am. Stat. Assoc. 83:414–425.

Emrich, L. J., P. A. Reese, and J. D. Kalbfleisch. 1981. COX MODEL A proportional hazards model analysis package for SPSS users. *In* Proc. 5th Annual SPSS Users and Coordinators Conf. SPSS Inc., San Francisco, CA.

Engelman, L. 1983. PLR 14.5 Stepwise logistic regression. Pages 330–344 *in* W. J. Dixon ed. BMDP Statistical Software. Univ. of Calif. Press, Berkeley.

Freeman, D. H., Jr. 1987. Applied Categorical Data Analysis. Dekker, New York. 318 pp.

Harrell, F. 1980a. The PHGLM Procedure, SAS Inst. Tech. Rep. S-109. SAS Institute Inc., Raleigh, NC.

Harrell, F. 1980b. The LOGIST Procedure, SAS Inst. Tech. Rep. S-110. SAS Institute Inc., Raleigh, NC.

Heisey, D. M. and T. K. Fuller. 1985. Evaluation of survival and cause-specific mortality rates using telemetry data. J. Wildl. Manage. 49:668–674.

Hensler, G. L. and J. S. Nichols. 1981. The Mayfield method of estimating nesting success: a model, estimators and simulation results. Wilson Bull. 93:42–53.

Hollander, M. and D. A. Wolfe. 1973. Nonparametric Statistical Methods. Wiley, New York. 503 pp.

Hopkins, A. 1983. P2L 19.2 Survival analysis with covariates—Cox models. Pages 576–594 *in* W. J. Dixon ed. BMDP Statistical Software. Univ. of Calif. Press, Berkeley.

Johnson, D. H. 1979. Estimating nest success: the Mayfield method and an alternative. Auk 96:651–661.

Kalbfleisch, J. D. and R. L. Prentice. 1980. The Statistical Analysis of Failure Time Data. Wiley, New York. 321 pp.

Kaplan, E. L. and P. Meier. 1958. Nonparametric estimation from incomplete observations. J. Am. Stat. Assoc. 53:457–481.

Kurzejeski, E. W., L. D. Vangilder, and J. B. Lewis. 1987. Survival of wild turkey hens in north Missouri. J. Wildl. Manage. 51:188–193.

Lee, E. T. 1980. Statistical Methods for Survival Data Analysis. Lifetime Learning Publications, Belmont, CA. 557 pp.

Mayfield, H. 1961. Nesting success calculated from exposure. Wilson Bull. 73:255–261.

Mayfield, H. 1975. Suggestions for calculating nest success. Wilson Bull. 87:456–466.

Miller, H. W. and D. H. Johnson. 1978. Interpreting the results of nesting studies. J. Wildl. Manage. 42:471–476.

Nelson, L. J., D. R. Anderson, and K. P. Burnham. 1980. The effect of band loss on estimates of annual survival. J. Field Ornithol. 51:30–38.

North, P. M. and R. M. Cormack. 1981. On Seber's method for estimating age-specific bird survival rates from ringing recoveries. Biometrics 37:103–112.

Pollock, K. H. and D. G. Raveling. 1982. Assumptions of modern band-recovery models, with emphasis on heterogeneous survival rates. J. Wildl. Manage. 46:88–98.

Pollock, K. H., S. R. Winterstein, C. M. Bunck, and P. D. Curtis. 1989a. Survival analysis in telemetry studies: the staggered entry design. J. Wildl. Manage. 53:7–15.

Pollock, K. H., S. R. Winterstein, and M. J. Conroy. 1989b. Estimation and analysis of survival distributions for radio-tagged animals. Biometrics 45:99–109.

Press, S. J. and S. Wilson. 1978. Choosing between logistic regression and discriminant analysis. J. Am. Stat. Assoc. 73:699–705.

Reese, P. A. 1982. FREND procedures. Issues 3:9–10.

SAS Institute Inc. 1985. SAS® Language Guide for Personal Computers, Version 6 Edition. SAS Institute Inc., Cary, NC. 429 pp.

Sokal, R. R. and F. J. Rohlf. 1981. Biometry, 2nd ed. Freeman, San Francisco, CA. 859 pp.

Trent, T. T. and O. J. Rongstad. 1974. Home range and survival of cottontail rabbits in southwestern Wisconsin. J. Wildl. Manage. 38:459–472.

White, G. C. 1983. Numerical estimation of survival rates from band recovery and biotelemetry data. J. Wildl. Manage. 47:716–728.

White, G. C., R. A. Garrott, R. M. Bartmann, L. H. Carpenter, and A. W. Alldredge. 1987. Survival of mule deer in northwest Colorado. J. Wildl. Manage. 51:852–859.

CHAPTER

10

Population Estimation

Population estimation is one of the major responsibilities of a wildlife manager. In this chapter, we discuss how radio-tracking methods can be used to assist in the estimation of population size. Telemetry techniques can be used to evaluate whether a study meets the assumptions of the statistical methodology used, as well as to provide better marking methods.

Capture–Recapture Estimation

General reviews of capture–recapture estimation have been provided by Cormack (1968), Seber (1973, 1982), Otis et al. (1978), and White et al. (1982). Hence, this material is not considered here. Rather, techniques are discussed where biotelemetry methodology can assist with capture–recapture methods.

One of the critical assumptions of capture–recapture techniques is that the number of marked animals in the population is known. Often, loss of marks or the death of marked animals causes this assumption to be violated. The major advantage in using radios in a capture–recapture study is that marked animals can be verified in the population prior to a recapture occasion. If radios are not used to determine the number of marked animals in the population and the number of marked animals changes due to deaths or emigration, the Jolly–Seber methodology is the proper approach (see Seber 1982 for an introduction to this technique). If neck collars (without radios) are used to mark the animals, the researcher has no way of determining the exact number of marked animals in the study area. The Jolly–Seber method allows the estimation of the number of marks, and thus compensates for this problem. However, the precision of the population estimate is considerably poorer than if radios are used.

As an illustration of the use of biotelemetry for capture–recapture population estimation, suppose that a sample of radio-marked deer has been introduced into a population where aerial surveys are to be used to obtain second, third, etc., samples to estimate population levels. The presence of the radioed animals can be verified before the aerial survey is performed. Hence, animals that have died or emigrated from the study area can be removed from the marked sample for the purposes of the aerial survey.

Define n_1 as the number of radio-marked deer, and n_i for $i = 2, 3, \ldots, k + 1$ as the total number of deer observed on each of the k aerial surveys. Also, define m_i for $i = 2, 3, \ldots, k + 1$ as the number of radio-marked deer observed on each of the k aerial surveys. Then the estimate of the population size from the ith survey is

$$\hat{N}_i = \frac{(n_1 + 1)(n_i + 1)}{(m_i + 1)} - 1 \tag{10.1}$$

which is the Lincoln–Petersen estimate derived by Chapman (1951) (also presented by Seber 1982:60). The estimated variance of $\hat{N}_i$ is defined (Seber 1970, 1982:60) as

$$\hat{\mathrm{Var}}(\hat{N}_i) = \frac{(n_1 + 1)(n_i + 1)(n_1 - m_i)(n_i - m_i)}{(m_i + 1)^2(m_i + 2)} \tag{10.2}$$

with a 95% confidence interval constructed as

$$\hat{N}_i \pm 1.96 \, \hat{\mathrm{Var}}(\hat{N}_i)^{1/2} \tag{10.3}$$

For example, suppose a radio-tracking session on Tuesday shows that 29 radio-collared animals (n_1) are present in the study area. We assume that all radios are working, and that all working radios are on live animals. On Wednesday an observer on foot sees 85 deer (n_2), of which 15 have radio collars (m_2). Then the population is estimated as

$$\frac{(29 + 1) \times (85 + 1)}{(15 + 1)} - 1 = 161$$

with a variance of

$$\hat{\mathrm{Var}}(\hat{N}) = \frac{(29 + 1) \times (85 + 1) \times (29 - 15) \times (85 - 15)}{(15 + 1)^2 \times (15 + 2)} = 581$$

A 95% confidence interval for this single estimate is

$$\hat{N} \pm 1.96 \times (581)^{1/2} = \hat{N} \pm 47.2$$

Because sightings of animals are used to generate the recaptures for the population estimate, the collars should not attract the observer to the marked animals. Thus, the radios are not used to find the marked animals as part of the population estimation procedure, but only to verify the number of marked animals available in the population. In general, one individual should count the marked animals in the population, while others should perform the sightings to generate the population estimation data.

Because unmarked animals sighted during a survey are not marked (radio-collared) for subsequent surveys, the Schnabel–Darroch multiple recapture protocol (model M_t) of Otis et al. (1978) or White et al. (1982) is not applicable for this situation. Rather, multiple applications of the Lincoln–Petersen estimator are computed where the initial marking (radio-collaring) of animals constitutes the first sample, and the numbers of marked and unmarked animals counted during the surveys provide the second sample. Each survey produces a new Lincoln–Petersen estimate.

Combining Lincoln–Petersen Estimates

The precision of the estimation technique can be improved by increasing the number of surveys to observe animals, that is, increasing k. Suppose that five different observers independently sight deer (or one observer does the sightings on five different occasions or days). Several procedures are available for combining the five estimates to achieve an overall estimate of the population size.

Estimators Based on a Mean One possibility is to take the mean of the five estimates, that is,

$$\hat{N} = \sum_{i=2}^{k+1} \frac{\hat{N}_i}{k}$$

Rice and Harder (1977) used such a mean and standard error of five independent estimates for the estimation of the population level of a geographically closed (fenced) population of white-tailed deer (*Odocoileus virginianus*).

The simple mean may be inefficient, because the variance of the individual estimates is not used to create a weighted average. Thus, another possible estimator is

$$\hat{N} = \frac{\sum_{i=2}^{k+1} W_i \hat{N}_i}{\sum_{i=2}^{k+1} W_i}$$

where the weight of each estimate is taken as

$$W_i = \frac{1}{\hat{\text{Var}}(\hat{N}_i)}$$

The variance of the overall estimate is

$$\hat{\text{Var}}(\hat{N}) = \frac{\sum_{i=2}^{k+1} W_i(\hat{N}_i - \hat{N})^2}{(n-1)\sum_{i=2}^{k+1} W_i}$$

Because $\hat{\text{Var}}(\hat{N}_i)$ includes the term for the estimate of $\hat{N}_i$, that is, $\hat{\text{Var}}(\hat{N}_i)$ is proportional to $\hat{N}_i$, the estimate and its variance are strongly correlated. Therefore, a bias would be expected. The values of $\hat{N}_i$ that overestimate the true N would be weighted less than those that underestimate the true N.

The distribution of $\hat{N}_i$ (Eq. 10.1) is known to be skewed, with larger values making the right tail of the distribution heavier than the left tail. Logarithmic transformation will correct this problem to some extent. Thus, another possible estimator is the geometric mean of the k surveys is

$$\hat{N} = \exp\left(\sum_{i=2}^{k+1} \frac{\log \hat{N}_i}{k}\right)$$
$$= \exp(\overline{\log \hat{N}})$$

Confidence intervals can be constructed on $\overline{\log \hat{N}}$ as

$$\overline{\log \hat{N}} \pm t_{(k-1)}\hat{\text{Var}}(\log \hat{N})^{1/2}$$

where the $\hat{\text{Var}}(\log \hat{N})$ is estimated as

$$\hat{\text{Var}}(\log \hat{N}) = \sum_{i=2}^{k+1} \frac{(\log \hat{N}_i - \overline{\log \hat{N}})^2}{k-1}$$

The upper and lower confidence interval endpoints can then be exponentiated to obtain confidence bounds on the value of $\hat{N}$.

In addition to the unweighted geometric mean, a weighted geometric mean and variance can be used to construct the estimate and the confidence interval, where the weight is

$$W_i = \frac{1}{\hat{\text{Var}}(\log \hat{N}_i)}$$

with:

$$\hat{\text{Var}}(\log \hat{N}_i) = \frac{\hat{\text{Var}}(\hat{N}_i)}{\hat{N}_i^2}$$

Median Another possibility is to combine the medians of the five separate estimates into one overall estimate of the population size. The advantage to using the median is that a "poor" estimate does not affect the overall estimate. That is, suppose that the weather turns bad after a survey is started, and the number of animals seen is quite low. The resulting estimate turns out to be considerably larger than the other four. The median of the five estimates is not biased because of this overestimate (Bartmann et al. 1987). Based on rank-order statistics, the largest and the smallest of these five values are a $1 - (2 \times 0.5^5) = 1 - 0.0625$ confidence interval, or approximately a 95% interval (see Conover 1971:110–115).

To provide the rank order statistics, the five estimates of N are ranked from the largest to the smallest. Then the probability is calculated of observing another estimate in the six intervals created by the five estimates. Table 10.1 illustrates the calculation of the probabilities of observing another value of $\hat{N}_i$

TABLE 10.1
Calculation of the Probabilities of Another Estimate Falling in Each of the Intervals Created by Five Ordered Estimates

Order statistic	Formula for interval	Value
	$\binom{k}{5} 0.5^5(1 - 0.5)^{k-5}$	0.03125
N_5		
	$\binom{k}{4} 0.5^4(1 - 0.5)^{k-4}$	0.15625
N_4		
	$\binom{k}{3} 0.5^3(1 - 0.5)^{k-3}$	0.31250
N_3		
	$\binom{k}{2} 0.5^2(1 - 0.5)^{k-2}$	0.31250
N_2		
	$\binom{k}{1} 0.5^1(1 - 0.5)^{k-1}$	0.15625
N_1		
	$\binom{k}{0} 0.5^0(1 - 0.5)^{k-0}$	0.03125

in the intervals created by the five order statistics. With only five estimates ($k = 5$), the probability of observing another estimate greater than the largest estimate (i.e., the largest-order statistic) is

$$\binom{k}{5} 0.5^5 (1 - 0.5)^{k-5} = 0.03125$$

where $\binom{k}{5}$ is a binomial coefficient equal to $k!/(5![k - 5]!)$. The exclamation points mean factorial, that is, $5! = 5 \times 4 \times 3 \times 2 \times 1$, with $0! = 1$. Likewise, the probability of observing another estimate less than the smallest estimate (i.e., the smallest-order statistic) is

$$\binom{k}{0} 0.5^0 (1 - 0.5)^{k-0} = 0.03125$$

Table 10.2 demonstrates the calculations for $k = 12$.

Joint Hypergeometric Maximum Likelihood Estimator Bartmann et al. (1987) suggest a third estimator derived using the maximum likelihood technique. The estimator suggested by Chapman (1951) for the common Lincoln–Petersen survey with only one sighting occasion is a maximum likelihood estimator based on the hypergeometric probability density function. Because the individual sighting occasions are independent, the product of the hypergeometric likelihood functions (Seber 1982:59) can be numerically optimized to estimate N. The maximum likelihood estimator is the value of $\hat{N}$ that maximizes the expression

$$\prod_{i=2}^{k+1} \frac{\binom{n_i}{m_i} \times \binom{\hat{N} - n_1}{n_i - m_i}}{\binom{\hat{N}}{n_i}}$$

where n_i and m_i are the total number of animals observed and the number of marked animals observed, respectively, on occasion i, $i = 2, \ldots, k + 1$. Computer optimization is required to find this value of $\hat{N}$. Bartmann et al. (1987) suggested a 95% confidence interval constructed directly from the likelihood, as suggested by Hudson (1971). Estimates of the lower and upper confidence bounds are the values of $\hat{N}$ that produce values of the log likelihood which are 2 units less than the value of the log likelihood at the maximum. The value "2" derives from the fact that -2 times the differences of two log likelihoods from nested models yields a statistic that follows the χ^2 distribution. In Fig. 10.1, for example, the maximum of the log likelihood is -1.48, so 2

10   Population Estimation

TABLE 10.2
Calculation of the Probabilities of Another Estimate Falling in Each of the Intervals Created by $k = 12$ Ordered Estimates

Order statistic	Formula for interval	Value
	$\binom{k}{12} 0.5^{12}(1 - 0.5)^{k-12}$	0.0002441
N_{12}		
	$\binom{k}{11} 0.5^{11}(1 - 0.5)^{k-11}$	0.0029297
N_{11}		
	$\binom{k}{10} 0.5^{10}(1 - 0.5)^{k-10}$	0.0161133
N_{10}		
	$\binom{k}{9} 0.5^{9}(1 - 0.5)^{k-9}$	0.0537109
N_9		
	$\binom{k}{8} 0.5^{8}(1 - 0.5)^{k-8}$	0.1208496
N_8		
	$\binom{k}{7} 0.5^{7}(1 - 0.5)^{k-7}$	0.1933594
N_7		
	$\binom{k}{6} 0.5^{6}(1 - 0.5)^{k-6}$	0.2255859
N_6		
	$\binom{k}{5} 0.5^{5}(1 - 0.5)^{k-5}$	0.1933594
N_5		
	$\binom{k}{4} 0.5^{4}(1 - 0.5)^{k-4}$	0.1208496
N_4		
	$\binom{k}{3} 0.5^{3}(1 - 0.5)^{k-3}$	0.0537109
N_3		
	$\binom{k}{2} 0.5^{2}(1 - 0.5)^{k-2}$	0.0161133
N_2		
	$\binom{k}{1} 0.5^{1}(1 - 0.5)^{k-1}$	0.0029297
N_1		
	$\binom{k}{0} 0.5^{0}(1 - 0.5)^{k-0}$	0.0002441

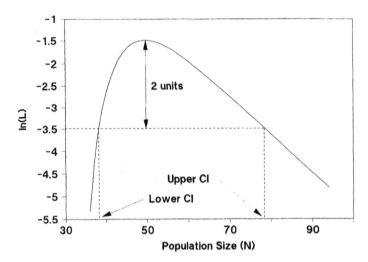

Figure 10.1 Example of the construction of a 95% confidence interval for the population size estimate ($\hat{N}$) from the likelihood function (solid line), using the procedure proposed by Hudson (1971).

units less is -3.48. Drawing a line across the log likelihood at this height and extending down to the x axis from the intersections with the function gives a lower confidence bound of 38 and an upper confidence bound of 79 for the number of individuals in the population. Advantages of this confidence interval are that a lower confidence bound is never estimated which is less than the minimum number of animals known to exist (i.e., the number of radio-collared animals present plus the largest number of uncollared animals seen on any of the surveys).

Example Calculations

Table 10.3 presents an example demonstrating the above calculations. Data were simulated with a true population of $N = 150$. Notice that the lower confidence bounds for the estimates are set to the minimum number of animals ($120 = 48 + 72$), because 48 radioed animals are known to be present and 72 unmarked animals were seen on the third survey (94 animals total minus 22 marked animals seen).

Assumptions

One important assumption for all of the estimators is that the recaptures (sightings) are obtained independently of the markings. Thus, researchers should not consistently observe animals at the trap sites where the radios were applied.

The best strategy is either to pick locations at random to observe animals or to use aerial techniques to observe the animals so that a large proportion of the population can be observed.

Each individual must have the same probability of capture or sighting as every other individual on a particular occasion. The Lincoln–Petersen estimator corresponds to model M_t of Otis et al. (1978) and White et al. (1982). Simulation results (Otis et al. 1978:26–27, 128) indicate that if heterogeneity of capture probabilities exists, then the estimator is biased low. That is, $\hat{N}$ is smaller than N. Most capture–recapture data suggest that heterogeneity of capture probabilities is present.

However, the simulations of Otis et al. (1978) of individual heterogeneity linked capture probabilities across occasions. That is, if an animal had a small capture probability on occasion 1, it also would have a small capture probability on occasion 2. Such a structure is not implicit in the resighting surveys described here. The fact that an animal has a radio is reasonably independent of its sighting probability on later resighting surveys. Sighting probabilities may differ among animals, but we do not expect sighting probabilities to be linked

TABLE 10.3
Example Demonstrating the Five Estimators of
N for $k = 5$ and $n_1 = 48$[a]

i	n_i	m_i	$\hat{N}_i$	SE($\hat{N}_i$)	95% CI
2	85	29	139.5	12.677	114.6–164.3
3	92	34	129.2	9.160	111.2–147.2
4	94	22	201.4	26.199	150.0–252.7
5	97	31	149.1	12.627	124.3–173.8
6	87	24	171.5	20.030	132.2–210.7

Estimator	$\hat{N}$	SE($\hat{N}$)	95% CI
Mean	158.1	12.879	122.4–193.9
Weighted mean	143.3	9.182	120.0–168.8
Geometric mean	156.1	—	125.5–194.3
Weighted geometric mean	146.9	—	121.3–178.0
Median	149.1	—	129.2–201.4
Joint hypergeometric maximum likelihood estimator	157	—	144–174

[a] That is, there are 48 ($= n_1$) radioed animals, and five separate resighting surveys were conducted. The number of animals sighted on each survey is n_i, m_i of them being marked animals. SE, standard error; CI, confidence interval.

to the probability of carrying a radio. Individual heterogeneity of sighting probabilities should not bias estimates of N (Seber 1970), but will cause the variance of the estimate to be underestimated.

In contrast to the above, the majority of estimates of mule deer from Bartmann et al. (1987) were smaller than the true population values (known number of animals in an enclosure), suggesting a negative bias. Thus, we caution users that heterogeneity may cause some bias in the estimates.

An additional assumption for the above techniques is that the k estimates are independent. That is, one observer cannot conduct resightings at the same water hole for 5 consecutive days and use these estimates to generate the estimate, because he probably saw the same individuals each day. Each animal in the population should have the same probability as any other animal of being sighted on a particular day. Thus, the importance of random locations for observation is again noted. If segments of the population have different sighting probabilities, both the point and interval estimates will be biased (Otis et al. 1978:26–27).

Preferred Estimator

Which of the six estimators is the best one to use? Bartmann et al. (1987) concluded that the joint hypergeometric maximum likelihood estimator was preferred over the arithmetic mean or the median, based on field studies using helicopter surveys to locate mule deer. The deer were confined in enclosures, so the true population was known. Although the accuracy of the maximum likelihood estimator did not appear to be better than the other two, the confidence interval length was shorter, while still providing good coverage.

To answer the question of which estimator to use, computer simulations are needed. The statistical analysis system (SAS) was used to perform 1000 replications of 12 cases: three proportions of marked animals (0.1, 0.2, and 0.3) times four sighting probabilities (0.2, 0.4, 0.6, and 0.8). Numbers of animals marked and sighted were generated with the random binomial function of SAS. The true population was $N = 150$, with $k = 5$ surveys. Six estimators were evaluated. For the four estimators based on means and the median, the lower confidence bound was set to the minimum number of animals when the estimated lower bound was less than this value. The SAS code used to perform the above simulations is included in Appendix 9 to demonstrate the computations necessary to calculate the joint hypergeometric maximum likelihood estimator and to provide programming details of the simulation. The simulation required approximately 50 hours of computer time on an IBM personal computer/AT with an 80287 coprocessor. The results are given in Table 10.4.

TABLE 10.4
Results of Monte Carlo Simulations to Examine the Bias and Confidence Interval Coverage of the Five Estimators for Combining Lincoln–Petersen Estimates.[a]

Probability of		Estimator	Average $\hat{N}$	Average CI length	Coverage (%)
Marking	Sighting				
0.1	0.2	Arithmetic mean	142.71	145.38	84
		Weighted arithmetic mean	101.63	70.67	30
		Geometric mean	132.56	137.69	86
		Weighted geometric mean	115.45	100.79	73
		Median	127.91	148.48	84
		Joint hypergeometric MLE	156.08	164.12	95
	0.4	Arithmetic mean	149.46	103.68	91
		Weighted arithmetic mean	122.49	61.21	56
		Geometric mean	143.85	99.34	93
		Weighted geometric mean	130.36	76.54	82
		Median	141.01	104.58	91
		Joint hypergeometric MLE	152.94	93.62	96
	0.6	Arithmetic mean	150.17	69.05	91
		Weighted arithmetic mean	132.09	44.20	66
		Geometric mean	146.75	67.19	92
		Weighted geometric mean	135.92	51.47	81
		Median	145.42	70.49	91
		Joint hypergeometric MLE	151.57	61.20	94
	0.8	Arithmetic mean	149.68	39.26	93
		Weighted arithmetic mean	137.79	20.01	61
		Geometric mean	149.11	40.53	94
		Weighted geometric mean	138.79	22.28	66
		Median	147.22	42.37	92
		Joint hypergeometric MLE	150.20	35.34	96
0.2	0.2	Arithmetic mean	151.18	116.53	91
		Weighted arithmetic mean	122.15	69.38	59
		Geometric mean	143.99	109.85	91
		Weighted geometric mean	131.20	85.95	85
		Median	141.39	117.20	91
		Joint hypergeometric MLE	155.29	102.71	95
	0.4	Arithmetic mean	150.75	70.71	94
		Weighted arithmetic mean	135.88	52.60	78
		Geometric mean	147.81	70.09	94
		Weighted geometric mean	139.97	60.56	90
		Median	146.47	70.53	93
		Joint hypergeometric MLE	151.95	58.19	95

(*continues*)

TABLE 10.4
(Continued)

Probability of Marking	Probability of Sighting	Estimator	Average $\hat{N}$	Average CI length	Coverage (%)
	0.6	Arithmetic mean	150.51	48.23	95
		Weighted arithmetic mean	141.22	38.70	85
		Geometric mean	149.32	46.44	94
		Weighted geometric mean	143.89	41.32	91
		Median	148.22	48.34	94
		Joint hypergeometric MLE	150.87	38.31	95
	0.8	Arithmetic mean	150.44	28.24	95
		Weighted arithmetic mean	144.72	21.81	85
		Geometric mean	149.51	28.32	95
		Weighted geometric mean	145.13	23.19	86
		Median	149.66	29.21	94
		Joint hypergeometric MLE	150.64	23.59	95
0.3	0.2	Arithmetic mean	149.20	87.38	91
		Weighted arithmetic mean	129.37	60.50	68
		Geometric mean	146.16	88.10	94
		Weighted geometric mean	136.15	71.92	88
		Median	143.43	86.72	90
		Joint hypergeometric MLE	151.35	72.92	95
	0.4	Arithmetic mean	149.89	53.46	94
		Weighted arithmetic mean	140.25	44.10	84
		Geometric mean	148.41	53.61	93
		Weighted geometric mean	143.13	48.20	91
		Median	147.70	53.15	93
		Joint hypergeometric MLE	150.60	43.62	94
	0.6	Arithmetic mean	150.02	36.47	96
		Weighted arithmetic mean	144.31	31.58	92
		Geometric mean	149.66	35.02	94
		Weighted geometric mean	146.33	32.42	92
		Median	148.91	36.27	95
		Joint hypergeometric MLE	150.31	28.97	95
	0.8	Arithmetic mean	150.07	21.32	95
		Weighted arithmetic mean	146.71	18.09	89
		Geometric mean	149.62	21.32	95
		Weighted geometric mean	147.07	18.99	92
		Median	149.58	21.84	94
		Joint hypergeometric MLE	150.16	17.80	94

The true population was 150, with $k = 5$ surveys. The expected confidence interval coverage was 95%.

Based on these simulations, the joint hypergeometric maximum likelihood estimator would be preferable to the other five estimators. This estimator was only slightly biased, and the confidence interval coverage was very close to the expected 95% level. Further, the length of the average confidence interval was generally smaller than the other five estimators when they had coverage around 95%. Often, confidence interval length of the joint hypergeometric maximum likelihood estimator was shorter than the other estimators, even when they did not have 95% coverage. Thus, the minimum variance property of the maximum likelihood estimator was demonstrated. The weighted estimators both performed poorly, indicating that the correlation between the estimate and the estimated variance was strong enough to seriously bias the weighted mean. The unweighted geometric mean was biased low, much more so than the unweighted arithmetic mean. Thus, the arithmetic mean would be preferred over the geometric mean. The median was also much more biased than the arithmetic mean, although average confidence interval length and confidence interval coverage were about the same as the arithmetic mean. Therefore, we conclude that if the joint hypergeometric maximum likelihood estimator cannot be used, the arithmetic mean would provide the next best alternative.

One caution about these simulations: Each of the five replicate surveys was given the same probability of sighting. That is, the five surveys were assumed to have equal effort (i.e., the same amount of flight time). Should this not be the case, the relative performances of the six estimators might change. From a small set of simulations using such conditions, J. Chafota (personal communication) has shown that the performance of the joint hypergeometric maximum likelihood estimator is nearly the same as shown here. However, additional simulations would be required to explore this situation further.

Sample Size Determination

A major problem with the proposed approach is that too few animals may be marked to generate reliable population estimates, and/or too few animals may be observed in the second sample. The simulations in Table 10.4 demonstrate how much more variable the estimates would be when low numbers of animals are marked and/or resighted. A smaller number of radio collars could be tolerated if a very large sample of observed animals was collected. The results in Table 10.4 provide information on the proportion of the population to mark and resight as a function of the expected confidence interval length for a population of 150 animals. Simulation of the proposed experiment can be performed using the SAS code in Appendix 9 to provide expected results and, hence, prevent expenditure of money on an effort likely to produce unreliable results. That is, if an adequate number of animals cannot be marked, or if an inadequate number of animals is expected to be resighted, then the project should not be started.

The sample size calculations of Jensen (1981) are useful, although they only apply to a single Lincoln–Petersen estimate. Jensen's formulas provide methods of calculating the number of animals to radio mark (n_1) and the number of animals to observe during the resighting survey (n_i) for the ith survey, given an initial estimate of the population size (number of animals), the cost of marking animals (C_1), the cost of resighting animals (C_i), and a desired confidence interval ($\pm B$) in units of numbers of animals. Then

$$n_1 = \left[\frac{4 N^3 C_i}{B^2 C_1}\right]^{1/2}$$

$$n_i = \left[\frac{4 N^3 C_1}{B^2 C_i}\right]^{1/2}$$

The variance of the mean of k independent surveys is

$$\sum_{i=2}^{k+1} \frac{\hat{\text{Var}}(\hat{N}_i)}{k^2}$$

Therefore, these formulas provide a method of determining expected confidence interval width as a function of cost for the arithmetic mean estimator. For the case of constant C_i, and hence constant n_i, the confidence interval width is $B/\sqrt{k}$. Based on the simulation results in Table 10.4, the results based on the arithmetic mean would probably provide a good approximation for use with the joint hypergeometric maximum likelihood estimator.

Rice and Harder (1977) presented graphs which have proved useful in designing surveys with repeated captures. However, these graphs only give a range of population marked from 0% to 25%, which, for many studies, is below the minimum level needed to obtain accurate estimates of small populations. For example, Bartmann et al. (1987) recommend that at least 45% of small populations (in their case, $N < 100$) should be marked before reliable estimates and confidence intervals can be obtained. This requirement may nullify the usefulness of this approach in many situations.

Another consideration of the techniques discussed here is the duration of the study being conducted. For short-term research studies the techniques may be appropriate, because precise results can be achieved even if the expense is large. In contrast, for long-term monitoring studies, the techniques may be too expensive and or too labor intensive to produce the quality of data necessary.

That is, less precise, but also less expensive, methods may suffice to detect trends in the population.

Line Transects

Line transect population estimation methods are a promising technique for estimating wildlife populations. A comprehensive review has been provided by Burnham et al. (1980). The two main assumptions of line transect methods are that animals that are on the line are *always* seen, and that animals do not move away from (or closer to) the line prior to the observer's seeing them. Radio-tracking can be used to check both of these assumptions.

It is a straightforward task to verify that a radio-marked animal standing on the line is not disturbed and moved away from the line prior to the observer seeing it. That an animal on or near the line is observed can also be verified by a second observer carrying a receiver.

Aerial Surveys

Aerial surveys are commonly used as a population estimation technique for big game populations (see Kufeld et al. 1980). The critical assumption of these surveys is that all of the animals in a quadrat are counted. Given that this is probably not possible, radioed animals can be used to estimate the proportion of animals not seen. An example of this methodology is provided by Biggins and Jackson (1982). In their experiment the radioed animals were used to estimate the proportion of animals observed from the air, in the same manner as in the capture–recapture experiments described earlier.

Summary

1. The main application of telemetry for population estimation is in capture–recapture studies, in which the radioed animals are the marked individuals in the population. Because of the radios, the number of marked animals is known exactly, so that the assumptions of the capture–recapture estimators are met.
2. Although the unweighted arithmetic mean of several Lincoln–Petersen estimates performs well, the joint hypergeometric estimator performs even better and is the estimator we recommend. Estimators based on weights constructed from the variance of the estimate do not perform

well for these problems because the weight is estimated and a strong correlation exists between the estimate and its weight, and this causes bias in the result.

References

Bartmann, R. M., G. C. White, L. H. Carpenter, and R. A. Garrott. 1987. Aerial mark-recapture estimates of confined mule deer in pinyon-juniper woodland. J. Wildl. Manage. 51:41–46.

Biggins, D. E. and M. R. Jackson. 1982. Biases in aerial surveys of mule deer. Pages 60–65 in R. D. Comer et al. eds. Issues and Technology in the Management of Impacted Western Wildlife. Thorne Ecological Institute, Boulder, CO.

Burnham, K. P., D. R. Anderson, and J. L. Laake. 1980. Estimation of density from line transect sampling of biological populations. Wildl. Monogr. 72:1–202.

Chapman, D. G. 1951. Some properties of the hypergeometric distribution with applications to zoological sample censuses. Univ. Calif. Berkeley, Publ. Stat. 1:131–160.

Conover, W. J. 1971. Practical Nonparametric Statistics. Wiley, New York. 462 pp.

Cormack, R. M. 1968. The statistics of capture-recapture methods. Oceanogr. Mar. Biol. Annu. Rev. 6:455–506.

Hudson, D. J. 1971. Interval estimation from the likelihood function. J. R. Stat. Soc. B 33:256–262.

Jensen, A. L. 1981. Sample sizes for single mark and single recapture experiments. Trans. Am. Fish. Soc. 110:455–458.

Kufeld, R. C., J. H. Olterman, and D. C. Bowden. 1980. A helicopter quadrat census for mule deer on Uncompahgre Plateau, Colorado. J. Wildl. Manage. 44:632–639.

Otis, D. L., K. P. Burnham, G. C. White, and D. R. Anderson. 1978. Statistical inference from capture data on closed animal populations. Wildl. Monogr. 62:1–135.

Rice, W. R. and J. D. Harder. 1977. Application of multiple aerial sampling to a mark-recapture census of white-tailed deer. J. Wildl. Manage. 41:197–206.

Seber, G. A. F. 1970. The effect of trap response on tag recapture estimates. Biometrics 26:13–22.

Seber, G. A. F. 1973. Estimation of Animal Abundance and Related Parameters. Griffin, London. 506 pp.

Seber, G. A. F. 1982. Estimation of Animal Abundance and Related Parameters, 2nd ed. Griffin, London. 654 pp.

White, G. C., D. R. Anderson, K. P. Burnham, and D. L. Otis. 1982. Capture-recapture and removal methods for sampling closed populations, LA-8787-NERP. Los Alamos Nat. Lab, Los Alamos, NM. 235 pp.

CHAPTER

11

Data Analysis System

Radio-tracking is one of the most commonly used techniques in wildlife management and research. As pointed out in earlier chapters, radio-tracking now provides tools to investigate all major aspects of wildlife sciences. Physiological parameters such as heart rate and internal and external body temperature can now be measured on telemetered animals. Electronic systems to continuously monitor the movements and physiological parameters of telemetered animals are feasible. As a result, tremendous amounts of data become available to the researcher, and hence, the use of a computer is required to process this data.

Since the beginning of wildlife radio-tracking, there has been a proliferation of computer programs to handle various phases of the analysis of radio-tracking data. However, most of these programs are specific to a local application. A general computer system is needed to handle all phases of the data analysis.

In this chapter the design and the beginning of the implementation of a system for the analysis of radio-tracking data are described. The general use of radio-tracking suggests that a data analysis system is needed to prevent duplication of effort and to help the researcher to fully analyze his/her data.

Need for Analysis System for Radio-Tracking Data

At this time no general package for data analysis is available for radio-tracking data. Most researchers who use computers to process radio-tracking data have one or more small programs to meet very specific needs (e.g., Kenward 1987). Several sets of programs are also available that provide a limited analysis of radio-tracking data, usually emphasizing one particular aspect of the analysis. For example, a set of programs available from James Dunn, University of Arkansas, Fayetteville, provides a home range estimation capability, utilizing the methods developed by Dunn and Gipson (1977) and by Dunn (1978a, b, 1979). The Cedar Creek analysis package is perhaps the most comprehensive package available, but still has several limitations, such as a lack of portability (CDC only), base map facilities, and quality graphics output.

The lack of a comprehensive analysis package also leads to a lack of standardized treatment of telemetry data. For example, data are not carefully checked for outliers or other errors because the software needed to do this is not available. Adequate graphics presentations are also frequently not made because of the lack of software. Researchers tend to perform the minimum analysis necessary to achieve their original goal. A comprehensive package would allow researchers to fully analyze their data with recent and relevant methods.

A general data analysis package would be cost effective because of the current duplication of software and the tendency not to thoroughly analyze the data collected. A data analysis package with general capabilities would do away with the need for each researcher to have his/her own program to do limited data analyses. Likewise, the general features of the package would allow the researcher to more fully analyze his/her data than if such a data analysis system were not available.

Development of radio-tracking systems to measure auxiliary variables such as heart rate, body temperature, and other physiological functions at the same time as the animal's location is determined provides a further need in data analysis. Future radio-tracking systems will allow researchers to measure an animal's Hutchinsonian niche. Means of analyzing such data sets are only now being considered by biologists. A system for data analysis that provides the necessary software to address these types of questions will be used heavily as more and more researchers use radio-tracking to address questions about an animal's physiological function in its natural environment.

The need for software to analyze maps is further justification for developing a data analysis system for radio-tracking data. The mapping and overlay statistical system (MOSS), developed by the Western Energy and Land Use Team, U. S. Fish and Wildlife Service, provides a sophisticated package for the analysis of maps, as do many other geographic information systems. However, methods to use a system such as MOSS to analyze an animal's movements

11 Data Analysis System

are not available. The thorough integration of a map analysis system with an analysis system for radio-tracking data would provide a powerful package for the interpretation of an animal's movements relative to habitat types, elevation, or other geographical features.

The recent development of high-quality graphics terminals and graphics systems is a further justification for a analysis system for radio-tracking data. Most researchers who analyze radio-tracking data do not understand graphics software packages well enough to fully utilize their capabilities. An analysis system that could produce quality graphics output on an interactive terminal would assist researchers in understanding their data. The capability to produce film or other types of hard-copy graphics would also assist a researcher in presenting information to other scientists.

Data Analysis System Design

In this section some general thoughts are presented concerning the capabilities of an analysis system for radio-tracking data. The comprehensive analysis of radio-tracking data requires at least the following capabilities: (1) conversion of azimuths to coordinates, (2) detection of outliers and errors, (3) editing of data, (4) query language, (5) base maps, (6) graphics, (7) area estimation, (8) hypothesis testing, and (9) pattern recognition.

Much of the radio-tracking data taken by researchers is through triangulation. Programs for the conversion of the bearings to point and area estimates of location should be included in the package. Procedures should be available to estimate the sampling variance of bearings taken on radios at known locations.

Several methods are available to detect outliers or errors in the data. First, the locations can be checked to see whether they are in a specified area. Second, the rate of movement between fixes can be estimated to see whether it is a reasonable value. Both of these approaches are utilized by the Cedar Creek program CHECKF to determine the validity of locations. These capabilities are incorporated in the BIOCHECK program described in Appendix 3.

If a bad fix is located, the user must be able to either delete or correct the location, and hence, an editing capability must be provided. Both interactive editing and files of replacement values would be useful. The capability to replace or delete locations from a plot at a graphics terminal would be especially useful.

The query language of statistical packages such as SAS is useful to select or reject groups of locations from the data file. All fixes for one animal or for one time period are common applications. Also, the ability to reject fixes based on various criteria is provided with the WHERE feature of SAS. Performing the same analysis on different parts of the data (i.e., by animal or by month) is provided with the BY capabilities of SAS.

The capability of analyzing data relative to a map is necessary. A common application would be to determine habitat utilization by telemetered animals. The ability to match animal movements to map characteristics would be necessary. A geographic information system should be incorporated into the analysis package for radio-tracking data.

Quality graphics presentations are another critical requirement of the analysis system for radio-tracking data. Both still plots and movies should be provided. Particularly important is the capability to plot movements on various base maps the user may enter. By far the most effective method of displaying an animal's movements is to present them on carefully selected maps.

Methods to estimate the area utilized by telemetered animals must be included in the analysis system for radio-tracking data. The general methods commonly used in the past include convex polygons (minimum area methods), center of activity methods (Dunn and Gipson 1977, Jennrich and Turner 1969, Hayne 1949), and grid cell methods (Siniff and Tester 1965). Extensions and updates discussed in Chapter 7 should be included in the analysis system.

Methods of testing hypotheses about an animal's utilization of space would be necessary. Multivariate ANOVA, Dunn's (1979) approach to testing for changes in home ranges, the MRPP procedure, and χ^2 methods based on grid counts would be needed in the data analysis system. A pattern recognition routine would also be needed to study activity records or to recognize patterns in the movements of individuals or groups of individuals.

This brief list of potential features of the data analysis system could be greatly extended by interaction with researchers who have radio-tracking data. These researchers could suggest many types of analyses that are not foreseeable at this time. Hence, an additional requirement of the system is the capability to easily add new analysis procedures.

Current Directions

We originally envisioned a separate data analysis system for radio-tracking data analysis. Later, our thinking changed to including the necessary programs as part of a larger statistical package such as SPSS (Nie et al. 1975, SPSS Inc. 1983).

Currently, we have chosen to shift our emphasis from the use of SPSS to SAS (SAS Institute Inc. 1985). Two reasons for this shift in direction are

(1) SAS is a more powerful data analysis system because of the extensive programming capabilities in the DATA step (demonstrated throughout this text) and (2) we expect SAS will be more able to include new PROCs than SPSS. Thus, throughout this text we supply SAS programs to solve the data analysis problems we discuss. Although in numerous cases stand-alone programs are also discussed, we believe that the need for these separate codes will disappear as the capabilities of SAS improve on personal computers.

Summary

An extensive data analysis system is needed for the analysis of all aspects of radio-tracking data. Currently, we believe that SAS is the best available system for the general analysis of radio-tracking data, but some of the stand-alone programs are still needed to fulfill all of the needs for this type of analysis. We expect that in the near future these stand-alone programs can be incorporated into the SAS environment.

References

Dunn, J. E. 1978a. Computer programs for the analysis of radio telemetry data in the study of home range, Stat. Lab. Tech. Rep. No. 7. Univ. of Arkansas, Fayetteville. 73 pp.

Dunn, J. E. 1978b. Optimal sampling in radio telemetry studies of home range. Pages 53–70 in H. H. Shugart, Jr. ed. Time Series and Ecological Processes. SIAM, Philadelphia, PA.

Dunn, J. E. 1979. A complete test of dynamic territorial interaction. Pages 159–169 in F. M. Long ed. Proc. 2nd Int. Conf. Wildl. Biotelemetry. Univ. of Wyoming, Laramie.

Dunn, J. E. and P. S. Gipson. 1977. Analysis of radio telemetry data in studies of home range. Biometrics 33:85–101.

Hayne, D. W. 1949. Calculation of size of home range. J. Mammal. 30:1–18.

Jennrich, R. I. and F. B. Turner. 1969. Measurement of non-circular home range. J. Theor. Biol. 232:227–237.

Kenward, R. E. 1987. Wildlife Radio Tagging. Academic Press, San Diego, CA. 222 pp.

Nie, N. H., C. H. Hull, J. G. Jenkins, K. Steinbrenner, and D. H. Bent. 1975. SPSS—Statistical Package for the Social Sciences, 2nd ed. McGraw-Hill, New York. 675 pp.

SAS Institute Inc. 1985. SAS® Language Guide for Personal Computers, Version 6 Edition. SAS Institute Inc., Cary, NC. 429 pp.

SPSS Inc. 1983. SPSS^x User's Guide. McGraw-Hill, New York. 806 pp.

Siniff, D. B. and J. R. Tester. 1965. Computer analysis of animal movement data obtained by telemetry. BioScience 15:104–108.

APPENDIX

1

Introduction

This appendix describes a set of programs implemented on the IBM PC and compatibles for the analysis of biotelemetry data. Program FIELDS formats a data file for input to the other programs. Program BIOCHECK checks a set of data for reasonable distances between consecutive locations, and a reasonable rate of movement between consecutive locations. Program BIOPLOT plots telemetry locations on the monitor of a PC. Program HOMER provides home range estimates for a variety of methods. The inputs for each of these programs are described in Appendices 2–5, respectively.

All the programs assume that biotelemetry data are entered from a specially formatted file, with x and y coordinates and time (t) for each case. In addition, an animal identification variable and age and sex codes are also often required or at least desirable. Hence, the basic biotelemetry observation should consist of the following variables:

1. XCOOR—x coordinate in meters (usually the UTM easterling coordinate).
2. YCOOR—y coordinate in meters (usually the UTM northerling coordinate).
3. TIME—calendar date and time in the form yymmdd.hhmm, where yy is the last two digits of the year, mm is the month (01–12), dd is the day of the month (01–31), the decimal (.) separates the day from the

fraction of a day, hh is hours in a 24-hour clock format (00–23), and mm the minutes (00–59).
4. ID—animal identification variable. Often this variable will be the frequency number, with either numeric or alphanumeric variables allowed. ID is also useful for creating subfiles, so that analyses can be conducted for each individual animal.
5. SEX—sex of the animal, usually coded as integer values but letters (i.e., M and F) can also be used.
6. AGE—age of the animal, again usually coded as integer values.

Graphic output is provided through Lotus 1-2-3 and Freelance. Lotus can be used to input files created in HOMER and will output plots that can be input to Freelance. Freelance is a powerful graphics editor, allowing the user to easily manipulate an image on the screen of the PC. Further, Freelance interfaces to a large number of graphics output devices, allowing the user freedom in both the type of product which can be prepared and in the use of whatever output device is available.

Conventions Used in this Appendix

The following conventions are observed in this appendix.

- Upper case words and letters, used in examples, indicate that you should type the word or letter exactly as shown.
- Lower case words and letters, used in syntax examples, indicate that you are to substitute a word or value of your choice.
- Brackets ([ ]) indicate optional elements.
- Braces ({ }) enclose lists from which one element is to be chosen.
- An ellipsis (. . .) indicates that the preceding item(s) can be repeated one or more times.

Example

Listings A1.1 to A1.3 illustrate the analysis of a typical set of biotelemetry data with the programs described in Appendices 2–5. These data are from an adult female mule deer tracked for two 12-hr periods per week over two summers on the Roan Plateau in northwestern Colorado. Observations were taken from a 3-tower triangulation system every 30 min, although some observations are missing because the bearings did not intersect closely enough to indicate a good location estimate. The data are analyzed as two subfiles, one for each

Appendix 1

summer. Listing A1.1 is the input file to FIELDS to prepare the input file for BIOPLOT, BIOCHECK, and HOMER. Listing A1.2 is the output file from BIOCHECK and Listing A1.3 is the output file from HOMER showing the results of each of the home range analyses. Figure A1.1 is the resulting plots produced from Lotus and Freelance. These plots have not been changed in Freelance, to represent the less sophisticated graphics produced by Lotus 1-2-3.

LISTING A1.1
Input file for Program FIELDS to create the input
for BIOPLOT, BIOCHECK, and HOMER.

The contents of the file FIELDS.DAT are:

Columns	Contents
1–11	date (*yymmdd*) followed by time (*hhmm*)
13–19	Animal identification
21	Animal age (1 = Fawn, 2 = Yearling, 3 = Adult)
23	Animal sex (1 = female, 2 = male)
25–31	*x* UTM coordinate
33–39	*y* UTM coordinate
41–47	Triangulation confidence ellipse size (ha)
49–57	Habitat code

The response to the input request from FIELDS to produce the output file was:

```
i=fields.dat o=biocheck.dat subfiles=2
1-11 13-19 21-21 23-23 25-31 33-39
^Z
```

so that the resulting file contains the TIME in field 1, ID in 2, AGE in 3, SEX in 4, XCOOR in 5, and YCOOR in 6.

```
840606.1219 Y84.210 3 1 0747137 4391208 0.22395 512US
840606.1351 Y84.210 3 1 0747137 4391208 0.22395 512US
840606.1405 Y84.210 3 1 0747137 4391208 0.22395 512US
840606.1438 Y84.210 3 1 0747131 4391198 0.22243 512US
840606.1740 Y84.210 3 1 0747022 4390855 0.36530 523SGHR
840606.1807 Y84.210 3 1 0747119 4390899 0.29264 523SGHR
840606.1838 Y84.210 3 1 0747136 4390919 0.29877 523SGHR
840606.2040 Y84.210 3 1 0747113 4390870 0.34206 523SGHR
840609.0017 Y84.210 3 1 0746985 4390993 0.19531 512US
```

Analysis of Wildlife Radio-Tracking Data

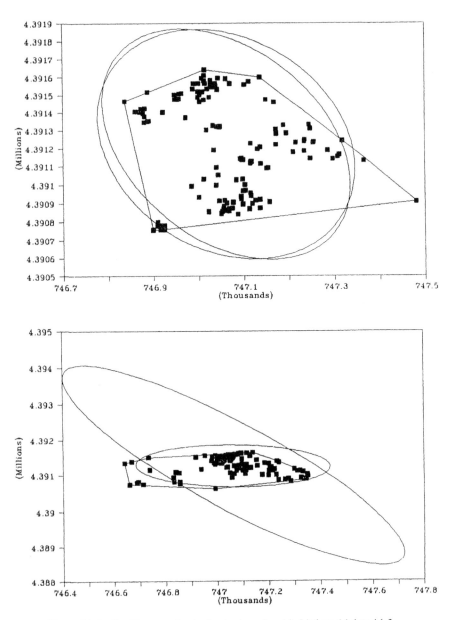

Figure A1.1 Graphic output for the 2 animals analyzed in Listings A1.1 to A1.3.

Appendix 1

```
840609.0107 Y84.210 3 1 0747095 4390924 0.27397 523SGHR
840609.0135 Y84.210 3 1 0747087 4390841 0.32381 523SGHR
840609.0203 Y84.210 3 1 0746924 4390760 0.37288 523SGHR
840609.0233 Y84.210 3 1 0746924 4390781 0.35308 523SGHR
840609.0304 Y84.210 3 1 0746916 4390780 0.35503 523SGHR
840609.0332 Y84.210 3 1 0746915 4390759 0.37424 523SGHR
840609.0408 Y84.210 3 1 0746899 4390757 0.37863 523SGHR
840609.0432 Y84.210 3 1 0746909 4390799 0.34135 532SG/US
840609.0505 Y84.210 3 1 0746908 4390779 0.35751 523SGHR
840609.1006 Y84.210 3 1 0747021 4391535 0.23127 411AS
840609.1034 Y84.210 3 1 0747000 4391520 0.21552 411AS
840609.1102 Y84.210 3 1 0747043 4391532 0.24247 411AS
840609.1131 Y84.210 3 1 0746989 4391536 0.22374 411AS
840623.1209 Y84.210 3 1 0746862 4391407 0.13478 512US
840623.1234 Y84.210 3 1 0746883 4391424 0.13966 411AS
840623.1303 Y84.210 3 1 0746874 4391422 0.14225 411AS
840623.1333 Y84.210 3 1 0746879 4391397 0.12046 512US
840623.1506 Y84.210 3 1 0746883 4391424 0.13966 411AS
840623.1612 Y84.210 3 1 0746838 4391466 0.19133 533SG/USHR
840623.1632 Y84.210 3 1 0746889 4391518 0.20852 533SG/USHR
840623.1903 Y84.210 3 1 0747150 4391473 0.27795 512US
840626.0136 Y84.210 3 1 0747053 4390856 0.30307 523SGHR
840626.0204 Y84.210 3 1 0747075 4390873 0.29073 523SGHR
840626.0235 Y84.210 3 1 0747068 4390871 0.28934 523SGHR
840626.0334 Y84.210 3 1 0747067 4390858 0.30063 523SGHR
840626.0408 Y84.210 3 1 0747056 4390883 0.28225 523SGHR
840626.0434 Y84.210 3 1 0747055 4390869 0.28804 523SGHR
840626.0505 Y84.210 3 1 0747098 4390895 0.29175 523SGHR
840626.0535 Y84.210 3 1 0747044 4391052 0.18183 512US
840626.0705 Y84.210 3 1 0747013 4391474 0.19240 411AS
840626.0740 Y84.210 3 1 0746973 4391373 0.21451 512US
840626.0809 Y84.210 3 1 0747025 4391489 0.35609 411AS
840626.0837 Y84.210 3 1 0746922 4391402 0.16287 512US
840626.0906 Y84.210 3 1 0746964 4391511 0.28030 411AS
840626.0936 Y84.210 3 1 0746958 4391480 0.25457 411AS
840626.1004 Y84.210 3 1 0746954 4391478 0.24874 411AS
840626.1036 Y84.210 3 1 0746958 4391508 0.27428 411AS
840626.1104 Y84.210 3 1 0746949 4391476 0.24326 411AS
840626.1135 Y84.210 3 1 0746949 4391476 0.24326 411AS
840709.1216 Y84.210 3 1 0747137 4391599 0.58024 522SG
840709.1239 Y84.210 3 1 0747103 4391559 0.28840 411AS
840709.1302 Y84.210 3 1 0747113 4391576 0.29562 411AS
```

```
840709.1333 Y84.210 3 1 0747009 4391596 0.26694 533SG/USHR
840709.1402 Y84.210 3 1 0746991 4391565 0.24410 533SG/USHR
840709.1433 Y84.210 3 1 0747028 4391582 0.28575 533SG/USHR
840709.1508 Y84.210 3 1 0747014 4391610 0.27279 533SG/USHR
840709.1533 Y84.210 3 1 0747014 4391580 0.26106 533SG/USHR
840709.1706 Y84.210 3 1 0747015 4391641 0.45009 533SG/USHR
840709.2005 Y84.210 3 1 0747365 4391130 0.35818 512US
840712.0251 Y84.210 3 1 0747136 4390873 0.33193 523SGHR
840712.0306 Y84.210 3 1 0747060 4390862 0.32548 523SGHR
840712.0336 Y84.210 3 1 0747050 4390845 0.32682 523SGHR
840712.0709 Y84.210 3 1 0747039 4390998 0.31061 512US
840712.1003 Y84.210 3 1 0747001 4390933 0.30871 512US
840712.1035 Y84.210 3 1 0747058 4390916 0.26210 512US
840712.1106 Y84.210 3 1 0747063 4390901 0.27245 523SGHR
840712.1131 Y84.210 3 1 0747098 4391031 0.22422 512US
840801.1936 Y84.210 3 1 0747034 4391192 0.12951 512US
840801.2034 Y84.210 3 1 0747247 4391142 0.30562 512US
840804.0704 Y84.210 3 1 0747172 4391306 0.25363 512US
840804.0743 Y84.210 3 1 0747071 4391564 0.26750 411AS
840804.0802 Y84.210 3 1 0746948 4391503 0.26326 411AS
840804.0903 Y84.210 3 1 0747016 4391566 0.26126 533SG/USHR
840804.0947 Y84.210 3 1 0747032 4391546 0.37753 411AS
840804.1007 Y84.210 3 1 0747028 4391570 0.37893 533SG/USHR
840804.1104 Y84.210 3 1 0747028 4391551 0.24346 411AS
840804.1130 Y84.210 3 1 0747038 4391564 0.25503 411AS
840817.1212 Y84.210 3 1 0746889 4391352 0.10407 512US
840817.1231 Y84.210 3 1 0746891 4391353 0.10501 512US
840817.1432 Y84.210 3 1 0746881 4391347 0.10042 512US
840817.1501 Y84.210 3 1 0746891 4391353 0.10501 512US
840817.1530 Y84.210 3 1 0746872 4391404 0.15332 512US
840817.1600 Y84.210 3 1 0746875 4391406 0.15431 512US
840817.1630 Y84.210 3 1 0746877 4391407 0.15541 512US
840817.1700 Y84.210 3 1 0746882 4391379 0.12951 512US
840817.1804 Y84.210 3 1 0747168 4391459 0.28538 512US
840817.1835 Y84.210 3 1 0747247 4391311 0.32602 512US
840817.1902 Y84.210 3 1 0747318 4391243 0.33100 512US
840817.1931 Y84.210 3 1 0747235 4391249 0.29070 512US
840817.2034 Y84.210 3 1 0747216 4391233 0.28026 512US
840817.2130 Y84.210 3 1 0747254 4391232 0.30405 512US
840817.2200 Y84.210 3 1 0747234 4391241 0.30175 512US
840817.2232 Y84.210 3 1 0747255 4391237 0.29984 512US
840820.0206 Y84.210 3 1 0747103 4391000 0.23936 512US
```

Appendix 1

```
840820.0336  Y84.210  3 1  0747084  4391026  0.21796  512US
840820.0404  Y84.210  3 1  0747115  4390956  0.27230  523SGHR
840820.0434  Y84.210  3 1  0747104  4390970  0.25466  523SGHR
840820.0503  Y84.210  3 1  0747099  4390969  0.25340  523SGHR
840820.0532  Y84.210  3 1  0747116  4390942  0.27283  523SGHR
840820.0603  Y84.210  3 1  0747094  4390939  0.26325  523SGHR
840820.0632  Y84.210  3 1  0747126  4390915  0.29040  523SGHR
840820.0704  Y84.210  3 1  0747157  4390907  0.30827  523SGHR
840820.0735  Y84.210  3 1  0747077  4390905  0.27607  523SGHR
840820.0808  Y84.210  3 1  0747046  4390913  0.26150  512US
840820.0838  Y84.210  3 1  0747012  4391031  0.18445  512US
840820.0905  Y84.210  3 1  0747040  4391095  0.15921  531SG/US
840820.0933  Y84.210  3 1  0747118  4391126  0.21275  512US
840820.1007  Y84.210  3 1  0747099  4391136  0.20519  512US
840820.1117  Y84.210  3 1  0747004  4391462  0.18231  411AS
840820.1137  Y84.210  3 1  0747039  4391549  0.39020  411AS
840828.1217  Y84.210  3 1  0747045  4391595  0.27908  533SG/USHR
840828.1241  Y84.210  3 1  0747062  4391592  0.28339  533SG/USHR
840828.1302  Y84.210  3 1  0746990  4391566  0.24223  533SG/USHR
840828.1331  Y84.210  3 1  0747002  4391566  0.24680  533SG/USHR
840828.1402  Y84.210  3 1  0746996  4391566  0.24433  533SG/USHR
840828.1429  Y84.210  3 1  0747010  4391519  0.21982  411AS
840828.1518  Y84.210  3 1  0747004  4391503  0.21325  411AS
840828.1533  Y84.210  3 1  0747000  4391489  0.19677  411AS
840828.1903  Y84.210  3 1  0747245  4391332  0.30634  512US
840828.2039  Y84.210  3 1  0747297  4391140  0.34422  512US
840921.1220  Y84.210  3 1  0747049  4391323  0.16265  512US
840921.1305  Y84.210  3 1  0747018  4391305  0.12297  512US
840921.1344  Y84.210  3 1  0747044  4391321  0.15171  512US
840921.1453  Y84.210  3 1  0747042  4391322  0.14940  512US
840921.1506  Y84.210  3 1  0747042  4391322  0.14940  512US
840921.1603  Y84.210  3 1  0747047  4391319  0.15418  512US
840921.1633  Y84.210  3 1  0747033  4391328  0.14096  512US
840921.2025  Y84.210  3 1  0747286  4391188  0.31320  512US
840921.2103  Y84.210  3 1  0747234  4391185  0.31030  512US
840924.0104  Y84.210  3 1  0747480  4390908  0.50656
840924.0204  Y84.210  3 1  0747311  4391163  0.53003  512US
840924.0354  Y84.210  3 1  0747308  4391153  0.33949  512US
840924.0445  Y84.210  3 1  0747172  4391222  0.28882  512US
840924.0547  Y84.210  3 1  0747195  4391286  0.26589  512US
840924.0603  Y84.210  3 1  0747173  4391283  0.25281  512US
840924.0650  Y84.210  3 1  0747189  4391329  0.26751  512US
```

```
840924.0703 Y84.210 3 1 0747210 4391179 0.27531 512US
840924.0754 Y84.210 3 1 0747115 4391118 0.21373 512US
840924.0807 Y84.210 3 1 0747138 4391115 0.23400 512US
840924.0846 Y84.210 3 1 0747093 4391142 0.19035 512US
840924.0902 Y84.210 3 1 0747096 4391139 0.19168 512US
840924.0951 Y84.210 3 1 0747096 4391144 0.19198 512US
840924.1004 Y84.210 3 1 0747116 4391128 0.21382 512US
840924.1048 Y84.210 3 1 0747155 4391093 0.24877 512US
840924.1102 Y84.210 3 1 0747151 4391090 0.24757 512US
840924.1130 Y84.210 3 1 0747116 4391230 0.20873 512US
850527.1228 Y85.210 3 1 0747113 4391156 0.20693 512US
850527.1238 Y85.210 3 1 0747106 4391180 0.20853 512US
850527.1309 Y85.210 3 1 0747099 4391146 0.19401 512US
850527.1336 Y85.210 3 1 0747099 4391146 0.19401 512US
850527.1406 Y85.210 3 1 0747101 4391144 0.19693 512US
850527.1436 Y85.210 3 1 0747112 4391156 0.20390 512US
850527.1636 Y85.210 3 1 0747076 4391239 0.18255 512US
850527.1709 Y85.210 3 1 0747131 4391226 0.22383 512US
850527.1732 Y85.210 3 1 0747130 4391218 0.22962 512US
850527.1803 Y85.210 3 1 0747042 4391230 0.14181 512US
850527.1834 Y85.210 3 1 0747129 4391216 0.22325 512US
850527.1942 Y85.210 3 1 0747312 4391121 0.59015 512US
850527.2037 Y85.210 3 1 0747349 4391039 0.40718 512US
850527.2105 Y85.210 3 1 0747351 4390987 0.45447 523SGHR
850527.2206 Y85.210 3 1 0747285 4390833 0.43153
850527.2304 Y85.210 3 1 0747263 4390883 0.39348 523SGHR
850527.2330 Y85.210 3 1 0747238 4390874 0.42347
850531.0022 Y85.210 3 1 0746714 4390749 0.51804
850531.0046 Y85.210 3 1 0746661 4390745 0.57573
850531.0108 Y85.210 3 1 0746689 4390800 0.49903
850531.0141 Y85.210 3 1 0746856 4390761 0.39376 532SG/US
850531.0309 Y85.210 3 1 0746856 4390822 0.35279 532SG/US
850531.0340 Y85.210 3 1 0746833 4390820 0.36149 532SG/US
850531.0435 Y85.210 3 1 0746695 4390829 0.47190
850531.0633 Y85.210 3 1 0746738 4391152 0.27731 531SG/US
850531.0703 Y85.210 3 1 0746643 4391347 0.41189
850531.0734 Y85.210 3 1 0746670 4391392 0.35590
850531.0905 Y85.210 3 1 0746734 4391506 0.29578 533SG/USHR
850622.1244 Y85.210 3 1 0746938 4391177 0.05858 531SG/US
850622.1307 Y85.210 3 1 0747075 4391133 0.18070 512US
850622.1345 Y85.210 3 1 0747066 4391047 0.20684 512US
850622.1402 Y85.210 3 1 0747050 4391075 0.18113 531SG/US
```

Appendix 1

```
850622.1551 Y85.210 3 1 0747202 4391137 0.27537 512US
850622.1905 Y85.210 3 1 0747240 4391367 0.30686 512US
850622.1934 Y85.210 3 1 0747232 4391403 0.31767 512US
850622.2004 Y85.210 3 1 0747200 4391335 0.27981 512US
850626.0037 Y85.210 3 1 0746854 4391074 0.15710 531SG/US
850626.0110 Y85.210 3 1 0746839 4391094 0.15373 531SG/US
850626.0133 Y85.210 3 1 0746842 4391094 0.15134 531SG/US
850626.0204 Y85.210 3 1 0746842 4391094 0.15134 531SG/US
850626.0232 Y85.210 3 1 0746842 4391094 0.15134 531SG/US
850626.0304 Y85.210 3 1 0746842 4391094 0.15134 531SG/US
850626.0333 Y85.210 3 1 0746842 4391094 0.15134 531SG/US
850626.0603 Y85.210 3 1 0746831 4390947 0.26164 532SG/US
850626.0707 Y85.210 3 1 0747156 4391273 0.24405 512US
850626.1034 Y85.210 3 1 0747016 4391505 0.21137 411AS
850626.1104 Y85.210 3 1 0747039 4391507 0.22425 411AS
850626.1132 Y85.210 3 1 0747052 4391536 0.24489 411AS
850707.1221 Y85.210 3 1 0747086 4391207 0.18493 512US
850707.1304 Y85.210 3 1 0747037 4391221 0.14045 512US
850707.1433 Y85.210 3 1 0747109 4391212 0.20375 512US
850707.1533 Y85.210 3 1 0747114 4391304 0.21498 512US
850707.1604 Y85.210 3 1 0747096 4391316 0.20242 512US
850707.1742 Y85.210 3 1 0747084 4391282 0.18893 512US
850707.1805 Y85.210 3 1 0747066 4391276 0.16992 512US
850707.1902 Y85.210 3 1 0747069 4391265 0.18638 512US
850707.1935 Y85.210 3 1 0747068 4391269 0.17336 512US
850707.2234 Y85.210 3 1 0747332 4391105 0.37995 512US
850707.2331 Y85.210 3 1 0747321 4391132 0.36483 512US
850711.0022 Y85.210 3 1 0746993 4391320 0.09978 512US
850711.0039 Y85.210 3 1 0747021 4391375 0.14717 512US
850711.0135 Y85.210 3 1 0747000 4391350 0.11759 512US
850711.0205 Y85.210 3 1 0747017 4391337 0.12781 512US
850711.0236 Y85.210 3 1 0747011 4391391 0.14744 512US
850711.0304 Y85.210 3 1 0747009 4391379 0.13851 512US
850711.0333 Y85.210 3 1 0747001 4391372 0.12907 512US
850711.0400 Y85.210 3 1 0746988 4391352 0.10928 512US
850711.0446 Y85.210 3 1 0746995 4391351 0.11402 512US
850711.0602 Y85.210 3 1 0747051 4391413 0.18408 512US
850711.0651 Y85.210 3 1 0746978 4391415 0.14100 512US
850711.0732 Y85.210 3 1 0747000 4391432 0.16236 512US
850711.0838 Y85.210 3 1 0747009 4391488 0.20126 411AS
850711.0903 Y85.210 3 1 0747063 4391535 0.24910 411AS
850711.0933 Y85.210 3 1 0746981 4391521 0.22051 411AS
```

```
850711.1006  Y85.210  3  1  0747032  4391534  0.23611  411AS
850711.1032  Y85.210  3  1  0747047  4391566  0.26083  411AS
850711.1103  Y85.210  3  1  0747039  4391550  0.24653  411AS
850711.1131  Y85.210  3  1  0747047  4391566  0.26083  411AS
850801.1246  Y85.210  3  1  0747008  4391549  0.25247  411AS
850801.1302  Y85.210  3  1  0747079  4391594  0.45781  533SG/USHR
850801.1333  Y85.210  3  1  0747064  4391562  0.43376  411AS
850801.1403  Y85.210  3  1  0747070  4391590  0.44245  411AS
850801.1433  Y85.210  3  1  0747083  4391621  0.46300  533SG/USHR
850801.1503  Y85.210  3  1  0747093  4391625  0.47801  533SG/USHR
850801.1533  Y85.210  3  1  0747138  4391646  0.54871  531SG/US
850801.1602  Y85.210  3  1  0747138  4391646  0.54871  531SG/US
850801.1631  Y85.210  3  1  0747115  4391635  0.51107  533SG/USHR
850801.1703  Y85.210  3  1  0747036  4391573  0.39008  533SG/USHR
850801.1832  Y85.210  3  1  0747127  4391551  0.32126  411AS
850805.0018  Y85.210  3  1  0747276  4390910  0.47826  523SGHR
850805.0204  Y85.210  3  1  0746993  4390624  0.44852  523SGHR
850805.0732  Y85.210  3  1  0746917  4391515  0.25654  411AS
850805.0804  Y85.210  3  1  0746952  4391564  0.24982  533SG/USHR
850805.0904  Y85.210  3  1  0747044  4391577  0.40194  411AS
850805.0932  Y85.210  3  1  0747052  4391581  0.41459  411AS
850805.1009  Y85.210  3  1  0746993  4391553  0.33238  533SG/USHR
850805.1103  Y85.210  3  1  0747038  4391520  0.22948  411AS
850805.1131  Y85.210  3  1  0747064  4391562  0.43376  411AS
850812.1247  Y85.210  3  1  0747060  4391483  0.22948  512US
850812.1314  Y85.210  3  1  0747049  4391456  0.21196  411AS
850812.1339  Y85.210  3  1  0747025  4391419  0.16940  512US
850812.1408  Y85.210  3  1  0747054  4391444  0.19975  512US
850812.1435  Y85.210  3  1  0747030  4391428  0.17625  512US
850812.1502  Y85.210  3  1  0747030  4391428  0.17625  512US
850812.1533  Y85.210  3  1  0747011  4391411  0.15619  512US
850812.1633  Y85.210  3  1  0747184  4391346  0.27640  512US
850812.1717  Y85.210  3  1  0747204  4391207  0.26969  512US
850812.1741  Y85.210  3  1  0747209  4391225  0.27660  512US
850812.1811  Y85.210  3  1  0747208  4391221  0.27396  512US
850812.1831  Y85.210  3  1  0747210  4391233  0.28690  512US
850812.1939  Y85.210  3  1  0747299  4391183  0.33138  512US
850812.2050  Y85.210  3  1  0747058  4390946  0.25337  512US
850812.2136  Y85.210  3  1  0747344  4390916  0.40248  523SGHR
850812.2202  Y85.210  3  1  0747325  4390928  0.38867  523SGHR
850812.2233  Y85.210  3  1  0747327  4390909  0.41209  523SGHR
850812.2304  Y85.210  3  1  0747337  4390892  0.45911  523SGHR
```

Appendix 1

```
850812.2330 Y85.210 3 1 0747345 4390896 0.44098 523SGHR
850816.0203 Y85.210 3 1 0747175 4391017 0.29515 512US
850816.0231 Y85.210 3 1 0747109 4391056 0.22535 512US
850816.0303 Y85.210 3 1 0747175 4391017 0.29515 512US
850816.0332 Y85.210 3 1 0747199 4391005 0.31964 523SGHR
850816.0402 Y85.210 3 1 0747230 4391021 0.31636 523SGHR
850816.0435 Y85.210 3 1 0747214 4391140 0.35470 512US
850816.0536 Y85.210 3 1 0747222 4391183 0.28366 512US
850816.0603 Y85.210 3 1 0747219 4391179 0.28505 512US
850816.0634 Y85.210 3 1 0747215 4391171 0.29577 512US
850816.0705 Y85.210 3 1 0747152 4391437 0.26632 512US
850816.0733 Y85.210 3 1 0747096 4391468 0.24127 512US
850816.0802 Y85.210 3 1 0747025 4391467 0.20379 411AS
850816.0832 Y85.210 3 1 0747002 4391492 0.20072 411AS
850816.0903 Y85.210 3 1 0747006 4391492 0.20165 411AS
850816.0932 Y85.210 3 1 0747011 4391492 0.20291 411AS
850816.1004 Y85.210 3 1 0747011 4391479 0.20017 411AS
850816.1031 Y85.210 3 1 0747005 4391519 0.21748 411AS
850816.1104 Y85.210 3 1 0747022 4391521 0.22413 411AS
850816.1130 Y85.210 3 1 0747004 4391488 0.19890 411AS
```

LISTING A1.2
Output from program BIOCHECK for the file created from FIELDS

The input in response to the interactive request from BIOCHECK was:

```
xcoor=5 ycoor=6 time=1 id=2 age=3 sex=4
xmin=746800 xmax=747600 ymin=4390600 ymax=4391700
speedlim=1 kmph   distance=1.5 km
^Z
```

```
Biotelemetry data analysis system for IBM-PC Version 1.00 Page 2
Analysis for animal Y84.210

    output from procedure biocheck for animal Y84.210

         utm xmin =     746800.      utm ymin =    0.439060E+07
         utm xmax =     747600.      utm ymax =    0.439170E+07
         speed limit =    1.00000    km/hr
         distance limit =    1.50000    km
```

Analysis of Wildlife Radio-Tracking Data

```
date   time    x            y            case    error message
                                         no.
-----------------------------------------------------------------
       number of cases checked =  144
       number of missing cases =    0
```

Biotelemetry data analysis system for IBM-PC Version 1.00 Page 2
Analysis for animal Y85.210

```
       output from procedure biocheck for animal Y85.210

              utm xmin =     746800.        utm ymin =    0.439060E+07
              utm xmax =     747600.        utm ymax =    0.439170E+07
              speed limit =    1.00000      km/hr
              distance limit =    1.50000   km

date   time    x            y            case    error message
                                         no.
-----------------------------------------------------------------
850531 0023  746714.0     4390749.        18  fix outside limits specified
850531 0047  746661.0     4390745.        19  fix outside limits specified
850531 0108  746689.0     4390800.        20  fix outside limits specified
850531 0436  746695.0     4390829.        24  fix outside limits specified
850531 0633  746738.0     4391152.        25  fix outside limits specified
850531 0704  746643.0     4391347.        26  fix outside limits specified
850531 0734  746670.0     4391392.        27  fix outside limits specified
850531 0906  746734.0     4391506.        28  fix outside limits specified

       number of cases checked =     135
       number of missing cases =       0
```

LISTING A1.3
Output from program HOMER for the file created from FIELDS.

The input in response to the interactive request from HOMER was:

```
xcoor=5 ycoor=6 time=1 id=2 age=3 sex=4
random dunntime=45 minutes
lotus ( jennrich polygon dunn data )
^Z
```

Appendix 1

Biotelemetry data analysis system for IBM-PC Version 1.00   Page 1
Analysis for animal Y84.210

    v a r i a b l e   s u m m a r y
    x coordinate - xcoor   '
    y coordinate - ycoor   '
    time   - time   '

    date and time on first case used-84/06/06 1220

    date and time on last case used -84/09/24 1131

    randomization test for time series correlation between locations

    mean distance between sequential locations is 93.864746
    variance of distance between sequential locations is 16089.986
    pr(mean distance this small)=0.8530

Biotelemetry data analysis system for IBM-PC Version 1.00   Page 2
Analysis for animal Y84.210

    home range by minimum area method 35.370899   hectares
                            (87.403404   acres)
    number of sides on minimum area polygon   8

    coordinates of convex polygon peripheral points
        case number   x   y
            57   747015.00   4391641.0
            49   747137.00   4391599.0
            87   747318.00   4391243.0
           129   747480.00   4390908.0
            12   746924.00   4390760.0
            16   746899.00   4390757.0
            28   746838.00   4391466.0
            29   746889.00   4391518.0

    area of home range rectangle 56.752800   hectares
                          (140.23924   acres)
    coordinates of lower left-hand corner of rectangle   x = 746838.00
                                                       y = 4390757.0
    maximum x distance   642.00000   meters
    maximum y distance   884.00000   meters

Analysis of Wildlife Radio-Tracking Data

jennrich-turner home range estimates

	all fixes	unique fixes	
mean x	747060.15	747062.49	
mean y	4391231.9	4391227.5	
variance x	13391.935	13356.237	
variance y	67729.813	69797.784	
covariance	-6662.9017	-6309.6823	
95% ellipse	55.270226	56.216297	(hectares)
lower 95% ci	47.197193	47.846691	(hectares)
upper 95% ci	65.619713	67.004585	(hectares)
95% ellipse	136.57571	138.91351	(acres)
lower 95% ci	116.62682	118.23176	(acres)
upper 95% ci	162.14986	165.57196	(acres)
no. fixes	144	138	

dunn home range estimates

	all fixes	
mean x	747092.70	
mean y	4391241.4	
variance x	11976.278	
variance y	61651.566	
covariance	-8397.3360	
95% ellipse	48.642860	(hectares)
95% ellipse	120.19913	(acres)

Biotelemetry data analysis system for IBM-PC Version 1.00   Page 3
Analysis for animal Y85.210

```
        v a r i a b l e   s u m m a r y
        x coordinate - xcoor    '
        y coordinate - ycoor    '
        time         - time     '
```

date and time on first case used - 85/05/27 1228

date and time on last case used - 85/08/16 1131

randomization test for time series correlation between locations

mean distance between sequential locations is 94.630013
variance of distance between sequential locations is 18585.908
pr(mean distance this small)=0.8620

Appendix 1

Biotelemetry data analysis system for IBM-PC Version 1.00   Page 4
Analysis for animal Y85.210

    home range by minimum area method 53.913849   hectares
                                  (133.22403   acres)
    number of sides on minimum area polygon   11

    coordinates of convex polygon peripheral points

case number	x	y
85	747138.00	4391646.0
35	747232.00	4391403.0
58	747332.00	4391105.0
13	747349.00	4391039.0
14	747351.00	4390987.0
116	747345.00	4390896.0
15	747285.00	4390833.0
91	746993.00	4390624.0
19	746661.00	4390745.0
26	746643.00	4391347.0
28	746734.00	4391506.0

    area of home range rectangle 72.357597   hectares
                             (178.79955   acres)
    coordinates of lower left-hand corner of rectangle   x = 746643.00
                                                                      y = 4390624.0

    maximum x distance   708.00000   meters
    maximum y distance   1022.0000   meters

    jennrich-turner home range estimates

	all fixes	unique fixes	
mean x	747060.44	747065.91	
mean y	4391266.8	4391266.2	
variance x	23558.547	23713.645	
variance y	59404.272	59816.533	
covariance	-1954.6028	-3231.7263	
95% ellipse	70.301758	70.612526	(hectares)
lower 95% ci	59.731445	59.623714	(hectares)
upper 95% ci	83.965752	84.964348	(hectares)
95% ellipse	173.71945	174.48737	(acres)
lower 95% ci	147.59962	147.33342	(acres)
upper 95% ci	207.48392	209.95149	(acres)
no. fixes	135	125	

291

```
dunn home range estimates
                        all fixes
         mean x         747077.24
         mean y         4391202.6
         variance x     79650.061
         variance y     1578430.0
         covariance     -314839.33
         95% ellipse    306.97925    (hectares)
         95% ellipse    758.56232    (acres)
```

APPENDIX

2

FIELDS: Radio-Tracking Data Preprocessor

Program FIELDS is used to prepare a master data file for processing by the programs BIOPLOT, BIOCHECK, and HOMER. It does 2 things: (1) selects the pertinent columns from the master data file as fields, separated by colons, and (2) splits the master file into subfiles by putting in subfile markers (a character string of 10 #'s) that are recognized by the data processing programs. The data processing programs expect to find up to 6 fields: (1) ID—animal's identification, (2) SEX—animal's sex, (3) AGE—animal's age, (4) TIME—date and time in the format yymmdd.hhmm, (5) XCOOR—UTM x coordinate, and (6) YCOOR—UTM y coordinate. The UTM coordinates are always used by each of the data analysis programs. Typically, the data file will be split into subfiles by animal, and possibly categories of age within animals.

Program FIELDS provides the details of specifying fields and subfiles when executed.

Limitations of Program FIELDS

The TIME field is too format specific—a variety of date and time formats should be supported by this software.

Availability

Program FIELDS is available on a PC diskette from the authors as a FORTRAN source code file and an executable file compiled with the Ryan–McFarland FORTRAN compiler.

APPENDIX

3

BIOCHECK: Radio-Tracking Data Checking

The BIOCHECK program provides a method of checking the (x, y, t) vectors of radio-tracking input data. Two basic methods of detecting errors in the data are provided. First, distances between locations are checked, with the rate of movement also considered. Second, the locations are checked to see if they are within a specified rectangle.

The input specifications to the BIOCHECK program are the following:

XCOOR=x_UTM_field YCOOR=y_UTM_field TIME=date_time_field
[AGE=age_field] [SEX=sex_field] [ID=animal_id_field]
[XMIN=minimum_x] [XMAX=maximum_x]
[YMIN=minimum_y] [YMAX=maximum_y]
[SPEEDLIM=speed_limit [mph|kmph|mps|m]]
[DISTANCE=maximum_distance_traveled [miles|km|m|feet]]
[MAXANIM=max_#_of_animals_or_subfiles]

XCOOR, YCOOR, and TIME parameters specify the field numbers of the x and y coordinates, and the time field, respectively. XCOOR and YCOOR variables are assumed to be in units of meters. The TIME variable is assumed to have the format yymmdd.hhmm, where yy is the last 2 digits of the year, mm is the month (01–12), dd is the day (01–31), hh is the hour (00–23), and mm is the minute (00–59).

AGE, SEX, and ID parameters specify the field numbers of the age, sex, and animal identification codes. These fields are checked to see that they do

295

not change, hence, that a subfile contains only one animal. If the contents of these fields do change, then an error message is printed and error checking continues. If one (or all) of these parameters is not specified, error checking is not provided for the missing field.

The XMIN, XMAX, YMIN, and YMAX parameters specify values for the rectangle to use in checking the geographic locations of fixes. That is, if an (x, y) pair is not inside of this rectangle, an error message is printed. Default values of these parameters are very large and very small values, so that all locations are inside the rectangle formed by the default values. Multiple values can be specified for each of these parameters to handle multiple subfiles in one run. The first value is used for the first subfile, the second value is used for the second subfile, and so on. If enough values are not specified to provide one for each subfile, then the last value specified is used for all remaining subfiles.

The SPEEDLIM parameter specifies the fastest rate of movement at which an animal is assumed to be able to move between fixes (locations). Rates of movement between 2 consecutive fixes (locations) greater than SPEEDLIM are flagged as errors. After the numerical value, a units value can be specified to interpret the speedlimit parameter. As with the XMIN parameter, multiple values can be specified to provide each subfile a different SPEEDLIM parameter value to be processed with BIOCHECK.

The DISTANCE parameter specifies the largest allowable distance between 2 consecutive fixes (locations) before an error message is printed. The units specified with this parameter allow units other than the default setting of km. Also, multiple values are matched with their respective subfiles as for XMIN. A synonym for DISTANCE is MAXDIST.

The MAXANIM parameter specifies the maximum number of animals to be processed in one subfile of data. The default value is 100, that is, 100 different values of the ID variable can be handled within a single subfile. If more than 100 values of the ID variable occur within a subfile, then the larger value must be specified for the MAXANIM parameter. This parameter should seldom, if ever, be used, because few users would ever have more than 100 radio-marked animals in a single data file.

Limitations of Program BIOCHECK

No methods are provided to test for locations inside an area that is not a feasible habitat. For example, the boundaries of a lake cannot be specified to check for locations of a terrestrial animal occurring in the lake. At this time, BIOCHECK only reads input data (i.e., locations) from the file named BIOCHECK.DAT. An additional parameter to specify the input data file should be included.

Output for Program BIOCHECK

Output of BIOCHECK consists of a summary of the errors located in the data for each animal checked, plus a summary of the parameter values used in the check. This output is listed on the screen, although the > symbol can be used on the execution line to redirect the output to a file.

Availability

Program BIOCHECK is available on a PC diskette from the authors as a FORTRAN source code file and an executable file compiled with the Ryan–McFarland FORTRAN compiler.

APPENDIX

4

BIOPLOT: Radio-Tracking Data Plotting and Editing

Program BIOPLOT is an Advanced BASIC (BASICA) program used to plot radio-tracking locations on a graphics monitor. A graphics board is required. Data are assumed to be in fields, as generated by Program FIELDS. BIOPLOT asks the user for the name of the data input file and the field numbers of the UTM x and y coordinate fields. Fields are assumed to be separated by colons. The user has the capability to delete bad locations from the plot, and hence from the output data file. In addition, the user can mark questionable locations for examination later in the output data file.

Limitations of Program BIOPLOT

BIOPLOT does not have an animation capability, i.e., to read in the data for one or more animals and display their movements on the screen as in a movie.

Availability

Program BIOPLOT is available from the authors as a BASICA source code file on a PC diskette.

APPENDIX

5

HOMER: Home Range Estimation

HOMER provides home range estimates from a set of (x, y, t) data. Available estimates are the minimum area polygon, Jennrich–Turner ellipse (Jennrich and Turner 1969), Dunn ellipse (Dunn and Gipson 1977), and a grid-cell estimator (Siniff and Tester 1965).

The general format of the HOMER calling sequence is

```
XCOOR=x_UTM_field YCOOR=y_UTM_field TIME=date_time_field
[AGE=age_field] [SEX=sex_field] [ID=animal_id_field]
[MAP=[fill|both]] [MAPSCALE] [REMOVE]
[XORIGIN=value(s)] [YORIGIN=value(s)]
[SQUARE=value(s) [mile|km|m|feet]] [GRIDS=#_grid_cells]
[VECTIME=value(s) [day|days|hour|hours|minute|minutes]]
[VECDIST=value(s) [mile|km|m|feet]] [FILL=value(s)]
[DUNNTIME=value [day|days|hour|hours|minute|minutes]]
[DUNN] [RESIDUAL] [LIKELIHO] [DIAGNOST]
[RANDOM]
[LOTUS [ (all jt jennrich polygon dunn grid grids unique data) ]]
[MAXANIM=max_#_animals] [MAXCASE=max_#_cases]
```

XCOOR, YCOOR, and TIME specify the variable names for x and y coordinates and time, respectively. These parameters are the only ones that must be specified to obtain home range estimates. The default estimates are the minimum area polygon and the Jennrich–Turner estimate. XCOOR and YCOOR are assumed to have units in meters and TIME in days. The TIME field must be in the format yymmdd.hhmm where yy is the last 2 digits of the year, mm is the month (01–12), dd is the day (01–31), hh is the hour (00–23), and mm is the minute (00–59). Specification of any of the parameters from MAP= to FILL= causes the estimation of home range with the grid cell method, i.e., simply specifying VECTIME=10 hours would produce output from the grid cell estimator. The MAP parameter causes the grid cell home range map to be printed. MAP without an equal sign and second parameter prints the grid cell map without any "filled" locations. MAP=FILL causes only the "filled" map to be printed, and MAP=BOTH causes both maps to be printed. The MAP-SCALE parameter causes the printed map to be scaled so that the x and y distances are approximately equal.

XORIGIN and YORIGIN parameters provide the x and y coordinates for the lower left corner of the grid cell map. If they are not specified, the data file is read twice, once to obtain values for XORIGIN and YORIGIN, and once to generate the home range estimates. Note that both XORIGIN and YORIGIN parameters allow more than one value. Multiple values are used for successive subfiles, with remaining subfiles set to the last value in the list. Thus,

```
XORIGIN=700000 800000 900000
```

sets the value of the left side of the grid cell map to 700000 for the first subfile, 800000 for the second subfile, and 900000 for the third and all remaining subfiles.

REMOVE is the wild fix (location) removal option. Generally it is not needed.

The SQUARE= parameter sets the size of the grid cell square used for home range estimation. GRIDS= parameter sets the number of grids in one dimension to be used for home range estimation. Thus, the total number of grid cells is this value squared. Both SQUARE and GRIDS allow multiple values for multiple subfiles. The defaults for these parameters are GRIDS=100 and SQUARE equal to min($y_{max} - y_{min}$, $x_{max} - x_{min}$)/GRIDS.

VECTIME parameter specifies the maximum amount of time allowed between fixes (locations) to perform vector fill, i.e., add additional locations along the line between 2 observed locations. The default value is zero, hence, no vector fill is performed. A positive value for vector fill causes a fix (location) to be generated in a cell that is crossed by the line connecting 2 consecutive locations within VECTIME time units. Although the TIME variable is assumed

Appendix 5

to have units of days, a value for VECTIME can be specified in hours or minutes by including the units after the value. Like the above parameters, VECTIME can have multiple values, one for each subfile.

The VECDIST parameter allows vector fill of grid cells where no locations were found. VECDIST sets the maximum distance of the vector along which vector fill can operate. If VECTIME = 0 is specified, then VECDIST is ignored. The default value is zero, hence, no vector fill takes place.

The FILL parameter specifies the number of grid cells that will be filled in, i.e., locations added, to eliminate patchiness in the map. The value of FILL is interpreted as the size of holes in the map that will be filled. Thus, FILL = 2 causes grid cells within 2 cells of filled cells to be filled, i.e., a fix (location) is generated for the empty cell. The default is zero, hence, no cells are filled. FILL also allows multiple values, one for each subfile.

The DUNNTIME parameter specifies the time interval between fixes (locations) within a burst for the Dunn home range estimator. Specification of this parameter causes the Dunn estimator to be calculated. The value specified is assumed to have units of days unless hours or minutes are added to the specification. The resulting value is used to divide the input data into bursts. Two consecutive locations within DUNNTIME units of each other are assumed to be in the same burst. Likewise, two consecutive locations separated by more than DUNNTIME units are separated into different bursts.

A potential danger is that a constant time interval between locations within bursts will not be maintained. The Dunn method assumes all locations within a burst are a fixed time unit apart, and that the time between bursts is large enough to assume that data in separate bursts are independent. HOMER *does not* check these assumptions. As an example, consider 4 consecutive locations with the time interval between them of 28, 29.5, and 30.5 minutes, respectively. If DUNNTIME = 30 minutes was specified, then the first 3 observations would be in one burst, and the last observation would be in a separate burst. Specifying DUNNTIME = 31 minutes would include all 4 locations in a single burst.

Four additional options may be specified with the DUNNTIME parameter to obtain additional output. These are mutually exclusive: DUNN, RESIDUAL, LIKELIHO, and DIAGNOST. Each produces output additional to the Dunn estimator itself.

To determine if significant autocorrelation exists through time, the RANDOM option causes HOMER to compute a randomization test. The mean distance between consecutive locations is computed. Then the locations are sorted into random order 1000 times, and the distribution of the average distance computed. The probability level of the observed average distance is computed from this empirical distribution. This option can be quite time consuming if there are a large number of observations per animal. Thus don't use it if you're in a hurry.

The LOTUS parameter specifies which data are to be written to output files for input to Lotus 1-2-3 for graphics. The options and their corresponding output are:

Parameter	Output	File name
DATA	Data Points	HOMER.LOC
UNIQUE	Unique data points	HOMER.UNQ
GRID or GRIDS	Frequencies in each grid cell	HOMER.GRD
POLYGON	Coordinates of polygon	HOMER.CON
JT or JENNRICH	200 points on ellipse	HOMER.ELP
DUNN	200 points on ellipse	HOMER.DUN
ALL	All of the above	As above

The UNIQUE option specifies that only the unique data be included in the file, i.e., 2 locations with identical coordinates would only occur in the output file as a single location. The user must incorporate these files into Lotus 1-2-3 with the File Import function of 1-2-3.

Limitations for Program HOMER

The harmonic mean estimator of Dixon and Chapman (1980) has not been implemented. The Fourier Series estimator of Anderson (1982) has not yet been implemented. The XORIGIN and YORIGIN values must be specified on the input, as the PC DOS version of HOMER does not do the 2 passes described above. HOMER reads its input from the file named HOMER.DAT in the local directory. It should read input from a user-specified file, or the standard input.

Output from Program HOMER

Output depends on the parameter specified. Generally the user will want to look at the output on the screen. However, by using the redirection capability of DOS (i.e., the > symbol) with the execution of HOMER (e.g., homer > temp.lis), the output can be saved to a file for printing later.

Appendix 5

Availability

Program HOMER is available on a PC diskette from the authors as a FORTRAN source code file and an executable file compiled with the Ryan–McFarland FORTRAN compiler.

References

Anderson, D. J. 1982. The home range: a new nonparametric estimation technique. Ecology 63:103–112.

Dixon, K. R. and J. A. Chapman. 1980. Harmonic mean measure of animal activity areas. Ecology 61:1040–1044.

Dunn, J. E. and P. S. Gipson. 1977. Analysis of radio telemetry data in studies of home range. Biometrics 33:85–101.

Jennrich, R. I. and F. B. Turner. 1969. Measurement of non-circular home range. J. Theor. Biol. 22:227–237.

Siniff, D. B. and J. R. Tester. 1965. Computer analysis of animal movement data obtained by telemetry. BioScience 15:104–108.

APPENDIX

6

PC SURVIV User's Manual Version 1.4

Program SURVIV (a stand alone program) is used to estimate survival rates from user specified cell probability functions. The program reads the user specified algebraic expressions and constructs a FORTRAN 77 subroutine that calculates the log-likelihood function. Numerical optimization is then used to estimate the unknown parameters. The program is available for PC-DOS and MS-DOS operating systems to run on an IBM PC or compatible. A math co-processor is required to run the program because of the large amount of numerical processing performed.

The syntax for each of the procedures in Program SURVIV is shown in Table A6.1. As the reader works through the examples illustrating these procedures, he should occasionally refer back to Table A6.1 to understand the information presented. Table A6.1 is a summary of each procedure and does not provide the detailed explanations provided in the examples.

Survival Rate Estimation

Three general tasks are required to estimate survival rates with the above procedure. First, the algebraic expressions must be read and the FORTRAN 77 subroutine constructed, PROC MODEL performs this task. Second, the likelihood function must be numerically maximized by varying the parameter estimates for one or more sets of constraints on the parameters, PROC ESTIMATE performs this task. Finally comparisons of parameter estimates by hypothesis

Table A6.1
Summary of command syntax for each of the
procedures and their modifiers in Program SURVIV

Modifiers enclosed in brackets indicate optional input. Program SURVIV allows comments enclosed between /* and */ symbols, and statements are delineated with semicolons.

```
PROC TITLE Any title information to be printed at top of each page;

PROC MODEL NPAR = # of S(I)'s in model;

        COHORT = # of animals in cohort;

                # observed: algebraic expression for cell probability;
                .
                .
                repeated for up to maximum number of cells per cohort;

        COHORT = repeated for up to maximum number of cohorts;

        LABELS;
                S(1) = identifying label;
                .
                .
                S(NPAR) = identifying label;

PROC ESTIMATE [NSIG = # of significant digits to find for estimates
                 (default is 5)]
              [MAXFN = maximum number of function evaluations (default
                 is 1000)]
              [NAME = name of submodel]
              [TRANSFORM = NONE or ABS (default is SIN transform)];
              [NOVAR (Do not print variance-covariance matrix, default
                 is to print matrix)];

        INITIAL;
                S(1) = value;
                .
                .
                S(NPAR) = value;
```

Appendix 6

Table A6.1
(Continued)

```
        or    ALL = value;

        or    RETAIN = modname /* Obtain initial estimates from output
                      of previous model named in ''modname'' */

        CONSTRAINTS /*Constraints of the form */;
              S(1) = S(2) /* S(1) is the same as S(2) */ ;
              S(1) < S(2) /* S(1) is less than S(2) */ ;
              S(1) > S(2) /* S(1) is greater than S(2) */;
              S(1) = value /* Fix parameter to constant */;

PROC TEST /* No modifiers */;

PROC BROWNIE [NSIG = # of significant digits to find for estimates
    (default is 5)]
        [MAXFN = maximum number of function evaluations (default
          is 1000)]
        [TRANSFORM = NONE or ABS (default is SIN transform)];
        [NOVAR (Do not print variance-covariance matrix, default is to
          print matrix)];
        /* Input to BROWNIE is exactly as described by Brownie et al.
        (1985:155-158) for programs ESTIMATE and BROWNIE, with the
        additional requirement that each card image must have a
        semicolon as the last non-blank character. */;
        /* Currently BROWNIE does not handle data of types 2 or 3 */;

PROC BMDPAR /* No modifiers */;

        INITIAL /* As in PROC ESTIMATE */;

        CONSTRAINTS /* As in PROC ESTIMATE */;

PROC IDENTIFIABLE
    [ADD = Constant to add for numerical derivative (Default is 0.05)]
    [MULTIPLY = Constant to multiply by for numerical derivative
      (Default is 1.05)];

        INITIAL /* As in PROC ESTIMATE */;

        CONSTRAINTS /* As in PROC ESTIMATE */;
```

(continues)

Table A6.1
(*Continued*)

```
PROC SIMULATE [NSIM = # of replications of each simulation (default
   is 5)]
  [DETAIL (print details of each simulation,
          with default of only simulation summaries)]
  [SEED = random number seed (default is 7654321)];
  [PARFILE = file name for simulated data output for later
          analysis (default is no data output)]
  [ERROR TEST (do not include simulations in data summaries
          with warnings about convergence - default to
          include questionable convergence simulations)];

          INITIAL /* As in PROC ESTIMATE */;

          CONSTRAINTS /* As in PROC ESTIMATE */;

PROC SAMPLE SIZE /* Notation follows Brownie et al. 1985:190-193 */;
     S-BAR = values of S /* Define average adult survival rates. */;
     S'-BAR = values of S' /* Define average young survival rates. */;
     F-BAR = values of f /* Define average adult recovery rate. */;
     CV(S) = values of CV of S /* Define coefficients of variation for
        values of S */;
     CV(S') = values of CV of S' /* Define coefficient of variation for
        values of S' */;
     K = number of years /* Define number of years to conduct banding
        effort. */;

PROC STOP /* Stops execution. */;
```

testing must be made to identify the model appropriate for the observed data. This process is performed by PROC TEST.

We will illustrate these procedures first with a simple example taken from Brownie et al. (1985), page 21. Ducks were banded for 3 years, and band recoveries summarized for 5 years. The input to run this model and the resulting output are shown in Fig. A6.1.

All input instructions read by Program SURVIV are echoed in the output with the identifying INPUT—to delineate them. Thus on page 1 of the output in Fig. A6.1, PROC TITLE is called with a title for the run. Statements are separated by semicolons. Next PROC MODEL is called, with the number of parameters to be estimated set to 7. Additional comments are specified to clarify the input. PROC MODEL processes the observed data and the associated cell probability statements. Thus, the next line of input is COHORT = 1603 with a comment to identify this cohort as the number of ducks banded in 1964. The COHORT card specifies the number of animals that are subject to mortality in the mortality categories that follow it. Therefore, the next input statement specifies that 127 duck bands were returned, and the expected cell probability is $S(3)$. That is, the observed multinomial cell value is 127, and the expected cell value is $1603 \times S(3)$, but the number of animals at risk of mortality is known from the COHORT statement, so only the expected cell probability is given. The colon (:) separates the observed number of recoveries and the expected cell probability.

```
SURVIV-Survival Rate Estimation with User Specified Cell Probabilities
12-Jun-89     13:46:31     Version 1.4(PC-DOS) June, 1989     Page 001
---------------------------------------------------------------------

This version of SURVIV can only handle input lines
of about 120 characters. Do not exceed this limitation.

     Dimension limitations for this run:
       Maximum number of parameters                36
       Maximum number of cohorts                   25
       Maximum number of classes within a cohort   10
       Maximum number of models for PROC TEST      20
     If your problem needs larger dimensions, reset the values
     in the MODELC include file and recompile the program.

INPUT --- PROC TITLE Banding Handbook Example from Page 21 of Brownie
INPUT --- et al. (1985);

     CPU time in seconds for last procedure was 0.33

INPUT --- PROC MODEL NPAR=7 /* Banding Handbook Page 21 */;

INPUT ---     COHORT = 1603 /* Banded 1964*/;
INPUT ---       127:S(3) /* Bands Recovered in 1964 */;
INPUT ---       44:S(1)*S(4) /* Bands Recovered in 1965 */;
INPUT ---       37:S(1)*S(2)*S(5) /* Bands Recovered in 1966 */;
```

Figure A6.1 Output from Program SURVIV for a banding analysis model from Brownie et al. (1985:21). (*Figure continues.*)

Analysis of Wildlife Radio-Tracking Data

```
SURVIV - Survival Rate Estimation with User Specified Cell Probabilities
12-Jun-89     13:46:31     Version 1.4(PC-DOS)     June, 1989     Page 002
Banding Handbook Example from Page 21 of Brownie et al. (1985)
-----------------------------------------------------------------

   INPUT ---    40:S(1)*S(2)*S(6) /* Bands Recovered in 1967 */;
   INPUT ---    17:S(1)*S(2)*S(7) /* Bands Recovered in 1968 */;

   INPUT ---    COHORT = 1595 /* Banded 1965 */;
   INPUT ---     62:S(4) /* Bands Recovered in 1965 */;
   INPUT ---     76:S(2)*S(5) /* Bands Recovered in 1966 */;
   INPUT ---     44:S(2)*S(6) /* Bands Recovered in 1967 */;
   INPUT ---     28:S(2)*S(7) /* Bands Recovered in 1968 */;

   INPUT ---    COHORT = 1157 /* Banded 1966*/;
   INPUT ---     82:S(5) /* Bands Recovered in 1966 */;
   INPUT ---     61:S(6) /* Bands Recovered in 1967 */;
   INPUT ---     24:S(7) /* Bands Recovered in 1968 */;

   INPUT ---    LABELS;
   INPUT ---     S(1)=Survival for 1964;
   INPUT ---     S(2)=Survival for 1965;
   INPUT ---     S(3)=Recovery for 1964;
   INPUT ---     S(4)=Recovery FOR 1965;
   INPUT ---     S(5)=Recovery FOR 1966;
   INPUT ---     S(6)=Survival for 1966 and Recovery 1967;
   INPUT ---     S(7)=Survival for 1966-67, Recovery 1968;

     CPU time in seconds for last procedure was    2.19

   INPUT --- PROC ESTIMATE NSIG=5 MAXFN=750 NAME=SV&RC_SAME /*
   INPUT --- Combination of Survival and Recovery Rates Constant by
   INPUT --- Time*/;

   INPUT ---    INITIAL;
   INPUT ---     ALL=0.05;

   INPUT ---    CONSTRAINTS;
   INPUT ---     S(1)=S(2);
   INPUT ---     S(3)=S(4);
   INPUT ---     S(3)=S(5);
```

Figure A6.1 (*Continued*)

Appendix 6

SURVIV - Survival Rate Estimation with User Specified Cell Probabilities
12-Jun-89   13:46:31   Version 1.4(PC-DOS)   June, 1989   Page 003
Banding Handbook Example from Page 21 of Brownie et al. (1985)

Number of parameters in model = 7

Number of parameters set equal = 3

Number of parameters fixed = 0

Number of parameters estimated = 4

I	Parameter	S(I)	Lower Bound	Upper Bound	Label
1	1	0.050	0.000	1.000	Survival for 1964
2	1	0.050	0.000	1.000	Survival for 1965
3	2	0.050	0.000	1.000	Recovery FOR 1964
4	2	0.050	0.000	1.000	Recovery FOR 1965
5	2	0.050	0.000	1.000	Recovery FOR 1966
6	3	0.050	0.000	1.000	Survival for 1966 and Recovery 1967
7	4	0.050	0.000	1.000	Survival for 1966-67, Recovery 1968

Final function value 2648.6431   (Error Return = 0)

Number of significant digits   7

Number of function evaluations   133

I	Parameter	S(I)	Standard Error	95% Confidence Interval Lower	Upper
1	1	0.64058338	0.34957855E-01	0.57206598	0.70910078
2	1	0.64058338	0.34957855E-01	0.57206598	0.70910078
3	2	0.60617526E-01	0.32176157E-02	0.54310999E-01	0.66924053E-01
4	2	0.60617526E-01	0.32176157E-02	0.54310999E-01	0.66924053E-01
5	2	0.60617526E-01	0.32176157E-02	0.54310999E-01	0.66924053E-01
6	3	0.51111879E-01	0.47218499E-02	0.41857053E-01	0.60366705E-01
7	4	0.24322205E-01	0.30891896E-02	0.18267393E-01	0.30377016E-01

(Figure continues.)

SURVIV - Survival Rate Estimation with User Specified Cell Probabilities
12-Jun-89   13:46:31   Version 1.4(PC-DOS) June, 1989   Page 004
Banding Handbook Example from Page 21 of Brownie et al. (1985)

Variance-Covariance matrix of estimates on diagonal and below,

Correlation matrix of estimates above diagonal.

```
   |    1          2          3          4          5          6
   |    7
---|----------------------------------------------------------------
   |
 1 |  0.12221E-02  1.0000    -0.50835   -0.50835   -0.50835   -0.47185
   | -0.34320
   |
 2 |  0.12221E-02  0.12221E-02 -0.50835  -0.50835   -0.50835   -0.47185
   | -0.34320
   |
 3 | -0.57180E-04 -0.57180E-04 0.10353E-04 1.0000    1.0000     0.19792
   |  0.14396
   |
 4 | -0.57180E-04 -0.57180E-04 0.10353E-04 0.10353E-04 1.0000    0.19792
   |  0.14396
   |
 5 | -0.57180E-04 -0.57180E-04 0.10353E-04 0.10353E-04 0.10353E-04 0.19792
   |  0.14396
   |
 6 | -0.77886E-04 -0.77886E-04 0.30071E-05 0.30071E-05 0.30071E-05 0.22296E-04
   |  0.13952
   |
 7 | -0.37063E-04 -0.37063E-04 0.14309E-05 0.14309E-05 0.14309E-05 0.20351E-05
   |  0.95431E-05
```

Cohort	Cell	Observed	Expected	Chi-square	Note
1	1	127	97.170	9.158	0 < P < 1
1	2	44	62.245	5.348	0 < P < 1
1	3	37	39.873	0.207	0 < P < 1
1	4	40	33.621	1.210	0 < P < 1
1	5	17	15.999	0.063	0 < P < 1
1	6	1338	1354.092	0.191	0 < P < 1
1 Cohort df= 5				16.177	P = 0.0064

Figure A6.1 (*Continued*)

Appendix 6

```
SURVIV - Survival Rate Estimation with User Specified Cell Probabilities
12-Jun-89      13:46:31        Version 1.4(PC-DOS) June, 1989      Page 005
Banding Handbook Example from Page 21 of Brownie et al. (1985)
------------------------------------------------------------------
```

2	1	62	96.685	12.443	0 < P < 1
2	2	76	61.935	3.194	0 < P < 1
2	3	44	52.223	1.295	0 < P < 1
2	4	28	24.851	0.399	0 < P < 1
2	5	1385	1359.307	0.486	0 < P < 1
2 Cohort df= 4				17.817	P = 0.0013
3	1	82	70.134	2.007	0 < P < 1
3	2	61	59.136	0.059	0 < P < 1
3	3	24	28.141	0.609	0 < P < 1
3	4	990	999.588	0.092	0 < P < 1
3 Cohort df= 3				2.767	P = 0.4289

```
------------------------------------------------------------------
  G Total (Degrees of freedom =    8)      38.092
  Pr(Larger Chi-square) = 0.0000
  With pooling, Degrees of freedom =   8  Pearson Chi-square =    36.761
  Pr(Larger Chi-square) = 0.0000

  Log-likelihood = -52.874994

    CPU time in seconds for last procedure was 3.74

INPUT --- PROC ESTIMATE NSIG=5 MAXFN=750 NAME=GENERAL/*General model*/;

INPUT ---     INITIAL;

INPUT ---     RETAIN=SV&RC_SAME ;

    Number of parameters in model   = 7

    Number of parameters set equal  = 0

    Number of parameters fixed      = 0

    Number of parameters estimated  = 7
```

(Figure continues.)

Analysis of Wildlife Radio-Tracking Data

```
SURVIV - Survival Rate Estimation with User Specified Cell Probabilities
12-Jun-89      13:46:31      Version 1.4(PC-DOS) June, 1989      Page 006
Banding Handbook Example from Page 21 of Brownie et al. (1985)
----------------------------------------------------------------
```

			Lower	Upper	
I	Parameter	S(I)	Bound	Bound	Label
--	------	----	---	---	--------------
1	1	0.641	0.000	1.000	Survival for 1964
2	2	0.641	0.000	1.000	Survival for 1965
3	3	0.061	0.000	1.000	Recovery FOR 1964
4	4	0.061	0.000	1.000	Recovery FOR 1965
5	5	0.061	0.000	1.000	Recovery FOR 1966
6	6	0.051	0.000	1.000	Survival for 1966 and Recovery 1967
7	7	0.024	0.000	1.000	Survival for 1966-67, Recovery 1968

Final function value 2632.5922   (Error Return =   0)

Number of significant digits   8

Number of function evaluations   190

				95% Confidence Interval	
I	Parameter	S(I)	Standard Error	Lower	Upper
--	------	-----------	-----------	-----------	-----------
1	1	0.65386329	0.67817609E-01	0.52094078	0.78678581
2	2	0.63432469	0.65052567E-01	0.50682166	0.76182773
3	3	0.79226450E-01	0.67459754E-02	0.66004338E-01	0.92448563E-01
4	4	0.40103773E-01	0.41474639E-02	0.31974743E-01	0.48232802E-01
5	5	0.68816791E-01	0.60804365E-02	0.56899135E-01	0.80734446E-01
6	6	0.51171460E-01	0.50073002E-02	0.41357151E-01	0.60985768E-01
7	7	0.24350557E-01	0.31908973E-02	0.18096398E-01	0.30604716E-01

Figure A6.1 (*Continued*)

Appendix 6

SURVIV - Survival Rate Estimation with User Specified Cell Probabilities
12-Jun-89   13:46:31   Version 1.4(PC-DOS) June, 1989   Page 007
Banding Handbook Example from Page 21 of Brownie et al. (1985)

Variance-Covariance matrix of estimates on diagonal and below,

Correlation matrix of estimates above diagonal.

```
  |    1           2           3           4           5           6
  |    7
--|------------------------------------------------------------------
  |
1 |  0.45992E-02-0.38874    -0.70638E-01-0.38549     0.58352E-12 0.68986E-12
  |  0.15635E-11
  |
2 | -0.17150E-02 0.42318E-02-0.24774E-12 0.11893    -0.56545    -0.51057
  | -0.38127
  |
3 | -0.32316E-04-0.10872E-15 0.45508E-04 0.85102E-12 0.46756E-12 0.68641E-12
  |  0.10534E-11
  |
4 | -0.10843E-03 0.32088E-04 0.23811E-16 0.17201E-04-0.15498E-13-0.24780E-13
  | -0.32248E-12
  |
5 |  0.24062E-15-0.22366E-03 0.19179E-16-0.39084E-18 0.36972E-04 0.30982
  |  0.23136
  |
6 |  0.23426E-15-0.16631E-03 0.23186E-16-0.51463E-18 0.94330E-05 0.25073E-04
  |  0.20891
  |
7 |  0.33835E-15-0.79142E-04 0.22675E-16-0.42678E-17 0.44888E-05 0.33378E-05
  |  0.10182E-04
```

Cohort	Cell	Observed	Expected	Chi-square	Note
1	1	127	127.000	0.000	0 < P < 1
1	2	44	42.034	0.092	0 < P < 1
1	3	37	45.754	1.675	0 < P < 1
1	4	40	34.022	1.050	0 < P < 1
1	5	17	16.190	0.041	0 < P < 1
1	6	1338	1338.000	0.000	0 < P < 1
1 Cohort df = 5				2.858	P = 0.7219

(Figure continues.)

SURVIV - Survival Rate Estimation with User Specified Cell Probabilities
12-Jun-89   13:46:31   Version 1.4(PC-DOS) June, 1989   Page 008
Banding Handbook Example from Page 21 of Brownie et al. (1985)

```
       2      1       62       63.966      0.060   0 < P < 1
       2      2       76       69.625      0.584   0 < P < 1
       2      3       44       51.773      1.167   0 < P < 1
       2      4       28       24.637      0.459   0 < P < 1
       2      5     1385     1385.000      0.000   0 < P < 1
       2 Cohort df = 4                     2.270   P = 0.6862
       3      1       82       79.621      0.071   0 < P < 1
       3      2       61       59.205      0.054   0 < P < 1
       3      3       24       28.174      0.618   0 < P < 1
       3      4      990      990.000      0.000   0 < P < 1
       3 Cohort df = 3                     0.744   P = 0.8629
```

G Total (Degrees of freedom =   5)   5.990
Pr(Larger Chi-square) = 0.3072
With pooling, Degrees of freedom =   5   Pearson Chi-square =   5.872
Pr(Larger Chi-square) = 0.3189

Log-likelihood = -36.824114

   CPU time in seconds for last procedure was   4.78

INPUT --- PROC ESTIMATE NSIG=5 MAXFN=750 NAME=SURVL_SAME /* 1964
INPUT --- Survival Equals 1965 Survival*/;

INPUT ---   INITIAL;

INPUT ---   RETAIN=GENERAL ;

INPUT ---   CONSTRAINTS;
INPUT ---   S(2)=S(1);

Figure A6.1   (*Continued*)

Appendix 6

SURVIV - Survival Rate Estimation with User Specified Cell Probabilities
12-Jun-89   13:46:31   Version 1.4(PC-DOS) June, 1989   Page 009
Banding Handbook Example from Page 21 of Brownie et al. (1985)

    Number of parameters in model   = 7

    Number of parameters set equal   = 1

    Number of parameters fixed   = 0

    Number of parameters estimated   = 6

                      Lower   Upper
 I   Parameter S(I) Bound   Bound Label
--   --------- ---- -----   ----- -----
 1   1   0.654 0.000   1.000 Survival for 1964
 2   1   0.654 0.000   1.000 Survival for 1965
 3   2   0.079 0.000   1.000 Recovery FOR 1964
 4   3   0.040 0.000   1.000 Recovery FOR 1965
 5   4   0.069 0.000   1.000 Recovery FOR 1966
 6   5   0.051 0.000   1.000 Survival for 1966 and Recovery 1967
 7   6   0.024 0.000   1.000 Survival for 1966-67, Recovery 1968

    Final function value 2632.6078   (Error Return = 0)

    Number of significant digits   9

    Number of function evaluations   140

                                              95% Confidence Interval
 I   Parameter   S(I)   Standard Error   Lower   Upper
--   --------- -----------   -------------- -----------   -----------
 1   1   0.64393595   0.36818149E-01 0.57177237   0.71609952
 2   1   0.64393595   0.36818149E-01 0.57177237   0.71609952
 3   2   0.79277320E-01 0.67417990E-02 0.66063394E-01   0.92491246E-01
 4   3   0.40325278E-01 0.39726728E-02 0.32538839E-01   0.48111717E-01
 5   4   0.68453790E-01 0.57011518E-02 0.57279533E-01   0.79628048E-01
 6   5   0.50901536E-01 0.47490351E-02 0.41593428E-01   0.60209645E-01
 7   6   0.24222110E-01 0.30925154E-02 0.18160780E-01   0.30283441E-01

(Figure continues.)

Analysis of Wildlife Radio-Tracking Data

```
SURVIV Survival Rate Estimation with User Specified Cell Probabilities
12-Jun-89     13:46:31      Version 1.4(PC-DOS) June, 1989      Page 010
Banding Handbook Example from Page 21 of Brownie et al. (1985)
-----------------------------------------------------------------
         Variance-Covariance matrix of estimates on diagonal and below,

             Correlation matrix of estimates above diagonal.

      |    1           2          3          4           5          6
      |    7
   ---|----------------------------------------------------------------
      |
   1  |  0.13556E-02  1.0000    -0.63570E-01-0.25020    -0.54625   -0.48762
      | -0.35633
      |
   2  |  0.13556E-02  0.13556E-02-0.63570E-01-0.25020   -0.54625   -0.48762
      | -0.35633
      |
   3  | -0.15780E-04-0.15780E-04 0.45452E-04-0.13455E-01 0.15415E-01 0.13761E-01
      |  0.10056E-01
      |
   4  | -0.36596E-04-0.36596E-04-0.36037E-06 0.15782E-04 0.11355     0.10136
      |  0.74072E-01
      |
   5  | -0.11466E-03-0.11466E-03 0.59250E-06 0.25718E-05 0.32503E-04 0.23273
      |  0.17007
      |
   6  | -0.85260E-04-0.85260E-04 0.44058E-06 0.19124E-05 0.63011E-05 0.22553E-04
      |  0.15181
      |
   7  | -0.40572E-04-0.40572E-04 0.20965E-06 0.91002E-06 0.29984E-05 0.22296E-05
      |  0.95637E-05

     Cohort  Cell   Observed  Expected  Chi-square  Note
     ----    ---    -----     -----     -------     --------
       1      1       127     127.082     0.000     0 < P < 1
       1      2        44      41.625     0.136     0 < P < 1
       1      3        37      45.501     1.588     0 < P < 1
       1      4        40      33.834     1.124     0 < P < 1
       1      5        17      16.100     0.050     0 < P < 1
       1      6      1338    1338.859     0.001     0 < P < 1
         1 Cohort df= 5                   2.898     P = 0.7157
```

Figure A6.1 (*Continued*)

Appendix 6

```
SURVIV - Survival Rate Estimation with User Specified Cell Probabilities
12-Jun-89    13:46:31      Version 1.4(PC-DOS) June, 1989      Page 011
              Banding Handbook Example from Page 21 of Brownie et al. (1985)
------------------------------------------------------------------------
        2      1       62        64.319     0.084    0 < P < 1
        2      2       76        70.307     0.461    0 < P < 1
        2      3       44        52.280     1.311    0 < P < 1
        2      4       28        24.878     0.392    0 < P < 1
        2      5     1385      1383.216     0.002    0 < P < 1
        2 Cohort df= 4                      2.250    P = 0.6899
        3      1       82        79.201     0.099    0 < P < 1
        3      2       61        58.893     0.075    0 < P < 1
        3      3       24        28.025     0.578    0 < P < 1
        3      4      990       990.881     0.001    0 < P < 1
        3 Cohort df= 3                      0.753    P = 0.8606
------------------------------------------------------------------------
G Total (Degrees of freedom =   6)       6.021
Pr(Larger Chi-square) = 0.4208
With pooling, Degrees of freedom =   6  Pearson Chi-square =      5.901
Pr(Larger Chi-square) = 0.4343

Log-likelihood = -36.839708

   CPU time in seconds for last procedure was    3.79

INPUT --- PROC ESTIMATE NSIG=5 MAXFN=750 NAME=REC_SAME /*Constant
INPUT --- Recovery RATES*/;

INPUT ---    INITIAL;

INPUT ---    RETAIN=GENERAL ;

INPUT ---    CONSTRAINTS;
INPUT ---       S(3)=S(4);
INPUT ---       S(3)=S(5);

   Number of parameters in model    =  7

   Number of parameters set equal   =  2

   Number of parameters fixed       =  0

   Number of parameters estimated   =  5
```

(Figure continues.)

Analysis of Wildlife Radio-Tracking Data

```
SURVIV - Survival Rate Estimation with User Specified Cell Probabilities
12-Jun-89    13:46:31      Version 1.4(PC-DOS) June, 1989      Page 012
Banding Handbook Example from Page 21 of Brownie et al. (1985)
-----------------------------------------------------------------
```

			Lower	Upper	
I	Parameter	S(I)	Bound	Bound	Label
1	1	0.654	0.000	1.000	Survival for 1964
2	2	0.634	0.000	1.000	Survival for 1965
3	3	0.079	0.000	1.000	Recovery FOR 1964
4	3	0.079	0.000	1.000	Recovery FOR 1965
5	3	0.079	0.000	1.000	Recovery FOR 1966
6	4	0.051	0.000	1.000	Survival for 1966 and Recovery 1967
7	5	0.024	0.000	1.000	Survival for 1966-67, Recovery 1968

Final function value 2647.1385   (Error Return = 0)

Number of significant digits   7

Number of function evaluations   131

				95% Confidence Interval	
I	Parameter	S(I)	Standard Error	Lower	Upper
1	1	0.56107125	0.53357772E-01	0.45649002	0.66565249
2	2	0.72176072	0.63359611E-01	0.59757588	0.84594556
3	3	0.60639881E-01	0.32256549E-02	0.54317597E-01	0.66962164E-01
4	3	0.60639881E-01	0.32256549E-02	0.54317597E-01	0.66962164E-01
5	3	0.60639881E-01	0.32256549E-02	0.54317597E-01	0.66962164E-01
6	4	0.48934496E-01	0.47092640E-02	0.39704339E-01	0.58164654E-01
7	5	0.23286071E-01	0.30247206E-02	0.17357618E-01	0.29214523E-01

Figure A6.1 (*Continued*)

Appendix 6

SURVIV - Survival Rate Estimation with User Specified Cell Probabilities
12-Jun-89   13:46:31   Version 1.4(PC-DOS) June, 1989   Page 013
Banding Handbook Example from Page 21 of Brownie et al. (1985)

         Variance-Covariance matrix of estimates on diagonal and below,
              Correlation matrix of estimates above diagonal.
    | 1   2   3   4   5   6
    | 7
 ---|---
    |
  1 | 0.28471E-02 -0.26502   -0.29049   -0.29049   -0.29049   -0.57322E-01
    | -0.42469E-01
    |
  2 | -0.89595E-03 0.40144E-02 -0.32750   -0.32750   -0.32750   -0.49398
    | -0.36598
    |
  3 | -0.49998E-04 -0.66933E-04 0.10405E-04 1.0000   1.0000   0.19733
    | 0.14620
    |
  4 | -0.49998E-04 -0.66933E-04 0.10405E-04 0.10405E-04 1.0000   0.19733
    | 0.14620
    |
  5 | -0.49998E-04 -0.66933E-04 0.10405E-04 0.10405E-04 0.10405E-04 0.19733
    | 0.14620
    |
  6 | -0.14404E-04 -0.14739E-03 0.29975E-05 0.29975E-05 0.29975E-05 0.22177E-04
    | 0.18810
    |
  7 | -0.68542E-05 -0.70138E-04 0.14264E-05 0.14264E-05 0.14264E-05 0.26793E-05
    | 0.91489E-05

  Cohort Cell   Observed Expected Chi-square Note
  ------ ----   -------- -------- ---------- --------
     1   1   127   97.206   9.132   0 < P < 1
     1   2   44   54.539   2.037   0 < P < 1
     1   3   37   39.364   0.142   0 < P < 1
     1   4   40   31.766   2.134   0 < P < 1
     1   5   17   15.116   0.235   0 < P < 1
     1   6   1338   1365.009   0.534   0 < P < 1
     1 Cohort df= 5   14.214   P = 0.0143

(*Figure continues.*)

Analysis of Wildlife Radio-Tracking Data

```
SURVIV - Survival Rate Estimation with User Specified Cell Probabilities
12-Jun-89     13:46:31       Version 1.4(PC-DOS) June, 1989      Page 014
Banding Handbook Example from Page 21 of Brownie et al. (1985)
---------------------------------------------------------------------
        2       1       62      96.721      12.464    0 < P < 1
        2       2       76      69.809       0.549    0 < P < 1
        2       3       44      56.334       2.700    0 < P < 1
        2       4       28      26.807       0.053    0 < P < 1
        2       5     1385    1345.329       1.170    0 < P < 1
        2 Cohort df= 4                      16.936    P = 0.0020
        3       1       82      70.160       1.998    0 < P < 1
        3       2       61      56.617       0.339    0 < P < 1
        3       3       24      26.942       0.321    0 < P < 1
        3       4      990    1003.280       0.176    0 < P < 1
        3 Cohort df= 3                       2.834    P = 0.4179
---------------------------------------------------------------------
    G Total (Degrees of freedom =   7)    35.083
    Pr(Larger Chi-square) = 0.0000
    With pooling, Degrees of freedom =   7   Pearson Chi-square =    33.985
    Pr(Larger Chi-square) = 0.0000


    Log-likelihood = -51.370480

    CPU time in seconds for last procedure was   3.46

INPUT --- PROC TEST;

        Submodel   Name        Log-likelihood  NDF   Akaike Inf. Criter.   G-O-F
        --------   -------     --------------  ---   -------------------   -----
           1       SV&RC_SAME   -52.874994      8         113.74999        0.0000
           2       GENERAL      -36.824114      5          87.648229       0.3072
           3       SURVL_SAME   -36.839708      6          85.679416       0.4208
           4       REC_SAME     -51.370480      7         112.74096        0.0000

                  Likelihood Ratio Tests Between Models

                General      Reduced                    Degrees    Pr(Larger
                Submodel     Submodel     Chi-square    Freedom    Chi-square)
                --------     --------     ----------    -------    -----------
                GENERAL      SV&RC_SAME     32.102         3         0.0000
                SURVL_SAME   SV&RC_SAME     32.071         2         0.0000
                REC_SAME     SV&RC_SAME      3.009         1         0.0828
```

Figure A6.1 (*Continued*)

Appendix 6

```
SURVIV - Survival Rate Estimation with User Specified Cell Probabilities
12-Jun-89    13:46:31        Version 1.4(PC-DOS) June, 1989      Page 015
Banding Handbook Example from Page 21 of Brownie et al. (1985)
------------------------------------------------------------------------
              GENERAL      SURVL_SAME     0.031       1     0.8598
              GENERAL      REC_SAME      29.093       2     0.0000
              SURVL_SAME   REC_SAME      29.062       1     0.0000

 * * WARNING * *   Sequence of models reinitialized to zero.

   CPU time in seconds for last procedure was    0.22

 INPUT --- PROC STOP /* End of this run. */;

   CPU time in minutes for this job was    0.31
         E X E C U T I O N    S U C C E S S F U L
```

Figure A6.1 (*Continued*)

The next statement continues to describe the fate of the 1603 ducks banded in 1964. Forty-four bands were recovered during the year 1965, with the expected cell probability being $S(1) \times S(4)$. The cell probability statements must be FORTRAN 77 readable, as these algebraic statements are used to construct the likelihood subroutine. The specification of observed recoveries and expected cell probabilities continues from the 1964 cohort through 1968. By this time, $127 + 44 + 37 + 40 + 17 = 265$ bands have been returned. Program SURVIV knows that the sum of the observed cells should add up to the number of animals specified on the proceeding cohort card. Because this is not the case, an additional cell is constructed with observed value equal to $1603 - 265 = 1338$ ducks never seen again. This cell has an expected cell probability equal to 1.0 minus the sum of the previous 5 cells. The sum of all the cell probabilities must be 1.0. Thus, if a duck was never seen again (i.e., the band is not reported), then it falls into this last cell.

PROC MODEL continues to process COHORT statements and their associated cell probability statements for the ducks banded in 1965 and 1966. Note that some of the parameters to be estimated occur in more than one cohort. This is reasonable because the survival rate of a duck banded in 1964 and one banded in 1965 should both be the same in 1966, assuming that the leg band is not changing the survival rate.

The LABELS statement is used to document the biological meaning of the parameters. Thus, $S(1)$ is shown to be the finite survival rate for the year 1964, i.e., S(1) = Survival for 1964. S(3) is the recovery rate for 1964. These parameters correspond to S_1 and f_1 in Brownie et al. (1985:21). S(6) and S(7) are products of survival and recovery rates which are not identifiable. These "nuisance" parameters must be in the model to estimate S(1) through S(5), but biologically are not of much value.

When PROC MODEL reaches the end of its input (signified by the PROC ESTIMATE statement), it constructs the FORTRAN 77 subroutine and writes it to a disk file. Program SURVIV then exits, the FORTRAN 77 compiler is called to compile the subroutine, the linker is called to place the binary code into the core image, and Program SURVIV is started up again; only this time, the compile run option is not set on the execute line so that Program SURVIV assumes that the version of the likelihood subroutine in the core image corresponds to the input to PROC MODEL. Thus PROC MODEL does not stop execution when the PROC ESTIMATE statement is reached, but passes control to PROC ESTIMATE.

PROC ESTIMATE then performs the numerical optimization of the likelihood function. Several parameters are specified on the statement. NSIG = 5 tells PROC ESTIMATE to estimate the parameters to 5 significant digits. MAXFN = 750 says that only 750 evaluations of the likelihood function are to be performed before an error is printed and optimization stopped. NAME = SV&RC_SAME gives this model a name (up to 10 characters) to identify it later in the PROC TEST output. All the parameters for PROC ESTIMATE have defaults, including NAME = SUBMODEL*ii*, where *ii* is the counter for the number of times that PROC ESTIMATE has been executed.

The INITIAL statement tells PROC ESTIMATE to initialize the parameter values in preparation for optimization. ALL = 0.05 sets all the values to 0.05, i.e., S(1) = 0.05, S(2) = 0.05, ..., S(7) = 0.05.

The next statement is used to estimate the model parameters with equal survival in 1964 and 1965, and equal recovery rates in 1964, 1965, and 1966. These equalities are set with the CONSTRAINTS statement. S(1) = S(2) sets survival rates for 1964 and 1965 equal. S(3) = S(4) and S(3) = S(5) sets the 1964, 1965, and 1966 recovery rates all equal to one another.

These constraints are reflected in the summary of initial conditions, because only 4 parameters are to be estimated. The column labeled PARAMETER in the summary table on page 003 of the output (Fig. A.6.1) indicates the relationship between S(1) and the parameter index used internally in SURVIV. Thus S(1) and S(2) are both parameter 1, and S(3), S(4), and S(5) are all parameter 2.

Following the input to PROC ESTIMATE, the summary of the initial con-

Appendix 6

ditions for the upcoming optimization is shown in the output. The initial value of each parameter, the lower bound, the upper bound, and the label are summarized in a table. After the table is completed, the results of the optimization procedure are printed. No errors are reported, the number of significant digits in the estimates is 7, and 133 function calls were required. The next table summarizes the parameter values, standard errors, and 95% confidence intervals. Next the variance–covariance and correlation matrix is printed for the parameter values. Finally, a table of the observed versus expected cell probabilities is printed, and the contribution of each cell to the overall chi-square goodness-of-fit statistic. If the model does not fit the data, this chi-square test will reject the null hypothesis. The NOTE column provides another method to check that the model was correctly optimized. The message $0 < P < 1$ indicates that the cell probability was in the interval [0–1]. Error messages are printed if the cell probability is not in the [0–1] interval.

In the summary of the estimation results, the values of $S(1)$ and $S(2)$ are identical, and likewise for $S(3)$, $S(4)$, and $S(5)$. Similarly, the variance–covariance matrix and the correlation matrix reflect that some of the parameters are constrained to be equal.

Three additional runs of PROC ESTIMATE are made in the sample output. First, all parameters are estimated individually (NAME = GENERAL). Then, the 1964 and 1965 survival rates are constrained, but not the recovery rates (NAME = SURVL_SAME). Finally, the 1964 and 1965 survival rates are not constrained, but the 1964, 1965, and 1966 recovery rates are all constrained to be equal (NAME = REC_SAME). Thus, a series of 4 models has been generated. The most general model allows every identifiable survival and recovery rate to be estimated. The other three models force various constraints on the estimates.

PROC TEST (page 014 of the output in Fig. A6.1) helps determine which of the four models is appropriate for the observed data. Each model is summarized in a table giving the log-likelihood function value, the degrees of freedom of the model, Akaike's Information Criterion (AIC) (Akaike 1973), and the chi-square goodness-of-fit significance probability. The AIC is computed as $-2*(\text{log-likelihood}) + 2*(\text{number of independently adjusted parameters})$. The smaller the value of AIC, the better the fit of the model to the observed data. In the example, only the GENERAL and SURVL_SAME models fit the observed data based on the goodness-of-fit test. Akaike's Information Criterion is the smallest for the SURVL_SAME model, suggesting that it is the appropriate model for the observed data. Both models with recoveries constrained equal for 1964, 1965, and 1966 reject the goodness-of-fit null hypothesis.

The next 6 lines provide the likelihood ratio test results of all pairs of models that do not have equal degrees of freedom. Thus, the first test suggests that the GENERAL model with 7 parameters fits the data much better than when survival and recovery rates are constrained to be equal by year, because the significance level of the chi-square statistic is less than 0.05. Tests of GENERAL versus either of the models with recovery rates constant by year strongly reject the null hypothesis, suggesting that constant recovery rates do not fit the observed data. However, the test of GENERAL versus SURVL_SAME does not reject ($P = 0.8598$), suggesting that survival may be constant across years.

Thus, the researcher would conclude from this analysis that constant recovery rates are not valid for the observed data, but constant survival rates may be.

If PROC ESTIMATE Fails

Chances that the numerical optimization procedure will not converge increase with: (1) the number of parameters, (2) with parameter estimates against their boundaries such as survival equal to 1, and (3) with model complexity. Generally the failure of the procedure to converge will be obvious: the number of significant digits will be less than what was requested, an error message will be printed, and/or some parameters will not have been changed from their initial values.

Generally, when PROC ESTIMATE does not converge properly, a warning is printed about the possible lack of parameter identifiability. When this warning is received, the user should verify whether each of the parameters can actually be estimated with the model and data input to SURVIV. We have not found a suitable numerical approach to determine parameter identifiability other than the error messages printed from PROC ESTIMATE. Occasionally, the numerical optimization procedure in PROC ESTIMATE will converge, but the variance–covariance matrix of the parameter estimates will not be positive-definite (meaning that the variance–covariance was not properly inverted). An error message will be printed in such cases. Lack of positive-definiteness of the variance–covariance matrix is a strong indication that the problem lacks identifiability.

Determination of parameter identifiability with numerical methods is difficult because of the limited numerical precision of any of the calculations performed. That is, a large problem (with many parameters to be estimated) will appear to lack identifiability, even though the parameters may actually be identifiable. We have found no suitable method to distinguish numerical problems in a large or poorly conditioned problem from the case where the parameters are really not identifiable. Thus, the user must carefully check the model input to SURVIV to verify that parameters are identifiable.

When PROC ESTIMATE does not produce reasonable answers, the user can try several things to correct the situation. The first is to change the initial parameter estimates to values closer to the true values, i.e., make better initial guesses. Observed data can be used to obtain these guesses rather than just setting the initial value to 0.5. Also, PROC ESTIMATE could be run with a constrained model to obtain better estimates, and use the INITIAL; RETAIN = ___ feature of PROC ESTIMATE to input the parameter estimates from this reduced model for the more complex model, as was done with GENERAL in Fig. A6.1.

Second, PROC ESTIMATE allows an absolute value transformation that may improve convergence of the algorithm when one or more parameters are at their upper or lower boundaries. For example, if $S(1) < 0.75$ is specified as a constraint, and the optimization procedure wants to make $S(1) > 0.75$, a convergence problem may result. By specifying TRANSFORM = ABS on the PROC ESTIMATE statement, the difficulties with constrained parameters may disappear.

Finally, another procedure is available in SURVIV to generate input files for the BMDPAR program of the BMDP statistical package (Dixon 1983). Two files are generated for input to BMDPAR. BMDFUN.FOR contains the FORTRAN source code for the BMDPAR FUN subroutine; and BMDINP.DAT contains the input to BMDPAR. PROC BMDPAR utilizes the same INITIAL and CONSTRAINT statements as PROC ESTIMATE and passes this information on to the BMDPAR program. One advantage of using the BMDPAR program to perform the optimization is the greater flexibility to specify inequality constraints. Once the BMDPAR input file has been generated, the user can modify it to provide constraints not handled by SURVIV. Thus, the user can modify the file of instructions to the BMDP-BMDPAR program to handle constraints such as survival $< (1 - \text{recovery})$. Verification of results from SURVIV with BMDPAR is also encouraged to guard against the possibility of a local maximum of the likelihood function being accepted as the global maximum.

Data Simulation and Sample Sizes

The preceding example illustrates the basic methods of data analysis for Program SURVIV. Two additional procedures help design survival estimation studies.

PROC SIMULATE performs Monte Carlo simulations for the model defined in the preceding PROC MODEL statements. An example utilizing a simple biotelemetry experiment is shown in Fig. A.6.2 and is also used in White (1983). The question to be answered in this simulation is whether 40 collars in year 1 and 50 collars in year 2 are enough effort to detect changes in survival rates of 0.1. Thus, the first PROC ESTIMATE simulates 4 survival rates: 0.6, 0.7, 0.5, and 0.6. The second PROC ESTIMATE uses the same data, but constrains S(1) = S(2) and S(3) = S(4). Thus, the PROC TEST procedure tests for constant survival in years 1 and 2 and in years 3 and 4.

```
SURVIV - Survival Rate Estimation with User Specified Cell Probabilities
12-Jun-89    14:33:40        Version 1.4(PC-DOS) June, 1989      Page 001
----------------------------------------------------------------------
This version of SURVIV can only handle input lines
of about 120 characters. Do not exceed this limitation.

       Dimension limitations for this run:
          Maximum number of parameters            36
          Maximum number of cohorts               25
          Maximum number of classes within a cohort  10
          Maximum number of models for PROC TEST  20
       If your problem needs larger dimensions, reset the values
       in the MODELC include file and recompile the program.



    INPUT --- PROC TITLE Small Radio-tracking Example for Simulation in
    INPUT --- User's Manual;

       CPU time in seconds for last procedure was     0.22



    INPUT --- PROC MODEL NPAR=4 /* Simple Radio-tracking Example */;

       INPUT ---    COHORT = 40 /* Number of animals collared in year 1 */;
       INPUT ---    0:(1.-S(1));
       INPUT ---    0:S(1)*(1.-S(2));
       INPUT ---    0:S(1)*S(2)*(1.-S(3));
       INPUT ---    40:S(1)*S(2)*S(3);
```

Figure A6.2 Example of estimating survival rates from biotelemetry data when the data are simulated with PROC SIMULATE. (*Figure continues.*)

Appendix 6

```
SURVIV - Survival Rate Estimation with User Specified Cell Probabilities
12-Jun-89    14:33:40      Version 1.4(PC-DOS) June, 1989     Page 002
Small Radio-tracking Example for Simulation in User's Manual
------------------------------------------------------------------
   INPUT ---    COHORT = 50 /* Number of animals collared in year 2 */;
   INPUT ---       0:(1.-S(2));
   INPUT ---       0:S(2)*(1.-S(3));
   INPUT ---       0:S(2)*S(3)*(1.-S(4));
   INPUT ---      50:S(2)*S(3)*S(4);

   INPUT ---    LABELS;
   INPUT ---       S(1)=Survival year 1;
   INPUT ---       S(2)=Survival year 2;
   INPUT ---       S(3)=Survival year 3;
   INPUT ---       S(4)=Survival year 4;

    CPU time in seconds for last procedure was    1.59

   INPUT --- PROC SIMULATE NSIM=200 SEED=4567555 PARFILE=SIMDATA ERROR
   INPUT --- TEST;

   INPUT ---    INITIAL;
   INPUT ---       S(1)=0.6;
   INPUT ---       S(2)=0.7;
   INPUT ---       S(3)=0.5;
   INPUT ---       S(4)=0.6;

    Number of parameters in model   = 4

    Number of parameters set equal  = 0

    Number of parameters fixed      = 0

    Number of parameters estimated  = 4

                         Lower    Upper
   I   Parameter  S(I)   Bound    Bound  Label
   --  ---------  -----  -----    -----  ----------------
   1       1      0.600  0.000    1.000  Survival year 1
   2       2      0.700  0.000    1.000  Survival year 2
   3       3      0.500  0.000    1.000  Survival year 3
   4       4      0.600  0.000    1.000  Survival year 4
```

(Figure continues.)

```
SURVIV-Survival Rate Estimation with User Specified Cell Probabilities
12-Jun-89    14:33:40       Version 1.4(PC-DOS) June, 1989    Page 003
Small Radio-tracking Example for Simulation in User's Manual
-----------------------------------------------------------------------

    INPUT ---  PROC ESTIMATE NAME=GENERAL /* All parameters individually
    INPUT ---  estimated */;

    INPUT ---      INITIAL;
    INPUT ---         S(1)=0.6;
    INPUT ---         S(2)=0.7;
    INPUT ---         S(3)=0.5;
    INPUT ---         S(4)=0.6;

       Number of parameters in model    =  4

       Number of parameters set equal   =  0

       Number of parameters fixed       =  0

       Number of parameters estimated   =  4


                              Lower    Upper
   I   Parameter   S(I)       Bound    Bound   Label
   --  ---------  -----       -----    -----   ---------------
   1       1      0.600       0.000    1.000   Survival year 1
   2       2      0.700       0.000    1.000   Survival year 2
   3       3      0.500       0.000    1.000   Survival year 3
   4       4      0.600       0.000    1.000   Survival year 4

    INPUT ---  PROC ESTIMATE NAME=CONSTRAIN /* Sets of 2 parameters
    INPUT ---  constrained equal */;

    INPUT ---      INITIAL;
    INPUT ---         S(1)=0.6;
    INPUT ---         S(2)=0.7;
    INPUT ---         S(3)=0.5;
    INPUT ---         S(4)=0.6;
```

Figure A6.2 (*Continued*)

Appendix 6

```
SURVIV - Survival Rate Estimation with User Specified Cell Probabilities
12-Jun-89    14:33:40       Version 1.4(PC-DOS) June, 1989    Page 004
Small Radio-tracking Example for Simulation in User's Manual
------------------------------------------------------------------------

    INPUT ---     CONSTRAINTS;
    INPUT ---        S(1)=S(2);
    INPUT ---        S(3)=S(4);

      Number of parameters in model    = 4

      Number of parameters set equal   = 2

      Number of parameters fixed       = 0

      Number of parameters estimated   = 2

                         Lower    Upper
  I    Parameter  S(I)   Bound    Bound  Label
  --   ---------  -----  -------  -----  ---------------
  1       1       0.600  0.000    1.000  Survival year 1
  2       1       0.600  0.000    1.000  Survival year 2
  3       2       0.500  0.000    1.000  Survival year 3
  4       2       0.500  0.000    1.000  Survival year 4

      Parameter estimates are written to the file SIMDATA
      with the order being Replicate Number, Model Number,
      and the parameters (S(I),SE(S(I))),I=1,NPAR)
      The format used is '(8F10.6)', which generates (NPAR*2+2)/8+1 = 2
      'CARDS' per model.

*** ERROR MESSAGE IER=130 FROM ROUTINE OPTMIZ
 * * WARNING * *   Rounding errors became dominant before parameters
                   estimated to NSIG digits for Model GENERAL
 * * WARNING * *   Check to be sure the parameters are identifiable,
                   but the problem may just be ill-conditioned.
```

(Figure continues.)

```
SURVIV - Survival Rate Estimation with User Specified Cell Probabilities
12-Jun-89   14:33:40      Version 1.4(PC-DOS) June, 1989      Page 005
Small Radio-tracking Example for Simulation in User's Manual
```

```
                  S I M U L A T I O N   R E S U L T S
              MODEL GENERAL     NSIM = 200   Converged = 198

        MEAN              SE              95% C. I.
   I    S(I)-Hat         S(I)-Hat       Lower        Upper
   -------------------------------------------------------------
   1  0.60391414    0.56356261E-02  0.59286831    0.61495997   Survival year 1
   2  0.70488091    0.33449553E-02  0.69832480    0.71143702   Survival year 2
   3  0.49442084    0.44410065E-02  0.48571647    0.50312521   Survival year 3
   4  0.59348927    0.87697724E-02  0.57630052    0.61067802   Survival year 4

        MEAN              SE              95% C. I.
   I   SE(S(I)-Hat)   (SE(S(I)-Hat)   Lower        Upper
   -------------------------------------------------------------
   1  0.76241201E-01 0.23528895E-03 0.75780035E-01 0.76702368E-01 Survival year 1
   2  0.52660168E-01 0.19771195E-03 0.52272652E-01 0.53047683E-01 Survival year 2
   3  0.68774219E-01 0.20534482E-03 0.68371743E-01 0.69176695E-01 Survival year 3
   4  0.11446660     0.86390657E-03 0.11277334     0.11615985     Survival year 4


   Results of Goodness of Fit Tests
      <0.01  <0.05  <0.10  <0.50
   -----------------------------
         1     11     20    100

                  S I M U L A T I O N   R E S U L T S
              MODEL CONSTRAIN   NSIM = 200   Converged = 199

        MEAN              SE              95% C. I.
   I    S(I)-Hat         S(I)-Hat       Lower        Upper
   -------------------------------------------------------------
   1  0.66939285    0.30297456E-02  0.66345454    0.67533115   Survival year 1
   2  0.66939285    0.30297456E-02  0.66345454    0.67533115   Survival year 2
   3  0.51826990    0.42738014E-02  0.50989325    0.52664655   Survival year 3
   4  0.51826990    0.42738014E-02  0.50989325    0.52664655   Survival year 4
```

Figure A6.2 (*Continued*)

Appendix 6

```
SURVIV - Survival Rate Estimation with User Specified Cell Probabilities
12-Jun-89    14:33:40        Version 1.4(PC-DOS) June, 1989    Page 006
Small Radio-tracking Example for Simulation in User's Manual
-------------------------------------------------------------------------
        MEAN           SE              95% C. I.
  I  SE(S(I)-Hat)  SE(S(I)-Hat)    Lower          Upper
-------------------------------------------------------------------------
  1 0.43344202E-01 0.12598464E-03 0.43097272E-01 0.43591132E-01 Survival year 1
  2 0.43344202E-01 0.12598464E-03 0.43097272E-01 0.43591132E-01 Survival year 2
  3 0.59965351E-01 0.19635173E-03 0.59580501E-01 0.60350200E-01 Survival year 3
  4 0.59965351E-01 0.19635173E-03 0.59580501E-01 0.60350200E-01 Survival year 4

  Results of Goodness of Fit Tests
    <0.01  <0.05  <0.10  <0.50
  ----------------------------
      8     28     47    136

            Likelihood Ratio Test Results
                        <0.01  <0.05  <0.10  <0.50  <1.00
  --------------------------------------------------------
  GENERAL   VS. CONSTRAIN   10    38     57    136    197

    CPU time in seconds for last procedure was  379.98

  INPUT --- PROC STOP /* End of this run. */;

    CPU time in minutes for this job was   6.36

          E X E C U T I O N   S U C C E S S F U L
```

Figure A6.2 *(Continued)*

Specification of DETAIL for **PROC SIMULATE** causes a detailed printout of each of the simulations. Thus, parameter estimates and the chi-square goodness-of-fit table are both printed. If a large number of simulations are performed, the output produced will be very large. Thus, DETAIL should only be used with the specification NSIM = 10 or less, i.e., only 10 simulations are performed.

The PARFILE = file_name_specification on PROC SIMULATE is used to name an output file to receive the estimates and associated standard errors for each of the parameters to be estimated for each of the PROC ESTIMATEs. This file might be used as input to a statistical package to perform a more thorough analysis of the simulations than the summary printed by PROC SIMULATE. The format of this file is documented in the output from PROC SIMULATE.

A random number seed, in order to simulate data with PROC SIMULATE, is provided in the SEED = value specification. Generally, the seed should be a 5 or 7 digit odd random integer to provide a proper random starting value. Specifying the same seed for two PROC SIMULATEs will generate identical data if the same model is used.

PROC SAMPLE SIZE is used to estimate the number of animals to band, based on the methods described in Brownie et al. (1985:190–193). The example shown in Fig. A.6.3 duplicates the example in Brownie et al. (1985:186–193).

```
SURVIV - Survival Rate Estimation with User Specified Cell Probabilities
12-Jun-89    14:00:26      Version 1.4(PC-DOS) June, 1989      Page 001
------------------------------------------------------------------------

This version of SURVIV can only handle input lines
of about 120 characters. Do not exceed this limitation.

      Dimension limitations for this run:
        Maximum number of parameters              36
        Maximum number of cohorts                 25
        Maximum number of classes within a cohort 10
        Maximum number of models for PROC TEST    20
      If your problem needs larger dimensions, reset the values
      in the MODELC include file and recompile the program.



INPUT --- PROC TITLE Sample Size Estimation for Banding Study;

      CPU time in seconds for last procedure was 0.27
```

Figure A6.3 Example of input to Program SURVIV for PROC SAMPLE SIZE and the output produced. (*Figure continues.*)

Appendix 6

```
SURVIV - Survival Rate Estimation with User Specified Cell Probabilities
12-Jun-89    14:00:26       Version 1.4(PC-DOS) June, 1989     Page 002
Sample Size Estimation for Banding Study
------------------------------------------------------------------------

 INPUT ---  PROC SAMPLE SIZE /* Sample size's for Brownie et al.
 INPUT ---  (1985:192) example */;

 INPUT ---     S-BAR=0.6;

 INPUT ---     S'-BAR=0.5;

 INPUT ---     F-BAR=0.07;

 INPUT ---     CV(S)=0.04;

 INPUT ---     CV(S')=0.04 0.07 0.03;

 INPUT ---     K=2 10 1;
```

SAMPLE SIZE ESTIMATION FOLLOWING BROWNIE ET AL. 1985:190-193

NOTATION	INTERPRETATION
S-BAR	Mean adult survival rate per year
S'-BAR	Mean young survival rate per year
F-BAR	Mean adult recovery rate per year
CV(S)	Coefficient of variation of mean adult survival rate
CV(S')	Coefficient of variation of mean young survival rate
K	Number of years to conduct banding effort
M	Number of young to band each year
N	Number of adults to band each year

S-BAR	S'-BAR	F-BAR	CV(S)	CV(S')	K	N	M
0.60	0.50	0.07	0.04	0.04	2	22560	27269
0.60	0.50	0.07	0.04	0.04	3	5698	16597
0.60	0.50	0.07	0.04	0.04	4	2554	15982
0.60	0.50	0.07	0.04	0.04	5	1462	
0.60	0.50	0.07	0.04	0.04	6	959	
0.60	0.50	0.07	0.04	0.04	7	686	

(Figure continues.)

```
SURVIV - Survival Rate Estimation with User Specified Cell Probabilities
12-Jun-89      14:00:26          Version 1.4(PC-DOS) June, 1989      Page 003
Sample Size Estimation for Banding Study
------------------------------------------------------------------------
       0.60      0.50      0.07      0.04      0.04       8       520
       0.60      0.50      0.07      0.04      0.04       9       411
       0.60      0.50      0.07      0.04      0.04      10       336
       0.60      0.50      0.07      0.04      0.07       2     22560     6395
       0.60      0.50      0.07      0.04      0.07       3      5698     2798
       0.60      0.50      0.07      0.04      0.07       4      2554     1740
       0.60      0.50      0.07      0.04      0.07       5      1462     1257
       0.60      0.50      0.07      0.04      0.07       6       959      987
       0.60      0.50      0.07      0.04      0.07       7       686      816
       0.60      0.50      0.07      0.04      0.07       8       520      700
       0.60      0.50      0.07      0.04      0.07       9       411      616
       0.60      0.50      0.07      0.04      0.07      10       336      553

       CPU time in seconds for last procedure was     0.82

INPUT --- PROC STOP /* End of this run. */;

       CPU time in minutes for this job was    0.02

            E X E C U T I O N   S U C C E S S F U L
```

Figure A6.3 (*Continued*)

The first number in each of the specification cards [S-BAR, S'-BAR, F-BAR, CV(S), CV(S'), and K] provides the value to be used for that parameter in the sample size calculation. However, because a range of parameter values are usually required, two additional values can be used to specify a range of values. Thus, S-BAR = 0.1 0.5 0.1 tells PROC SAMPLE SIZE to use a range of S from 0.1 to 0.5 in increments of 0.1. Note that specifying ranges for all 5 input parameters can result in many pages of sample sizes to be output.

Additional Procedures

PROC STOP is used to exit the program and is provided to stop execution in a data file when additional models are included in the file, but it cannot be executed in this particular run because the wrong PROC MODEL has been loaded into memory. PROC STOP allows the stacking of multiple models in an input file, but only runs the first one. A text editor can be used to move models from within the input file to the top.

PROC BROWNIE will generate output of the type from PROC ESTIMATE from input to the programs ESTIMATE and BROWNIE described in Brownie et al. (1985:155–158). To use SURVIV on this input, a semicolon must be put at the end of each card image; i.e., each card of input to the Brownie et al. programs is treated as a statement in SURVIV. PROC BROWNIE generates the appropriate COHORT and cell probability cards for input to PROC MODEL, lists these cards on the output, and then runs PROC MODEL to generate the estimation subroutine. Once the estimation subroutine is loaded, PROC BROWNIE calls PROC ESTIMATE to generate estimates and likelihood function values for the models appropriate to the type of input. Thus, PROC BROWNIE can be viewed as a combination of PROC MODEL and PROC ESTIMATE statements. PROC BROWNIE will generate each of the models analyzed in the Brownie et al. programs, and if PROC TEST is used, likelihood ratio test between models can be generated. The big advantage of using SURVIV to analyze these models is that additional hypotheses can be tested by using PROC ESTIMATE to analyze models not specified in the Brownie et al. programs. For example, to test if survival is constant for years 1970–75 but different from years 1976–1980, the appropriate constraint cards can be used. The definitions of the $S(I)$'s must be read from a previous run of PROC BROWNIE unless the user is very familiar with the coding scheme used and can construct the tests without viewing the coding scheme.

PROC IDENTIFIABLE is useful in locating which pair of parameters are most highly correlated. The input specifications ADD and MULTIPLY allow the user to determine how the parameter value is changed to calculate the numerical derivatives used in this process. Generally PROC IDENTIFIABLE is a last resort for determining the nonidentifiable parameters in a model.

File Names and Execution Time Parameters

Program SURVIV uses five files: to read input from, to write output on, to construct the FORTRAN 77 subroutine on, to construct the BMDPAR FUN subroutine on, and to construct the BMDPAR input on. The default file names are SURVIN.DAT, SUPVLP.DAT, EST.FOR, BMDFUN.FOR, and BMDINP.DAT. For DOS, the default input and output files are the keyboard and monitor, respectively.

When SURVIV is started, the program asks for the user to enter the execution time parameters from the keyboard. The input file name can be modified at this time by specifying I = filename. Likewise, the output from Program SURVIV can be placed on file XYZ by specifying L = XYZ in response to the startup question. The subroutine output file and the BMDPAR files can only be changed by modifying the program.

Besides the I and L parameters that are specified at execution time, three additional parameters can be specified. COMPILE RUN tells PROC MODEL to write the estimation subroutine to disk and stop execution, that is, the correct model has not yet been replaced in the memory. If COMPILE RUN is not specified, execution proceeds, assuming that the correct model is in memory. NOECHO causes the input lines to not be printed in the output. This option can save some paper when the output is printed, but it is not advisable to use because of the difficulty in locating errors in the input. LINES = # specifies the number of output lines to print per page, with the default LINES = 60.

Installation Guide for DOS Microcomputers

SURVIV is distributed by the authors on two 5 1/4 inch DSDD floppy diskettes. The first disk contains the source code with separate files for each subroutine. The second floppy contains the object code generated with the Ryan–McFarland FORTRAN 77 compiler. To run SURVIV, at least 512k of RAM is required, plus a hard disk. For most problems, the size of the SURVIV.EXE file exceeds 360k, thus the hard disk requirement. A batch file (SURVIVE.BAT) is included on the second floppy disk to execute SURVIV, including the compile run, to create EST.FOR, to execute the Ryan–McFarland compiler to generate EST.OBJ, to re-create SURVIV.EXE, and to run SURVIV again to perform the numerical estimation phase of the process. The batch file is interactive, and tells the user what input to enter. Any DOS editor can be used to create the input file. Example input is included in the SURVIN file on the floppy diskettes.

The program has been compiled with the Ryan–McFarland Fortran 77 compiler, Version 2.4. Thus, the user must have this compiler available to execute the program as distributed, because the EST.FOR routine must be compiled with the same compiler as used to compile the rest of the program. Other compilers that are Fortran 77 compatible could be used if the entire program is re-compiled.

Appendix 6

Setting Larger Program Dimensions

Program SURVIV has been set up with flexible array dimensions through the use of FORTRAN 77 PARAMETER statements. All the critical dimensions that would need to be changed for larger problems are in the INCLUDE file MODELC. These PARAMETERS are well documented with comment statements. Values which can be changed are MAXPAR (maximum number of parameters that can be estimated), MAXCHT (maximum number of COHORTS allowed in a model), MXCLAS (maximum number of probability cells within a cohort), MAXMOD (maximum number of models that can be generated with PROC ESTIMATE and then tested with PROC TEST), and MAXCHS (the maximum size of the character string holding the input buffer). DOS does not allow character strings larger than 255 characters, so MAXCHS is set to 255. This limitation means that statements longer than about 120 characters are not allowed. Also, cell probability expressions in PROC MODEL may not exceed 255 characters in the DOS version of SURVIV.

References

Akaike, H. 1973. Information theory and an extension of the maximum likelihood principle. Pages 267–281 *in* B. Petrov and F. Czakil Proc. 2nd Int. Symp. Inf. Theory. Akadémiaí Kiadó, Budapest.

Brownie, C., D. R. Anderson, K. P. Burnham, and D. S. Robson. 1985. Statistical inference from band recovery data—A handbook, 2nd ed., Res. Publ. No. 131. U.S. Fish and Wildl. Serv., Washington, D.C. 305 pp.

Dixon, W. J. (ed.). 1983. BMDP Statistical Software 1983. Univ. of Calif. Press, Berkeley. 734 pp.

White, G. C. 1983. Numerical estimation of survival rates from band recovery and biotelemetry data. J. Wildl. Manage. 47:716–728.

APPENDIX

7

SAS Home Range Estimation Procedures

LISTING A7.1
SAS code to determine the coordinates defining the minimum area polygon, calculate the area, plot the data and polygon, and test the data against a uniform distribution within the polygon

```
*  ----------------------------------------------------------  *
| SAS procedure to determine minimum area polygon from set of  |
| biotelemetry data, calculate area, and plot polygon with     |
| original data.                                               |
|                                                              |
| A SAS data set is processed -- this data set should have the |
| following variables defined:                                 |
|  id - Animal id used for BY statement processing             |
|  x  - x coordinate                                           |
|  y  - y coordinate                                           |
|                                                              |
| Definitions of variables used in this program:               |
|                                                              |
|  xcoor   - 1 dimension array of x coordinates                |
|  ycoor   - 1 dimension array of y coordinates                |
|  n       - number of coordinates in xcoor and ycoor          |
|  npoly   - number of points in polygon, sorted in clockwise  |
|            order to top of coor on return (0 means error)    |
|  area    - area in polygon                                   |
|  wsq     - Value of Cramer-von Mises goodness of fit statistic |
|  gofprob - Interpolated probability level of wsq             |
*  ----------------------------------------------------------  *;
```

```
title 'Minimum Area Polygon Calculation';
libname library 'c:';
data datafile;
   set library.deer;   /* IMPORTANT -- Specify SAS data file here */
   year=year(date);
   keep id year x y;
proc sort; by year id;
data polygon;
   array xcoor{300} x1-x300;
   array ycoor{300} y1-y300;
   array z{300};
   array wcrit{5} w01 w025 w05 w10 w15;
   retain x1-x300 y1-y300 n;
   set datafile; by year id;
   label area='Minimum convex polygon area in units**2'
         n='Number of telemetry locations'
         npoly='Number of corners in convex polygon'
         wsq='W**2 gof test'
         gofprob='Probability level W**2';
   keep id x y area n npoly wsq gofprob;
   if first.id then n=0;
   n=n+1;
   if n > 300 then do;
      put 'ERROR -- More than 300 observations for' id=;
      abort return 300;
      end;
   xcoor{n}=x; ycoor{n}=y;
   if last.id then do;
        /* find maximum y value */
      imin=1;
      do i=2 to n;
         if ycoor{i} < ycoor{imin} then imin=i; end;
      /* put minimum y coordinate to position n+1 */
      xcoor{n+1}=xcoor{imin}; ycoor{n+1}=ycoor{imin};
      M=0; minangle=0;
      /* Start loop to find the rest of the corners */
      do until (imin=n+1);
        M=M+1;
        temp=xcoor{M}; xcoor{M}=xcoor{imin}; xcoor{imin}=temp;
        temp=ycoor{M}; ycoor{M}=ycoor{imin}; ycoor{imin}=temp;
        imin=n+1; v=minangle; minangle=360;
```

Appendix 7

```
        do i=M+1 to n+1;
            link thetacal;
            if theta > v then do;
                if theta < minangle then do;
                    imin=i; minangle=theta; end;
                end;
            end;
        end;
        npoly=M;
        area=0;
        if npoly >= 3 then do;
            area=xcoor{1}*(ycoor{npoly}-ycoor{2})+
                 xcoor{npoly}*(ycoor{npoly-1}-ycoor{1});
            do i=2 to npoly-1;
                area=area+xcoor{i}*(ycoor{i-1}-ycoor{i+1});
                end;
            area=area*(-0.5);
            end;
        else area=.;
* ----------------------------------------------------- *
| SAS procedure to test for bivariate uniform data.     |
|                                                       |
|   Reference: Samuel, M. D. and E. O. Garton. 1985. Home |
|     range: a weighted normal estimate and tests of under- |
|     lying assumptions. J. Wildl. Manage. 49(2):513-519.   |
|                                                       |
* ----------------------------------------------------- *;
        * Determine the rectangle that fits around the polygon;
        xu=xcoor{1}; xl=xcoor{1}; yu=ycoor{1}; yl=ycoor{1};
        do i=2 to npoly;
            xu=max(xu,xcoor{i});
            xl=min(xl,xcoor{i});
            yu=max(yu,ycoor{i});
            yl=min(yl,ycoor{i});
            end;
        seed=7654321; pi=4*atan(1);
        dbar=0;
        do i=1 to n;
            * Generate a random coordinate that is inside the polygon;
newval:     x=ranuni(seed)*(xu-xl)+xl;
            y=ranuni(seed)*(yu-yl)+yl;
```

```
   link inside;
   if not in then goto newval;
   distmin=sqrt((xcoor{1}-x)**2+(ycoor{1}-y)**2);
   do j=2 to n;
      distmin=min(distmin,sqrt((xcoor{j}-x)**2+(ycoor{j}-y)**2));
      end;
   * Calculate z{i} values for Cramer-von Mises test;
   z{i}=pi*distmin*distmin;
   dbar=dbar+z{i};
   end;
 dbar=dbar/n;
 do i=1 to n;
    z{i}=1-exp(-z{i}/dbar);
    end;
 link zsort;  /* Sort z array into ascending order */
 * Calculate critical values of this test.;
 den=1.0+0.6/n;
 w01=0.337/den;
 w025=0.273/den;
 w05=0.224/den;
 w10=0.177/den;
 w15=0.149/den;
 put @15 '****************' id= '****************';
 put @10 'Critical values of W**2 statistic';
 put @14 '(Listed in output as WSQ)';
 put @10 '----------------------------------';
 put @15 'W(0.01)   = ' w01;
 put @15 'W(0.025)  = ' w025;
 put @15 'W(0.05)   = ' w05;
 put @15 'W(0.10)   = ' w10;
 put @15 'W(0.15)   = ' w15;
 link w2calc; wsq=wvalue; gofprob=prob;
 * Output polygon vertices to file;
 do i=1 to npoly;
    x=xcoor{i};
    y=ycoor{i};
    output;
    end;
  end;  /* end of do for last.id if */
return;
```

Appendix 7

```
*---------------------------------------------*
|    This subroutine determines whether a point lies    |
|    inside or outside of a polygon (regular or irregular, |
|    including concave and re-entrant).                 |
|    Written by Melvin L. Prueitt. Version Aug. 25,1980. |
| npoly - Number of vertices of the polygon.            |
| x     - x coordinate of the point in question.        |
| y     - y coordinate of the point in question.        |
| xcoor, ycoor are arrays giving the coordinates of the |
|         polygon vertices.                             |
| in    - returned to the caller as a 1 if the point (x,y) |
|         is inside the polygon, 0 if outside. points on the |
|         boundary are considered to be outside, but rounding |
|         errors sometimes cause boundary points to be inside.|
*---------------------------------------------*;
inside: in=0;
        iii = npoly;
 s100:f = ycoor{1}-ycoor{iii};
        if f = 0 then do;
            iii = iii-1;
            if iii > 1 then goto s100;
            else goto s600;
            end;
        else
            do iii=1 to npoly;
                jjj = mod(iii,npoly)+1;
                if (y-ycoor{iii})*(y-ycoor{jjj}) > 0 then goto s500;
                if y = ycoor{jjj} then goto s500;
                if y ^= ycoor{iii} then goto s400;
                if f*(ycoor{jjj}-y) > 0 then goto s500;
 s400:          if x > (y-ycoor{iii})*(xcoor{jjj}-xcoor{iii})/
                    (ycoor{jjj}-ycoor{iii})+xcoor{iii} then in = 1 - in;
 s500:          if ycoor{jjj} ^= ycoor{iii} then f = ycoor{jjj}-ycoor{iii};
            end;
s600: return;
* Sort z{i} vector;
zsort:do i=1 to n-1;
        do j=i to n;
            if z{i} > z{j} then do;
                wvalue=z{i}; z{i}=z{j}; z{j}=wvalue; end;
            end;
        end;
    return;
```

```
* Calculate W**2 statistic;
w2calc: wvalue=0;
     do i=1 to n;
        wvalue=wvalue+(z{i}-(2*i-1)/(2*n))**2+(1/(12*n));
        end;
     link interp;
     return;
* Interpolate probability from critical values;
interp:if wvalue > wcrit{1} then prob=0.01;
     else if wvalue > wcrit{2} then
        prob=0.01+((wvalue-wcrit{1})/(wcrit{2}-wcrit{1}))*(0.025-0.01);
     else if wvalue > wcrit{3} then
        prob=0.025+((wvalue-wcrit{2})/(wcrit{3}-wcrit{2}))*(0.05-0.025);
     else if wvalue > wcrit{4} then
        prob=0.05+((wvalue-wcrit{3})/(wcrit{4}-wcrit{3}))*(0.10-0.05);
     else if wvalue > wcrit{5} then
        prob=0.10+((wvalue-wcrit{4})/(wcrit{5}-wcrit{4}))*(0.15-0.10);
     else prob=0.15;
     return;
* Determine theta -- a number between 0 and 360 that is not the ;
* angle made by point at M and point at i with the horizontal   ;
* but which has the same order properties as the true angle     ;
thetacal: dx=xcoor{i}-xcoor{M}; ax=abs(dx);
          dy=ycoor{i}-ycoor{M}; ay=abs(dy);
          if dx=0 & dy=0 then theta=0;
             else theta=dy/(ax+ay);
          if dx<0 then theta=2-theta;
             else if dy<0 then theta=4+theta;
          theta=theta*90;
     return;
proc print data=polygon;
*_____*
|  Code to plot polygon and merge with original data  |
* _____*;
data plots;
   retain firstx firsty lastx lasty;
   retain interval 25; /* Number of plot intervals along a side */
   keep x y id;
   set polygon; by id;
   if first.id then do;
      firstx=x;  firsty=y;
      lastx=x;   lasty=y;
      end;
```

```
     else do;
        xnow=x; ynow=y;
        link line;
        lastx=xnow; lasty=ynow;
        end;
     if last.id then do;
        xnow=firstx; ynow=firsty;
        link line;
        end;
     return;
line: xinc=(xnow-lastx)/interval;
      yinc=(ynow-lasty)/interval;
      do i=1 to interval;
         x=lastx+xinc*i;
         y=lasty+yinc*i;
         output;
         end;
      return;
data both;
   set datafile (in=indata) plots; by id;
   if indata=1 then plotsym='+';
   else plotsym='*';
proc sort; by id;
proc plot data=both;
   plot y*x=plotsym;
   by id;
run;
```

LISTING A7.2
SAS code to calculate the bivariate normal home range estimate (Jennrich and Turner 1969), plot the ellipse and data, and test the data against the bivariate normal distribution

```
* ---------------------------------------------- *
| SAS procedure to estimate and plot 95% Jennrich-Turner ellipse |
| home range estimate, and perform a chi-square goodness of fit  |
| test. The procedure uses PROC CORR to calculate the data sum-  |
| mary statistics needed, and uses a data step to calculate the  |
| home range and 95% confidence interval on this estimate.       |
| Then a second data step is used to calculate the points of     |
| the ellipse for plotting. This data set is then combined and   |
| sorted with the original data to produce plots of the ellipse  |
| on the original data. Finally, a goodness of fit test is       |
| performed in a separate data step.                             |
|                                                                |
| A SAS data set is processed -- this data set should have the   |
| following variables defined:                                   |
|   id   - Animal id used for BY statement processing            |
|   x    - x coordinate                                          |
|   y    - y coordinate                                          |
|   time - SAS date and time variable for this location          |
* ---------------------------------------------- *;
libname library 'c:';
data datafile;
    set library.deer;  /* <-- IMPORTANT -- specify SAS data file to
                                            process */
title 'Jennrich-Turner Home Range Estimation';
proc sort; by id;
proc corr cov noprint out=hrmss data=datafile; var x y; by id;
* ---------------------------------------------- *
|   Data step to calculate the home range estimate               |
* ---------------------------------------------- *;
data jt; /* Data set JT contains the sufficient statistics to
            calculate the home range, and the estimate and CI. */
    set hrmss; by id;
    drop _type_ _name_ x y;
    retain xmean ymean xvar yvar xycov corr area lci uci n;
    if _type_ = 'COV' & _name_ = 'X' then do;
        xycov=y; xvar=x; end;
    if _type_ = 'COV' & _name_ = 'Y' then yvar=y;
    if _type_ = 'CORR' & _name_ = 'X' then corr=y;
```

Appendix 7

```
   if _type_ = 'MEAN' then do;
      xmean=x; ymean=y; end;
   if _type_ = 'N' then n=min(x,y);
   if last.id then do;
      /* Use n-2 denominator in estimates of var and cov */
/*    yvar=yvar*(n-1)/(n-2);
      xvar=xvar*(n-1)/(n-2);
      xycov=xycov*(n-1)/(n-2);    */
      area=(4*atan(1))*sqrt(xvar*yvar-xycov**2)*cinv(0.95,2);
      lci=(2*n-4)*area/cinv(0.975,2*n-4);
      uci=(2*n-4)*area/cinv(0.025,2*n-4);
      output;
      end;
*----------------------------------------------------------*
| Code to calculate a chi-square goodness of fit of the    |
| bivariate normal distribution based on the number of     |
| locations between similar concentric ellipses.           |
*----------------------------------------------------------*;
data gof;
   retain ncells 10;
   array observ{10} observ1-observ10;
   retain observ1-observ10 xmean xsd ymean ysd cost sint;
   keep id n chisq gofprob observ1-observ10;
   label chisq='Chi-square gof statistic'
         gofprob='Chi-square gof probability';
   merge jt datafile; by id;
   if first.id then do;
      ncells=min(int(n/5),10);
      do i=1 to ncells;
         observ{i}=0;
         end;
      do i=ncells+1 to 10;
         observ{i}=.;
         end;
   end;
   chi2=(((x-xmean)**2)*yvar-2*(x-xmean)*(y-ymean)*xycov+
        ((y-ymean)**2)*xvar)/(xvar*yvar-(xycov**2));
   d=sqrt(chi2);
/* put x= y= chi2= d=;   */
   i=1;
   do while ( chi2 > cinv(i/ncells,2));
      i=i+1; if i = ncells then goto endloop;
      end;
```

351

```
endloop: observ{i}=observ{i}+1;
   if last.id then do;
      expect=n/ncells;
      chisq=0;
      do i=1 to ncells;
         chisq=chisq+(((observ{i}-expect)**2)/expect);
         end;
      gofprob=1-probchi(chisq,ncells-1);
      output;
   end;
data combine;
   merge jt gof;by id;
   drop observ1-observ10;
proc print;
*_____*
| Code to plot ellipse with original data    |
* _____ *;
data ellipse;
   set jt;
   keep id x y;
   pi=4*atan(1);
   R=sqrt((xvar+yvar)**2-4*(xvar*yvar-xycov**2));
   theta=atan(-2*xycov/(yvar-xvar-R));
   a=sqrt((yvar+xvar+R)*cinv(0.95,2)/2);
   b=sqrt((yvar+xvar-R)*cinv(0.95,2)/2);
   cost=cos(theta);
   sint=sin(theta);
   do psi=0 to 360 by 2;
      psir=psi*pi/180;
      acosp=a*cos(psir);
      bcosp=b*sin(psir);
      x=xmean+acosp*cost-bcosp*sint;
      y=ymean+acosp*sint+bcosp*cost;
      output;
      end;
  data plots;
     set datafile(in=indata) ellipse;
     if indata=1 then plotsym='+';
     else plotsym='*';
  proc sort; by id;
  proc plot; plot y*x=plotsym; by id;
  run;
```

LISTING A7.3

SAS code to estimate the Samuel-Garton weighted ellipse estimator (Samuel and Garton 1985), plot both the unweighted and weighted ellipses, and compute a goodness-of-fit test of the data to the bivariate normal distribution

```
* ---------------------------------------------------- *
| SAS procedure to determine weighted bivariate normal ellipse   |
| estimate of home range, goodness of fit tests, and plot the    |
| weighted and unweighted ellipses on top of the original data.  |
|                                                                |
|   Reference: Samuel, M. D. and E. O. Garton.  1985.  Home      |
|      range: a weighted normal estimate and tests of under-     |
|      lying assumtpions.  J. Wildl. Manage. 49(2):513-519.      |
|                                                                |
| A SAS data set is processed -- this data set must have the     |
| following variables defined:                                   |
|   id - Animal id used for BY statement processing              |
|   x  - x coordinate                                            |
|   y  - y coordinate                                            |
|                                                                |
| Definitions of variables used in this program:                 |
|                                                                |
|   xcoor   - 1 dimension array of x coordinates                 |
|   ycoor   - 1 dimension array of y coordinates                 |
|   n       - number of coordinates in xcoor and ycoor           |
|   xmean   - mean of xcoor                                      |
|   ymean   - mean of ycoor                                      |
|   xvar    - variance of xcoor                                  |
|   yvar    - variance of ycoor                                  |
|   xycov   - covariance of xcoor and ycoor                      |
|   area    - area in Jennrich-Turner 95% ellipse                |
|   lci     - lower 95% CI on area                               |
|   uci     - upper 95% CI on area                               |
|   wsq     - Value of Cramer-von Mises goodness of fit statistic|
|   gofprob - Interpolated probability level of wsq              |
|   wxmean  - weighted mean of xcoor                             |
|   wymean  - weighted mean of ycoor                             |
|   wxvar   - weighted variance of xcoor                         |
|   wyvar   - weighted variance of ycoor                         |
|   wxycov  - weighted covariance of xcoor and ycoor             |
|   warea   - area of robust Jennrich-Turner 95% ellipse         |
|   wlci    - lower 95% CI on warea                              |
|   wuci    - upper 95% CI on warea                              |
|   wwsq    - Value of Cramer-von Mises goodness of fit statistic|
|             for weighted estimates                             |
|   wgofprob- Interpolated probability level of wsq              |
* ---------------------------------------------------- *;
```

```
title 'Weighted Bivariate Normal Home Range Estimation';
libname library 'c:';
data datafile;
   keep id x y;
   set library.deer;   /* IMPORTANT -- specify SAS data file */
proc sort; by id;
data wellipse;
   array xcoor{300} x1-x300;
   array ycoor{300} y1-y300;
   array z{300};
   array ctable{5,5};
   array wcrit{5} w01 w025 w05 w10 w15;
   drop x1-x300 y1-y300 w wtot i j x y wvalue prob dbar n1 delta
      oxmean oymean oxvar oyvar oxycov d d2 oarea iter
      w01 w025 w05 w10 w15 den;
   retain x1-x300 y1-y300 n;
   label xmean='x mean JT ellipse'
         ymean='y mean JT ellipse'
         xvar='x variance JT ellipse'
         yvar='y variance JT ellipse'
         xycov='covariance JT ellipse'
         area='area JT ellipse'
         wsq='W**2 gof test'
         wxmean='x mean robust JT ellipse'
         wymean='y mean robust JT ellipse'
         wxvar='x variance robust JT ellipse'
         wyvar='y variance robust JT ellipse'
         wxycov='covariance robust JT ellipse'
         warea='area robust JT ellipse';
   set datafile; by id;
   if first.id then n=0;
   n=n+1;
   if n > 300 then do;
      put 'ERROR -- More than 300 observations for' id=;
      abort return 300;
      end;
   xcoor{n}=x; ycoor{n}=y;
   if last.id then do;
      xmean=0; ymean=0;
      do i=1 to n;
         xmean=xmean+xcoor{i};
         ymean=ymean+ycoor{i};
         end;
```

Appendix 7

```
xmean=xmean/n;
ymean=ymean/n;
xvar=0; yvar=0; xycov=0;
do i=1 to n;
    xvar=xvar+(xcoor{i}-xmean)**2;
    yvar=yvar+(ycoor{i}-ymean)**2;
    xycov=xycov+(xcoor{i}-xmean)*(ycoor{i}-ymean);
    end;
xvar=xvar/(n-2);
yvar=yvar/(n-2);
xycov=xycov/(n-2);
area=(4*atan(1))*sqrt(max(0,xvar*yvar-xycov**2))*cinv(0.95,2);
lci=(2*n-4)*area/cinv(0.975,2*n-4);
uci=(2*n-4)*area/cinv(0.025,2*n-4);
oxmean=xmean; oymean=ymean; oxvar=xvar; oyvar=yvar;
oxycov=xycov;
* Calculate z{i} values for Cramer-von Mises test;
dbar=0;
do i=1 to n;
    link dcalc;
    z{i}=d2;
    dbar=dbar+z{i};
    end;
dbar=dbar/n;
do i=1 to n;
    z{i}=probchi(z{i},2);   /* Why not same as next?? */
    z{i}=1-exp(-z{i}/dbar);
    end;
link zsort; /* Sort z array into ascending order */
* Calculate critical values of this test.;
den=1.0+0.6/n;
w01=0.337/den;
w025=0.273/den;
w05=0.224/den;
w10=0.177/den;
w15=0.149/den;
put @15 '****************' id= '****************';
put @10 'Critical values of W**2 statistic';
put @14 '(Listed in output as WSQ)';
put @10 '--------------------------------';
put @15 'W(0.01)   = ' w01;
put @15 'W(0.025)  = ' w025;
put @15 'W(0.05)   = ' w05;
```

355

```
put @15 'W(0.10)  = ' w10;
put @15 'W(0.15)  = ' w15;
link w2calc; wsq=wvalue; gofprob=prob;
oarea=area; warea=0.9*area; iter=0;
do while (abs(warea/oarea-1) > 0.00001);
   wxmean=0; wymean=0; wtot=0; oarea=warea; iter=iter+1;
   do i=1 to n;
      link dcalc;
      * put iter= i= d= w= ;
      z{i}=w;
      wxmean=wxmean+w*xcoor{i};
      wymean=wymean+w*ycoor{i};
      wtot=wtot+w;
      end;
   wxmean=wxmean/wtot;
   wymean=wymean/wtot;
   wtot=0; wyvar=0; wxvar=0; wxycov=0;
   do i=1 to n;
      w=z{i};
      wxvar=wxvar+w*w*(xcoor{i}-wxmean)**2;
      wyvar=wyvar+w*w*(ycoor{i}-wymean)**2;
      wxycov=wxycov+w*w*(xcoor{i}-wxmean)*(ycoor{i}-wymean);
      wtot=wtot+w*w;
      end;
   wxvar=wxvar/wtot;
   wyvar=wyvar/wtot;
   wxycov=wxycov/wtot;
   warea=(4*atan(1))*sqrt(max(0,wxvar*wyvar-wxycov**2))*cinv(0.95,2);
   * put wxmean= wymean= wxvar= wyvar= wxycov= warea=;
   oxmean=wxmean; oymean=wymean; oxvar=wxvar; oyvar=wyvar;
   oxycov=wxycov;
   end;
wlci=(2*n-4)*warea/cinv(0.975,2*n-4);
wuci=(2*n-4)*warea/cinv(0.025,2*n-4);
*Perform goodness of fit test on weighted d2 values;
do i=1 to n;
   link dcalc;
   z{i}=probchi(d2,1.96);
   end;
link zsort;
```

Appendix 7

```
* Construct interpolation table for probability level of wwsq;
ctable{1,1}=0.3003; ctable{1,2}=0.3370; ctable{1,3}=0.3022;
ctable{1,4}=0.3015; ctable{1,5}=0.3351;
ctable{2,1}=0.2452; ctable{2,2}=0.2967; ctable{2,3}=0.2588;
ctable{2,4}=0.2519; ctable{2,5}=0.2701;
ctable{3,1}=0.2142; ctable{3,2}=0.2278; ctable{3,3}=0.2173;
ctable{3,4}=0.2060; ctable{3,5}=0.2224;
ctable{4,1}=0.1711; ctable{4,2}=0.1737; ctable{4,3}=0.1662;
ctable{4,4}=0.1707; ctable{4,5}=0.1760;
ctable{5,1}=0.1496; ctable{5,2}=0.1518; ctable{5,3}=0.1398;
ctable{5,4}=0.1485; ctable{5,5}=0.1495;
* Interpolate value from table;
if n > 200 then do;
   n1=5; delta=0; end;
else if n <= 30 then do;
   n1=1; delta=0; end;
else if n<= 50 then do;
   n1=2; delta=(n-50)/(50-30); end;
else if n<= 70 then do;
   n1=3; delta=(n-70)/(70-50); end;
else if n<= 100 then do;
   n1=4; delta=(n-100)/(100-70); end;
else if n<= 200 then do;
   n1=4; delta=(n-200)/(200-100); end;
do i=1 to 5;
   if n1 > 1 then
      wcrit{i}=ctable{i,n1}+(delta*(ctable{i,n1}-ctable{i,n1-1}));
   else
      wcrit{i}=ctable{i,n1};
   end;
put @10 'Critical values for robust W**2 statistic';
put @15 '(Listed as WWSQ on output)';
put @10 '----------------------------------------';
put @15 'W(0.01)   = ' w01;
put @15 'W(0.025)  = ' w025;
put @15 'W(0.05)   = ' w05;
put @15 'W(0.10)   = ' w10;
put @15 'W(0.15)   = ' w15;
link w2calc; wwsq=wvalue; wgofprob=prob;
output;
end;
return;
```

```
* Calculate Malanobis d statistic (also d**2 and w);
dcalc: d2=(((xcoor{i}-oxmean)**2)*oyvar
       -2*(xcoor{i}-oxmean)*(ycoor{i}-oymean)*oxycov+
       ((ycoor{i}-oymean)**2)*oxvar)
       /(oxvar*oyvar-(oxycov**2));
       d=sqrt(max(0,d2));
       if d > 1.97 then w=1.97/d;
                   else w=1.;
    return;
* Sort z{i} vector;
zsort:do i=1 to n-1;
       do j=i to n;
         if z{i} > z{j} then do;
            wvalue=z{i}; z{i}=z{j}; z{j}=wvalue; end;
         end;
       end;
    return;
* Calculate W**2 statistic;
w2calc: wvalue=0;
       do i=1 to n;
         wvalue=wvalue+(z{i}-(2*i-1)/(2*n))**2+(1/(12*n));
         end;
       link interp;
    return;
* Interpolate probability from critical values;
interp:if wvalue > wcrit{1} then prob=0.01;
       else if wvalue > wcrit{2} then
         prob=0.01+((wvalue-wcrit{1})/(wcrit{2}-wcrit{1}))*(0.025-0.01);
       else if wvalue > wcrit{3} then
         prob=0.025+((wvalue-wcrit{2})/(wcrit{3}-wcrit{2}))*(0.05-0.025);
       else if wvalue > wcrit{4} then
         prob=0.05+((wvalue-wcrit{3})/(wcrit{4}-wcrit{3}))*(0.10-0.05);
       else if wvalue > wcrit{5} then
         prob=0.10+((wvalue-wcrit{4})/(wcrit{5}-wcrit{4}))*(0.15-0.10);
       else prob=0.15;
    return;
proc print;
*---------------------------------*
| Code to plot both ellipses with original data  |
*---------------------------------*;
```

Appendix 7

```
data ellipse;
   set wellipse;
   keep id x y plotsym;
   pi=4*atan(1);
   xxmean=xmean; yymean=ymean; xxvar=xvar; yyvar=yvar;
   xxyycov=xycov; plotsym='U';
   do i=1 to 2;
      R=sqrt(max(0,(xxvar+yyvar)**2-
         4*(xxvar*yyvar-xxyycov**2)));
      theta=atan(-2*xxyycov/(yyvar-xxvar-R));
      a=sqrt(max(0,(yyvar+xxvar+R)*cinv(0.95,2)/2));
      b=sqrt(max(0,(yyvar+xxvar-R)*cinv(0.95,2)/2));
      cost=cos(theta);
      sint=sin(theta);
      do psi=0 to 360 by 3;
         psir=psi*pi/180;
         acosp=a*cos(psir);
         bcosp=b*sin(psir);
         x=xxmean+acosp*cost-bcosp*sint;
         y=yymean+acosp*sint+bcosp*cost;
         output;
         end;
      xxmean=wxmean; yymean=wymean; xxvar=wxvar; yyvar=wyvar;
      xxyycov=wxycov; plotsym='W';
      end;
data plots;
   set datafile(in=indata) ellipse;
   if indata=1 then plotsym='+';
proc sort; by id;
proc plot; plot y*x=plotsym; by id;
run;
```

APPENDIX

8

Survival Estimation Computer Listings

LISTING A8.1
Program SURVIV input for elk data in Table 9.2

```
proc title Los Alamos Elk Biotelemetry Data Survival Analysis 2 year
batteries;
proc model npar=36 /* Survival on a calendar year basis starting 1 March
*/;
   cohort = 0 /* Adult males collared in East Jemez in 1978 */;
      0:(1.-S(1)) /* Recovered in 1978 */;
      0:S(1)*(1.-S(2)) /* Recovered in 1979 */;
      0:S(1)*S(2) /* Lived until 1980 */;
   cohort = 1 /* Adult males collared in East Jemez in 1979 */;
      0:(1.-S(2)) /* Recovered in 1979 */;
      1:S(2)*(1.-S(3)) /* Recovered in 1980 -- 216.720 */;
      0:S(2)*S(3) /* Lived until 1981 */;
   cohort = 1 /* Adult males collared in East Jemez in 1980 */;
      0:(1.-S(3)) /* Recovered in 1980 */;
      0:S(3)*(1.-S(4)) /* Recovered in 1981 */;
      1:S(3)*S(4) /* Lived until 1982 -- 191.315 */;
   cohort = 1 /* Adult males collared in West Jemez in 1980 */;
      1:(1.-S(5)) /* Recovered in 1980 -- 191.864 */;
      0:S(5)*(1.-S(6)) /* Recovered in 1981 */;
      0:S(5)*S(6) /* Lived until 1982 */;
```

cohort = 1 /* Yearling males collared in East Jemez in 1978 */;
   1:(1.-S(7)) /* Recovered in 1978 -- 216.588 */;
   0:S(7)*(1.-S(2)) /* Recovered in 1979 */;
   0:S(7)*S(2) /* Lived until 1980 */;
cohort = 1 /* Yearling males collared in East Jemez in 1979 */;
   1:(1.-S(8)) /* Recovered in 1979 -- 216.602 */;
   0:S(8)*(1.-S(3)) /* Recovered in 1980 */;
   0:S(8)*S(3) /* Lived until 1981 */;
cohort = 2 /* Yearling males collared in East Jemez in 1980 */;
   1:(1.-S(9)) /* Recovered in 1980 -- 191.815 */;
   1:S(9)*(1.-S(4)) /* Recovered in 1981 -- 191.588 */;
   0:S(9)*S(4) /* Lived until 1982 */;
cohort = 5 /* Yearling males collared in West Jemez in 1980 */;
   0:(1.-S(11)) /* Recovered in 1980 */;
   4:S(11)*(1.-S(6)) /* Recovered in 1981 -- 191.065, 191.130, 191.752,
                                             191.775 */;
   1:S(11)*S(6) /* Lived until 1982 -- 191.046 */;
cohort = 5 /* Male calves collared in East Jemez in 1978 */;
   1:(1.-S(13)) /* Recovered in 1978 -- 216.734 */;
   1:S(13)*(1.-S(8)) /* Recovered in 1979 -- 216.202 */;
   3:S(13)*S(8) /* Lived until 1980 -- 216.161, 216.771, 216.820 */;
cohort = 2 /* Male calves collared in East Jemez in 1979 */;
   0:(1.-S(14)) /* Recovered in 1979 */;
   1:S(14)*(1.-S(9)) /* Recovered in 1980 -- 216.173 */;
   1:S(14)*S(9) /* Lived until 1981 -- 216.405 */;
cohort = 1 /* Male calves collared in East Jemez in 1980 */;
   0:(1.-S(15)) /* Recovered in 1980 */;
   0:S(15)*(1.-S(10)) /* Recovered in 1981 */;
   1:S(15)*S(10) /* Lived until 1982 -- 191.373 */;
cohort = 0 /* Male calves collared in West Jemez in 1980 */;
   0:(1.-S(17)) /* Recovered in 1980 */;
   0:S(17)*(1.-S(12)) /* Recovered in 1981 */;
   0:S(17)*S(12) /* Lived until 1982 */;
cohort = 8 /* Adult females collared in East Jemez in 1978 */;
   0:(1.-S(19)) /* Recovered in 1978 */;
   0:S(19)*(1.-S(20)) /* Recovered in 1979 */;
   8:S(19)*S(20) /* Lived until 1980 -- 216.760, 216.280, 216.292,
                                        216.615, 216.639, 216.691,
                                        216.843, 216.880 */;
cohort = 4 /* Adult females collared in East Jemez in 1979 */;
   2:(1.-S(20)) /* Recovered in 1979 -- 216.388, 216.703 */;
   0:S(20)*(1.-S(21)) /* Recovered in 1980 */;
   2:S(20)*S(21) /* Lived until 1981 -- 216.447, 216.375 */;

Appendix 8

```
   cohort = 7 /* Adult females collared in East Jemez in 1980 */;
     1:(1.-S(21)) /* Recovered in 1980 -- 191.339 */;
     1:S(21)*(1.-S(22)) /* Recovered in 1981 -- 191.235 */;
     5:S(21)*S(22) /* Lived until 1982 -- 191.166, 191.187,
191.608, 191.687, 191.882 */;
   cohort = 5 /* Adult females collared in West Jemez in 1980 */;
     0:(1.-S(23)) /* Recovered in 1980 */;
     0:S(23)*(1.-S(24)) /* Recovered in 1981 */;
     5:S(23)*S(24) /* Lived until 1982 -- 191.397, 191.459, 191.564,
                                           191.629, 191.730 */;
   cohort = 0 /* Yearling females collared in East Jemez in 1978 */;
     0:(1.-S(25)) /* Recovered in 1978 */;
     0:S(25)*(1.-S(20)) /* Recovered in 1979 */;
     0:S(25)*S(20) /* Lived until 1980 */;
   cohort = 0 /* Yearling females collared in East Jemez in 1979 */;
     0:(1.-S(26)) /* Recovered in 1979 */;
     0:S(26)*(1.-S(21)) /* Recovered in 1980 */;
     0:S(26)*S(21) /* Lived until 1981 */;
   cohort = 4 /* Yearling females collared in East Jemez in 1980 */;
     0:(1.-S(27)) /* Recovered in 1980 */;
     1:S(27)*(1.-S(22)) /* Recovered in 1981 -- 191.207 */;
     3:S(27)*S(22) /* Lived until 1982 -- 191.017, 191.270, 191.482 */;
   cohort = 0 /* Yearling females collared in West Jemez in 1980 */;
     0:(1.-S(29)) /* Recovered in 1980 */;
     0:S(29)*(1.-S(24)) /* Recovered in 1981 */;
     0:S(29)*S(24) /* Lived until 1982 */;
   cohort = 2 /* Female calves collared in East Jemez in 1978 */;
     0:(1.-S(31)) /* Recovered in 1978 */;
     0:S(31)*(1.-S(26)) /* Recovered in 1979 */;
     2:S(31)*S(26) /* Lived until 1980 -- 216.313, 216.790 */;
   cohort = 4 /* Female calves collared in East Jemez in 1979 */;
     0:(1.-S(32)) /* Recovered in 1979 */;
     0:S(32)*(1.-S(27)) /* Recovered in 1980 */;
     4:S(32)*S(27) /* Lived until 1981 -- 216.262, 216.323(191.423),
                                           216.214, 216.565 */;
   cohort = 2 /* Female calves collared in East Jemez in 1980 */;
     0:(1.-S(33)) /* Recovered in 1980 */;
     0:S(33)*(1.-S(28)) /* Recovered in 1981 */;
     2:S(33)*S(28) /* Lived until 1982 -- 191.710, 191.920 */;
   cohort = 2 /* Female calves collared in West Jemez in 1980 */;
     0:(1.-S(35)) /* Recovered in 1980 */;
     1:S(35)*(1.-S(30)) /* Recovered in 1981 -- 191.501 */;
     1:S(35)*S(30) /* Lived until 1982 -- 191.523 */;
```

```
   labels;
      S(1) =East Jemez adult male survival 1978;
      S(2) =East Jemez adult male survival 1979;
      S(3) =East Jemez adult male survival 1980;
      S(4) =East Jemez adult male survival 1981;
      S(5) =West Jemez adult male survival 1980;
      S(6) =West Jemez adult male survival 1981;
      S(7) =East Jemez yearling male survival 1978;
      S(8) =East Jemez yearling male survival 1979;
      S(9) =East Jemez yearling male survival 1980;
      S(10)=East Jemez yearling male survival 1981;
      S(11)=West Jemez yearling male survival 1980;
      S(12)=West Jemez yearling male survival 1981;
      S(13)=East Jemez male calf survival 1978;
      S(14)=East Jemez male calf survival 1979;
      S(15)=East Jemez male calf survival 1980;
      S(16)=East Jemez male calf survival 1981;
      S(17)=West Jemez male calf survival 1980;
      S(18)=West Jemez male calf survival 1981;
      S(19)=East Jemez adult female survival 1978;
      S(20)=East Jemez adult female survival 1979;
      S(21)=East Jemez adult female survival 1980;
      S(22)=East Jemez adult female survival 1981;
      S(23)=West Jemez adult female survival 1980;
      S(24)=West Jemez adult female survival 1981;
      S(25)=East Jemez yearling female survival 1978;
      S(26)=East Jemez yearling female survival 1979;
      S(27)=East Jemez yearling female survival 1980;
      S(28)=East Jemez yearling female survival 1981;
      S(29)=West Jemez yearling female survival 1980;
      S(30)=West Jemez yearling female survival 1981;
      S(31)=East Jemez female calf survival 1978;
      S(32)=East Jemez female calf survival 1979;
      S(33)=East Jemez female calf survival 1980;
      S(34)=East Jemez female calf survival 1981;
      S(35)=West Jemez female calf survival 1980;
      S(36)=West Jemez female calf survival 1981;
proc estimate nsig=5 maxfn=1500 novar name=All_same /* Most reduced
   model */;
      initial;   S(1)=0.75;
      constraints;    S(1)=S(2);    S(1)=S(3);    S(1)=S(4);    S(1)=S(5);
         S(1)=S(6);   S(1)=S(7);   S(1)=S(8);   S(1)=S(9);   S(1)=S(10);
```

Appendix 8

```
    S(1)=S(11);   S(1)=S(12);   S(1)=S(13);   S(1)=S(14);   S(1)=S(15);
    S(1)=S(16);   S(1)=S(17);   S(1)=S(18);   S(1)=S(19);   S(1)=S(20);
    S(1)=S(21);   S(1)=S(22);   S(1)=S(23);   S(1)=S(24);   S(1)=S(25);
    S(1)=S(26);   S(1)=S(27);   S(1)=S(28);   S(1)=S(29);   S(1)=S(30);
    S(1)=S(31);   S(1)=S(32);   S(1)=S(33);   S(1)=S(34);   S(1)=S(35);
    S(1)=S(36);
proc estimate nsig=5 maxfn=1500 novar name=Hunt&Nonh /* Hunted and
nonhunted */;
    initial;   S(1)=0.50;   S(13)=0.90;
    constraints;   S(1)=S(2);   S(1)=S(3);   S(1)=S(4);   S(1)=S(5);
        S(1)=S(6);   S(1)=S(7);   S(1)=S(8);   S(1)=S(9);   S(1)=S(10);
        S(1)=S(11);  S(1)=S(12);                S(13)=S(14); S(13)=S(15);
        S(13)=S(16); S(13)=S(17); S(13)=S(18); S(13)=S(19); S(13)=S(20);
        S(13)=S(21); S(13)=S(22); S(13)=S(23); S(13)=S(24); S(13)=S(25);
        S(13)=S(26); S(13)=S(27); S(13)=S(28); S(13)=S(29); S(13)=S(30);
        S(13)=S(31); S(13)=S(32); S(13)=S(33); S(13)=S(34); S(13)=S(35);
        S(13)=S(36);
proc estimate nsig=5 maxfn=1500 novar name=Yr&Ar_Same /* Years and areas
the same */;
    initial;   retain=Hunt&Nonh;
    constraints;   S(1)=S(2);   S(1)=S(3);   S(1)=S(4);
        S(1)=S(5);   S(1)=S(6);   S(7)=S(8);   S(7)=S(9);   S(7)=S(10);
        S(7)=S(11);  S(7)=S(12);                S(13)=S(14); S(13)=S(15);
        S(13)=S(16); S(13)=S(17); S(13)=S(18);
        S(19)=S(20); S(19)=S(21); S(19)=S(22); S(19)=S(23); S(19)=S(24);
        S(25)=S(26); S(25)=S(27); S(25)=S(28); S(25)=S(29); S(25)=S(30);
        S(31)=S(32); S(31)=S(33); S(31)=S(34); S(31)=S(35); S(31)=S(36);
proc estimate nsig=5 maxfn=1500 novar name=Year_same /* Age and sex
classes constant through time */;
    initial;   retain=Yr&Ar_Same;
    constraints;   S(1)=S(2);   S(1)=S(3);   S(1)=S(4);
        S(5)=S(6);   S(7)=S(8);   S(7)=S(9);   S(7)=S(10);
        S(11)=S(12); S(13)=S(14); S(13)=S(15); S(13)=S(16);
        S(17)=1;     S(18)=1;     S(19)=S(20); S(19)=S(21); S(19)=S(22);
        S(23)=S(24); S(25)=S(26); S(25)=S(27); S(25)=S(28);
        S(29)=S(30); S(31)=S(32); S(31)=S(33); S(31)=S(34);
        S(35)=S(36);
proc estimate nsig=5 maxfn=1500 novar name=Area_same /* Age and sex
classes separate with two areas combined */;
    initial;   retain=Yr&Ar_Same;
    constraints;   S(3)=S(5);   S(4)=S(6);   S(9)=S(11);
        S(10)=S(12); S(15)=S(17); S(21)=S(23); S(22)=S(24);
```

```
    S(27)=S(29);   S(28)=S(30);  S(33)=S(35);
    S(1)=1.;       S(16)=1.;     S(18)=1.;     S(25)=1.;
    S(34)=1.;      S(36)=1.;
proc estimate nsig=5 maxfn=1500 novar name=General /* Most general model
 supported by the data */;
   initial;   retain=Year_same;
   constraints;   S(1)=1.;   S(12)=1.;   S(16)=1.;   S(17)=1.;
      S(18)=1.;   S(25)=1.;   S(29)=1.;   S(34)=1.;   S(36)=1.;
proc test;
proc stop /* End of the analysis */;
```

APPENDIX

9

SAS Monte Carlo Simulation of Capture-Recapture

LISTING A9.1
SAS code to simulate the Monte Carlo results presented in Chapter 10

```
*----------------------------------------------------*
| SAS procedure to simulate capture-recapture estimation . |
| Six estimators are simulated:                      |
|     1) unweighted arithmetic mean                  |
|     2) arithmetic mean weighted by 1/variance[N(i)] |
|     3) unweighted geometric mean                   |
|     4) geometric mean weighted by 1/variance[log N(i)] |
|     5) median                                      |
|     6) joint hypergeometric maximum likelihood     |
|                                                    |
| Variables that may be changed:                     |
|    noccas   - number of sighting occasions         |
|    ntrue    - true population size                 |
|    probcap  - probability of capture and marking with radio |
|    probsite - probability of sighting on an occasion |
|    nreps    - number of replications to simulate   |
*----------------------------------------------------*;
libname library 'c:';
data library.noremark;
   /* Random number to begin pseudo-random number string */
   seed1=0;
```

```
/* Arrays used for generating data */
array resight{5}  m2-m6;
array unmarked{5} u2-u6;
array chapest{5}  cr1-cr5;
array varchap{5}  varcr1-varcr5;
array logchap{5}  logcr1-logcr5;
array varlogc{5}  logvar1-logvar5;
/* Arrays used for summarizing data with the 6 estimators */
array nhat{6}      eavest ewavest elavest elwavest medest hypest;
array variance{6} eavvar ewavvar elavvar elwavvar medvar hypvar;
array se{6}        eavse  ewavse  elavse  elwavse  medse  hypse;
array uci{6}       eavuci ewavuci elavuci elwavuci meduci hypuci;
array lci{6}       eavlci ewavlci elavlci elwavlci medlci hyplci;
array range{6}     eavrng ewavrng elavrng elwavrng medrng hyprng;
array cover{6}     eavcov ewavcov elavcov elwavcov medcov hypcov;
/* Array used for median estimator */
array tempest{5} tmpest1-tmpest5;
/* Array used for hypergeometric estimator */
array savlik{1000};
label noccas='Number of sighting occasions'
      ntrue='True population'
      probcap='Probability of capture and marking'
      probsite='Probability of sighting on an occasion'
      eavest='Unweighted arithmetic average'
      ewavest='Weighted arithmetic average'
      elavest='Unweighted geometric average'
      elwavest='Weighted geometric average'
      medest='Median'
      hypest='Joint hypergeometric ML'
      eavran='Unweighted arithmetic average CI length'
      ewavran='Weighted arithmetic average CI length'
      elavran='Unweighted geometric average CI length'
      elwavran='Weighted geometric average CI length'
      medran='Median CI length'
      hypran='Joint hypergeometric ML CI length'
      eavcov='Unweighted arithmetic average coverage'
      ewavcov='Weighted arithmetic average coverage'
      elavcov='Unweighted geometric average coverage'
      elwavcov='Weighted geometric average coverage'
      medcov='Median coverage'
      hypcov='Joint hypergeometric ML coverage';
```

Appendix 9

```
/* Set parameter values */
noccas=5;
ntrue=150;
do nreps=1 to 50;
   do probcap=0.1 to 0.3 by 0.1;
      do probsite=0.2 to 0.8 by 0.2;
      /* Generate data for this replication */
      /* Number of marked animals in population */
      nmarked=ranbin(seed1,ntrue,probcap);
      /* Generate each of the resighting occasions */
      do i=1 to noccas;
         /* Number of marked animals sighted */
         resight{i}=ranbin(seed1,nmarked,probsite);
         /* Number of unmarked animals sighted */
         unmarked{i}=ranbin(seed1,ntrue-nmarked,probsite);
         /* Total number of animals sighted */
         sighted=resight{i}+unmarked{i};
         /* Chapman estimate for this occasion */
         chapest{i}=(((nmarked+1)*(sighted+1))/
                  (resight{i}+1)) - 1;
         logchap{i}=log(chapest{i});
         /* Variance of Chapman estimate for weights */
         varchap{i}=((nmarked+1)*(sighted+1)*(nmarked-resight{i})*
                  (sighted-resight{i}))/((resight{i}+1)*
                  (resight{i}+1)*(resight{i}+2));
         if varchap{i} <= 1 then varchap{i}=1;
         varlogc{i}=varchap{i}/chapest{i}**2;
      end;
      /* Minimum number known alive */
      mna=nmarked+max(of u2-u6);
      /* Find estimate and confidence intervals
         of each of the estimators */
      /* t statistics for all the estimates based on a mean */
      t=tinv(0.975,noccas-1);
      /* Find unweighted arithmetic mean */
      eavest=mean(of cr1-cr5);
      eavvar=var(of cr1-cr5)/noccas;
      eavse=sqrt(eavvar);
      eavlci=eavest-t*eavse;
      eavuci=eavest+t*eavse;
      /* Find weighted arithmetic mean */
      ewavest=0.;sumwgt=0.;
```

```
do i=1 to noccas;
   ewavest = ewavest+chapest{i}/varchap{i};
   sumwgt = sumwgt+1/varchap{i};
   end;
ewavest = ewavest/sumwgt;
ewavvar = 0;
do i=1 to noccas;
   ewavvar = ewavvar+(chapest{i}-ewavest)**2/varchap{i};
   end;
ewavvar = ewavvar/((noccas-1)*sumwgt);
ewavse = sqrt(ewavvar);
ewavlci = ewavest-t*ewavse;
ewavuci = ewavest+t*ewavse;
/* Find unweighted geometric mean */
elavest = mean(of logcr1-logcr5);
elavvar = var(of logcr1-logcr5)/noccas;
elavse = sqrt(elavvar);
elavlci = exp(elavest-t*elavse);
elavuci = exp(elavest+t*elavse);
elavest = exp(elavest);
/* Find weighted geometric mean */
elwavest = 0.; sumwgt = 0.;
do i=1 to noccas;
   elwavest = elwavest+logchap{i}/varlogc{i};
   sumwgt = sumwgt+1/varlogc{i};
   end;
elwavest = elwavest/sumwgt;
elwavvar = 0.;
do i=1 to noccas;
   elwavvar = elwavvar+(((logchap{i}-elwavest)**2)/varlogc{i});
   end;
elwavvar = elwavvar/((noccas-1)*sumwgt);
elwavse = sqrt(elwavvar);
elwavlci = exp(elwavest-t*elwavse);
elwavuci = exp(elwavest+t*elwavse);
elwavest = exp(elwavest);
/* Find median of estimates */
do i=1 to noccas;
   tempest{i} = chapest{i};
   end;
```

Appendix 9

```
   do i=1 to noccas-1; do j=i to noccas;
      if (tempest{i}>tempest{j}) then do;
         temp=tempest{i}; tempest{i}=tempest{j};
         tempest{j}=temp; end;
      end; end;
   medest=tempest{3};
                           /* These statements dependent on */
   meduci=tempest{5};      /* the number of occasions       */
   medlci=tempest{1};      /* contained in noccas           */
   /* Find joint hypergeometric estimate */
   nhatmp=mna+1;
   initial=mna;
   hyperg=1.;
   do i=1 to noccas;
      if resight{i} > 0 then
         hyperg=hyperg*(probhypr(nhatmp,nmarked,
            resight{i}+unmarked{i},resight{i})-
            probhypr(nhatmp,nmarked,
            resight{i}+unmarked{i},resight{i}-1));
      else
         hyperg=hyperg*probhypr(nhatmp,nmarked,
            resight{i}+unmarked{i},resight{i});
      end;
   savlik{nhatmp-initial}=log(hyperg);
   like=hyperg-1;
   do while(like < hyperg);
      nhatmp=nhatmp+1;
      like=hyperg;
      do i=1 to noccas;
         /* A recursive algorithm is used to speed up
            the calculation of a new value of the
            likelihood for each value of nhatmp       */
         hyperg=hyperg*(nhatmp-nmarked)*(nhatmp-resight{i}-
            unmarked{i}) /((nhatmp-nmarked-unmarked{i})*nhatmp);
         end;
      savlik{nhatmp-initial}=log(hyperg);
      end;
   hypest=nhatmp-1;
   maxlik=savlik{hypest-initial};
   do nhatmp=hypest-1 to initial+1 by -1;
      if maxlik-savlik{nhatmp-initial} > 2 then go to endloop1
      end;
   nhatmp=mna;
```

```
endloop1:hyplci=nhatmp;
        nhatmp=hypest+1;
        if maxlik-savlik{nhatmp-initial} > 2 then go to endloop2;
        do until (maxlik-log(hyperg)>2);
            nhatmp=nhatmp+1;
            do i=1 to noccas;
                hyperg=hyperg*(nhatmp-nmarked)
                    *(nhatmp-resight{i}-unmarked{i})
                    /((nhatmp-nmarked-unmarked{i})*nhatmp);
            end;
        end;
endloop2:hypuci=nhatmp;
        /* Coverage and confidence interval range calculations */
        do j=1 to 6;
            if lci{j} < mna then lci{j}=mna;
            range{j}=uci{j}-lci{j};
            cover{j}=0;
            if uci{j} > ntrue & lci{j} < ntrue then cover{j}=1
            end;
        output;
        end;
      end;
   end;
drop nmarked m2-m6 u2-u6 cr1-cr5 varcr1-varcr5;
drop logcr1-logcr5 logvar1-logvar5;
drop eavvar ewavvar elavvar elwavvar medvar hypvar;
drop eavse ewavse elavse elwavse medse hypse;
drop eavuci ewavuci elavuci elwavuci meduci hypuci;
drop eavlci ewavlci elavlci elwavlci medlci hyplci;
drop nhatmp tmpest1-tmpest5 seed1 i j nreps sighted;
drop sumwgt maxlik initial temp like hyperg t;
proc append base=library.sim5; run;
```

Index

Accuracy
   aerial tracking, 43–45
   in radio-tracking studies, 21–22
     habitat analysis, 200–201
   triangulation system design
     directional bearings, 80–90
     testing of, 79–80
Aerial surveys
   population estimations, 269
   tracking techniques, 42–45
   transmitter and tower locations with, 83, 85
Age, animal survival rates, 242–243
AGE variable, 278
Algorithms, for map coordinate systems, 4, 6
Andrews estimator
   data quality control and censoring, 72–74
   triangulation location, 69–72
     system design, computer simulation, 94
Animal associations, migration and dispersal studies, 137–140
Animal identification, data entry procedures, 7–8
Animal movement, home range estimation, 145–147
Animated graphics, simple animal movements, 119–121

Antenna-pointer orientation
   hand-held antennas, 92–94
   transmitter testing, 87–89
   vehicle-mounted systems, 91–92
AREAS system, habitat availability measurement, 184
Availability of habitat
   analysis of, 183–185
   known values for, 197
   Marcum–Loftsgaarden analysis, 191–192

Band recovery data, survival rates, 207–208
Bayesian approach, triangulation location techniques, 61–62
Bearing directions
   mobile triangulation tracking systems, 90–94
   precision, confidence ellipse size and, 103–104
   system design, 79–80
Behavioral changes
   effects of tagging on, 27–28
   migration and dispersal studies, 137–140
Bias
   animal survival rates, 224–226
   directional bearings, 80–90

Index

Binomial estimator
  animal survival rates, 208–213
    Kaplan–Meier product limit estimator, 232–242
    MICROMORT program, 231–232
BIOCHECK program, 273
  data error detection, 114
  input file to FIELDS program, 279–287
  output, 297
    to FIELDS program, 287–288
  program limitations, 296–297
  radio-tracking data checking, 295–297
BIOPLOT program
  data error detection, 114
  FIELDS program input file, 279–287
  limitations of, 299
  radio-tracking data plotting and editing, 299
Biotelemetry, capture–recapture estimates, 255–262
Birds, instrumentation effects in, 36–38
Bivariate estimates
  home range models
    Dunn estimator, 161–162
    Jennrich–Turner estimator, 155–160
    multiple ellipse, 161
    normality testing, 162–166
    weighted normal estimator, 160–161
  SAS definition codes, 350–352
BMDP statistical package
  animal survival rates
    Cox's model, 245
    logistic regression, 248
  habitat preference testing, 192–193
  SURVIV program and, 329
Bonferonni statistics
  confidence intervals, habitat preference analysis, 189
  habitat analysis and, sample size, 202–203

Capture–recapture population estimation, 255–269
  Lincoln–Petersen estimates, 257–262
    joint hypergeometric maximum likelihood estimates, 260–262

mean-based estimates, 257–259
median combinations, 259–260
Monte Carlo simulation, SAS codes, 367–372
Cause-and-effect relationships, migration and dispersal studies, 123–124
Cedar Creek data analysis systems, 272–274
Censoring
  animal survival rates, 249
  triangulation location studies, 72–74
Center-weighted distribution, home range estimation, 175
CHECKF program, 273
Chi-square analysis
  animal survival rates
    binomial distribution, 209, 212–213
    medical analysis, 248–249
    SURVIV Program, 218–220
  habitat preference analysis, 186–191
  home range estimation, 164–166
  noncentrality parameter, 34
  SAS program and, 33–34
  three-tower triangulation location, 69
Circular coordinate mapping system, 2
Circular normal distribution, home range estimation, 161
Circular scatter diagram
  migration and dispersal studies, 127–128
  SAS code, 138–133
Cluster analysis, migration and dispersal studies, 139–140
Compass bearings, triangulation location with, 49–50
Computer programs
  home range estimation, 173–174
  tracking data entry for, 6–10
Computer simulation
  simple animal movement
    animated graphics, 119–121
    still graphics, 115–117
  triangulation system design, 94–110
Confidence ellipse
  size and shape for two-, three- and four-tower systems, 108–110
  tower bearing standard deviation and number of towers, 102–103

Index

triangulation system design, area definition, 94-99
Confidence intervals
  animal survival rates, 208-209
    till death, Kaplan-Meier product limit estimator, 235-239
Correlational studies, defined, 15
Costs in radio-tracking studies, 21-22
  of satellite tracking, 46
Cox's regression method, animal survival rates, 244-245
Cramer-von Mises statistic, home range estimation, 155
Critical habitat, 198-200
Cumulative distribution function, 229-230

Data analysis
  current research trends, 274-275
  software development for, 271-273
  system design, 273-274
Data identification and error detection, 113-115
Data quality control, triangulation location studies, 72-74
Data simulation, SURVIV program and, 329-338
DATA step program, triangulation systems testing, 86-87
Database entry, computer analysis of tracking data, 6-10
DC80 program, home range estimation, 159
  harmonic mean, 171-172
Descriptive studies, defined, 14-15
Directional bearings
  bias and precision in, 80-82
  mobile tracking systems timing problems, 93-94
  triangulation system design accuracy, 80-90
Discriminant analysis, animal survival rates, 248
Dispersal, see Migration and dispersal tracking techniques
Dixon-Chapman home range estimator, 170-172
  extensions of, 174

Dunn's home range estimator, 161-163
  extensions of, 174
Dynamic territorial interaction, migration and dispersal studies, 138-139

Elevational gradients, triangulation system design and, 104-105
Endpoint criteria, migration and dispersal studies, 122
Environmental parameters
  animal survival rates till death, 228-230
  migration and dispersal studies, 123
  radio-tracking studies, 21-22
Error rates
  directional bearings, bias and precision, 80-82
  mobile tracking systems, 93-94
  nontriangulation tracking techniques
    aerial tracking, 43-45
    homing techniques, 42
    satellite tracking, 45-47
  radio-tracking locations, habitat analysis, 200-201
  satellite tracking systems, 46-47
  simple animal movements, 113-115
  triangulation tracking techniques
    three or more receiving stations, 59-69
    two-tower systems, 52-59

F statistic multiplier, home range estimation with, 158-160
"Fence post" transmitter testing, 87
Fidelity, see Site fidelity
Field conditions, radio-tracking studies, 21-22
FIELDS
  BIOCHECK output, 287-288
  HOMER output, 288-292
  input file, 279-287
  limitations, 293
  radio-tracking data preprocessor, 293
File names, SURVIV program, 339-340

Index

Fitness, animal survival rates, 226
Five-tower locations, confidence ellipse area, 95, 98
Fixed-location towers
 in radio-tracking studies, 22
 triangulation location with, 48
Fourier transform
 extensions of, 174
 home range estimation, series smoothing, 168–170
Four-tower triangulation system
 confidence ellipse
  area, 95, 98
  size and shape, 108, 110
 testing procedures, 83–85
Freelance Plus software, simple animal movements, 115–119
Friedman test, habitat preference testing, 194–196

G statistic
 animal survival rates, 218–220
 habitat preference analysis, 188–189
Geographic information systems (GIS)
 habitat availability measurement, 184
 simple animal movements, 117–118
 triangulation system design, 105–106
Goodness-of-fit measurements, *see* Chi-square analysis
Graphics, data analysis with, 272–274
Greenwood's formula, animal survival rates till death, 234–235
Grid cell counting
 habitat availability measurement, 184
 habitat utilization, 185–186
 home range estimation, 166–168

Habitat analysis
 animal differences, 196–197
 availability, 183–185
 chi-square testing, 186–191
 critical habitat, 198–200
 defined, 183

Friedman test, 194–196
 Heisey's analysis, 192–193
 home range estimation, 201
 Johnson's analysis, 193–194
 Marcum–Loftsgaarden analysis, 191–192
 preference, 186
 radio-tracking location accuracy, 200–201
 sample size, 202–203
 test selection, 197–198
 utilization, 185–186
Hand-held antenna
 antenna-pointer orientation, 92–94
 triangulation location with, 49
Hand-held receiving systems, 21–22
Hardware requirements
 location estimations and, 41
 triangulation location estimates with, 72–74
Harmonic mean, home range estimation, 170–172
Hazard function, animal survival rates
 Cox's model, 244–245
 medical analysis, 243
 MICROMORT program, 231–232
 smoothing in medical analysis, 243
 till death, 226–230
Heisey's analysis, habitat preference testing, 192–193
Heterogeneity of capture probabilities, 263–264
Heterogenous survival rates, animal survival studies, 225–226
Histogram, migration and dispersal studies, 129–132
Home range estimation
 bivariate normal models, 155–166
  Dunn estimator, 161–162
  Jennrich–Turner estimator, 155–160
  multiple ellipses, 161
  testing normality, 162–166
  weighted bivariate normal estimator, 160–161
 computer programs, 173–174
 defined, 145–146
 evaluation of, 174–175
 extension of, 174

376

Index

habitat analysis and, 201
HOMER program, 301–305
independence of observations, 147–148
limits of, 178–179
minimum convex polygon, 148–155
nonparametric approaches, 166–173
    Fourier series smoothing, 168–170
    grid cell counts, 166–168
    harmonic mean, 170–172
    limits of, 172–173
preferred methods, 178
similarity of ranges, 175–177
HOMER program
    DUNNTIME parameter, 303
    FIELDS program input file, 279–287
    FILL parameter, 303
    home range estimation with, 159, 301–305
    limitations, 304
    LOTUS parameter, 304
    output from, 304
        to FIELDS program, 288–292
    RANDOM option, 303–304
    REMOVE parameter, 302
    SQUARE= parameter, 302
    UNIQUE option, 304
    variable names, 301–302
    VECDIST parameter, 303
    VECTIME parameter, 302–303
    XORIGIN and YORIGIN parameters, 302
Homing techniques for animal tracking, 42
Hotelling's T^2 test, 134–135
Huber estimator
    data quality control and censoring, 72–74
    triangulation location, 69–72
        computer simulation, 94
Hunted animals, animal survival rates, 225

ID animal identification variable, 278
Independent observation
    home range estimation, 147–148
    radio-tracking studies, 20–21
INITIAL statement (SURVIV program), 326
Instantaneous survival rates, SURVIV program, 222

Instrumentation
    effects on animals, 28–29, 35–38
    guidelines for using, 37–38
    migration and dispersal studies, 124–125

Jennrich–Turner home range estimator
    comparisons with Dunn estimator, 162–163
    extensions of, 174
    Fourier series smoothing and, 169–170
    home range estimation, 155–160
Johnson's analysis, habitat preference testing, 193–194
Joint hypergeometric maximum likelihood estimator, 260–262
Joint maximum likelihood estimator, 264–267
Jolly–Seber estimation method, capture–recapture estimates, 256
Julian calendar, entry problems with, 7–8

Kaplan–Meier product limit estimator
    animal survival rates, 208
    till death, 232–242

LABEL command (SURVIV program), 325–326
LANDSAT data, habitat availability measurement, 184–185
Lenth's maximum likelihood estimation
    mobile triangulation tracking systems, 91
    performance of, 69–72
    triangulation system design
        computer simulation, 94–110
        three or more receiving stations, 59–69
Lincoln–Petersen estimates
    capture–recapture estimates, 257–262
        joint hypergeometric maximum likelihood estimator, 260–262
        mean-based estimators, 257–259
Line transect population estimations, 269
Linear models
    SURVIV program estimates, 222–223
    triangulation system design, 107–108

Index

Location estimations
  nontriangulation techniques, 42–47
  triangulation techniques, 47–74
Logistic regression, animal survival rates, 245–248
Long-term studies, capture–recapture estimates, 268–269
LORAN-C system for aerial tracking, 44–45
Los Alamos National Laboratory computer network, 120–121
Lotus 1-2-3 software
  HOMER program and, 304
  simple animal movements, 115–119

Mahalanobis distance, home range estimation with, 160–161
Mainframe computers, simple animal movements, 120–121
Manipulative studies, 15–16
Mann–Whitney test, migration and dispersal studies, 134–135
Map coordinate systems, 2–6
  circular, 2
  township–range land-mapping system, 6
  Universal Transverse Mercator (UTM) system, 2–6
Mapping and overlay statistical system (MOSS)
  data analysis with, 272–273
  habitat availability measurement, 184
  triangulation system design, 105–106
Marcum–Loftsgaarden analysis, 191–192
Marine species, satellite tracking of, 46
Markovian process, home range estimation, 161–162
Maximum likelihood estimation
  animal survival rates, 207–208
    logistic regression, 245–248
    till death, MICROMORT program, 230–232
  capture–recapture estimates, 260–262
  data quality control and censoring, 72–74
  triangulation location techniques, three or more receiving stations, 59–69
Mayfield estimator for nest survival, 207

McPAAL program
  Fourier series smoothing, 169–170
  home range estimation, 159
    advantages of, 174
    harmonic mean, 172
Mean-based estimators, capture–recapture estimates, 257–259
Median-based estimators, capture–recapture estimates, 259–260
Medical analysis, animal survival rates, 242–251
  censoring, 249
  chi-square analysis, 248–249
  Cox's model, 244–245
  discriminant analysis, 248
  example analysis, 249–251
  logistic regression, 245–248
MICROMORT program
  animal survival rates, 208
  till death, 230–232
Migration and dispersal tracking techniques, 121–133
Minimum area polygon
  home range estimation
    concavity, 153–154
    convexity, 148–155
  SAS definition codes, 343–349
Mobile receiving systems, 21–22
Mobile triangulation systems design and testing, 90–94
Monte Carlo simulations
  animal survival rates, 215
  capture–recapture estimates, 264–267
    SAS codes, 367–372
  habitat analysis, 194
  home range estimation, 174–175
Movement patterns, migration and dispersal studies, 121–122
MRPP test
  home range estimation, 177, 178–179
  migration and dispersal studies
    sample size, 140–141
    site fidelity measurements, 135–136
Multinomial distributions, animal survival rates, 213–215
Multiple ellipse, home range estimation, 161

Index

Multivariate analysis (MOVA), migration and dispersal studies, 134–135
Multivariate Ornstein–Uhlenbeck stochastic process (MOU), 161–162

NIMBUS satellites, animal tracking with, 45–46
Nonparametric techniques, home range estimation, 166–173
 Fourier series smoothing, 168–170
 grid cell counts, 166–168
 harmonic mean, 170–172
 limits of, 172–173
Nontriangulation location techniques, 42–47
 aerial tracking, 42–45
 homing-in, 42
 satellite tracking, 45–47
Numerical optimization, animal survival rates, 213–215

"Outliers," elimination of, home range estimation, 151–155

Parameter estimation, animal survival rates, 213–215
"Partial logistic regression," animal survival rates, 248
Pen studies, critical habitat analysis, 199–200
Perturbation studies, critical habitat, 198–200
Physiological parameters, radio-tracking studies, 21–22
Planimeter, habitat availability measurement, 184
Polaroid Palette film recorder, simple animal movements, still graphics, 117–119
Polygon mapping
 habitat availability measurement, 184
 habitat utilization, 185–186
Pooled data, habitat preference analysis, 190–191
Population estimation
 aerial surveys, 269
 capture–recapture estimation, 255–269
  assumptions, 262–264
  example calculations, 262
  Lincoln–Petersen estimates, 257–262
  preferred estimator, 264–267
  sample size, 267–269
 line transects, 269
Precision estimates
 directional bearings, 80–90
 triangulation system designs, 101
Predation rates, transmitter effects on vulnerability and, 30–31
PREFER program, habitat preference testing, 194
Preference of habitat
 animal differences and, 196–198
 chi-square analysis, 186–191
 Friedman testing, 194–196
 Heisey's analysis, 192–193
 Johnson's analysis, 193–194
 Marcum–Loftsgaarden analysis, 191–192
 test selection, 197–198
Probability density function, 226–230
Probability distribution, migration and dispersal studies, 140–141
Probability models, home range estimation, 165–166
"Probability polygons," home range estimation, 152–155
PROC BROWNIE command (SURVIV program), 339
PROC CATMOD command (SURVIV program), 248
PROC ESTIMATE command (SURVIV program)
 failure of, 328–329
 survival rate estimation, 307, 326–327
PROC IDENTIFIABLE command (SURVIV program), 339
PROC MODEL command (SURVIV program), survival rate estimation, 307
 COHORT statements, 325–326
 sample program, 311–325
PROC PLOT command (SURVIV program), 106
PROC SAMPLE SIZE (SURVIV program), 336–338

379

Index

PROC SIMULATE command (SURVIV program), 329–338
PROC STOP command (SURVIV program), 338–339
PROC TEST command (SURVIV program), 310, 327–328

Quade test, habitat preference testing, 196
Quattro software, simple animal movements, 115–119

Radio-tracking studies
  accuracy of, habitat analysis, 200–201
  correlational studies, 15
  decriptive studies, 14–15
  design of
    field considerations, 21–22
    sampling and statistical considerations, 19–21
    scientific method, 13–16
    treatments, controls, and replicates, 17–19
  manipulative studies, 15–16
Random sampling, radio-tracking studies, 20
Random walk models, migration and dispersal studies, 140–141
Rank-order statistics, capture–recapture estimates, 259–261
Rayleigh scattering, migration and dispersal studies, 132
Receiving system design, in radio-tracking studies, 22
Reflex database package, data error detection, 114–115
Replication techniques, radio-tracking studies, 18–19
"Reverse LORAN," triangulation location with, 47–48

Sample size
  capture–recapture estimates, 267–269
  habitat analysis and, 202–203
  home range estimation, 178
  bivariate normality testing, 165–166
  minimum convex polygon, 149–155
  size inflation, 147–148
  migration and dispersal studies, 140–141
  SURVIV program and, 329–338
Sampling techniques
  home range estimation, 146
  migration and dispersal studies, 121–122
  radio-tracking studies, 19–21
Samuel–Garton weighted ellipse estimator, 353–359
Satellite tracking techniques, 45–47
Schnabel–Darroch multiple recapture protocol, 257
Scientific method, radio-tracking studies and, 13–16
Seasonality, migration and dispersal studies, 122–123
SESAME user assumptions, 223–226
  survival estimation input, 361–366
  user's bulletin board, 10
SEX variable, 278
Shape criterion, home range estimation, 178
Short-term studies, capture–recapture estimates, 268–269
Sighting probabilities of animals, capture–recapture estimates, 263–264
Signal absorption, transmitter testing, 87
Signal attenuation, transmitter testing, 87
Signal reflection
  nonreflected bearings and, 106–108
  triangulation system design, 104–105
Simple animal movements
  animal, 137–140
  animated graphics, 119–121
  data error identification and correction, 113–115
  fidelity measurement, 133–137
  migration and dispersal, 121–133
  sample size, 140–141
  still graphics, 115–119
Site fidelity
  home range estimation, 175–177
  migration and dispersal studies, 133–137
  telemetry techniques, 134
Six-tower confidence ellipse area, 95, 98

Index

Social behavior of animals, 137–140
Software designs, radio-tracking data analysis and, 7–10
Spatial controls, radio-tracking studies, 17–19
SPSS program, animal survival rates
   Cox's model, 245
   logistic regression, 248
Standard deviation, directional bearings bias and precision, 82–90
Static territorial interaction, 137–138
Statistical analysis
   migration and dispersal studies, 125–133
   radio-tracking studies, 19–21
Statistical Analysis System (SAS)
   animal survival rates, 249–251
      Cox's model, 245
      logistic regression, 248
      till death, 235–239
   bearings calculation code, 86–87
   capabilities of, 274–275
   capture–recapture estimates, 264–267
   DATA step programming language, 8–9
   GRAPH program, 117–119
   habitat analysis and preference testing, 197
      sample size, 202–203
   home range estimation
      chi-square testing, 164–166
      procedures for, 343–359
      uniform distribution, 154–155
   Lenth's maximum likelihood estimation, 64–68
   migration and dispersal studies, 126–132
   Monte Carlo simulation, capture–recapture analysis, 367–372
   PC system, 8
   procedure and output in chi-square testing, 31, 33–34
   query languages, 273–274
   sample size and, 31–32
   triangulation error polygon construction, 54–57
   von Mises distribution parameters, 59–61
Statistical independence, home range estimation, 147–148
Still graphics, simple animal movements, 115–119

SURVIV program
   animal survival rates, 208, 215–226
   COHORT card, 216–217
   complex applications, 222–223
   example of, 220–222
   PROC ESTIMATE, 217–218
   PROC MODEL statement, 216–217
   PROC SAMPLE, 218
   PROC SIMULATE, 218
   PROC TEST, 218
   user assumptions, 223–226
   survival estimation input, 361–366
   user's manual, 307–341
      command syntax summary, 308–310
      data simulation and sample sizes, 329–338
      DOS installation guide, 340
      failure of PROC ESTIMATE, 328–329
      file names and execution time parameters, 339–340
      output model, 311–325
      PROC BROWNIE command, 339
      PROC IDENTIFIABLE command, 339
      PROC STOP command, 338–339
      program enlargement, 341
      survival rate estimation, 307–328
Survival rate estimation
   binomial distribution, 208–213
   computer listings, 361–366
   defined, 207–208
   medical survival analysis, 242–251
      censoring, 249
      chi-square analysis, 248–249
      Cox's model, 244–245
      discriminant analysis, 248
      example analysis, 249–251
      logistic regression, 245–248
   parameter estimation by numerical methods, 213–215
   SURVIV program, 208, 215–226
      complex applications, 222–223
      example, 220–222
      user assumptions in, 223–226
      user guidelines, 307–328
   time until death methods, 226–242
      Kaplan–Meier method, 232–242

Index

Survival rate estimation—time until death methods (*continued*)
   MICROMORT program, 230–232
   transmitter failures, 223–224
Survivorship function
   animal survival rates till death, 226–230
   Kaplan–Meier product limit estimator, 235, 240–241
"Symbol plot," simple animal movements, 115

Tagging techniques, effects on animals, 27–38
   experiment designs, 28–35
   transmitter effects, 35–38
Telemetry
   data analysis systems, 274
   data standardization lacking, 272
   radio-tracking studies, 21–22
Temporal controls, radio-tracking studies, 17–19
Three-tower triangulation location
   confidence ellipse
     area, 95, 97
     size and shape, 108–109
   errors in, 72–74
   triangulation location techniques, 61–64
Time series randomization test, 178
TIME variable, 277–278
Timing
   animal survival rates
     death and, 226–230
     medical analysis, 243
   home range estimation, 146–148
     nonparametric methods, 172–173
   migration and dispersal studies, 122–124
   animal association analysis, 139–140
TIROS/ARGOS satellite system, animal tracking with, 45–46
Tower locations, triangulation system design, 99, 101
Township–range land-mapping system, 6
Transmitter attachment, animal behavior and, 27–29
Transmitter collar recovery, 215–221
Transmitter failures, animal survival rates, 223–224

Transmitters, effects on animals, 35–38
TRIANG program, triangulation tracking techniques
   maximum likelihood estimations, 63–64
   system design, 106
Triangulation location techniques
   confidence area construction, 53–54
   data quality control and censoring, 72–74
   Lenth's estimates, 69–72
   mobile, design and testing, 90–94
   system design
     computer simulations, 94–110
     data analysis system design and, 272–273
     design and testing of, 79–110
     directional bearings accuracy, 80–90
     mobile systems, 90–94
     tower location guidelines, 99, 101
     three receiving stations, 58–69
     two receiving stations, 47–58
Trigonometry, triangulation testing procedures, 83, 85–90, 97–99
Two-tower triangulation system, 53–59
   confidence ellipse
     area, 95, 97
     size and shape, 108–109

Uniform distribution, home range estimation, 154–155
Universal Polar Stereographic (UPS) projection, 4
Universal Transverse Mercator (UTM) mapping system, 2–6
   LORAN-C system and, 44
   zone designations, 2–4
U.S. Geological Survey maps, 4–6
Utilization distribution, home range estimation, 146, 176–177
Utilization of habitat, analysis of, 185–186

Variables in wildlife tracking data, 6–8
Variance–covariance matrix, home range estimation, 160–161
"Vector plot," simple animal movements, 115–116

Index

Vehicle-mounted receiving systems
   antenna-pointer orientation, 91–92
   triangulation location with, 49
von Mises distribution, triangulation location techniques, 59–64

Weighted bivariate normal estimator, home range estimation with, 160–161
Weighted mean, home range estimation with, 160–161
World Data Bank I and II, simple animal movement mapping, 118

WORLD program, simple animal movement mapping, 118–119
WRIS system, habitat availability measurement, 184

x, y, and t variables, data entry procedures, 6–7
XCOOR variable, 227

YCOOR variable, 277